# RSME Springer Series

Volume 14

As of 2015, RSME - Real Sociedad Matemática Española - and Springer cooperate in order to publish works by authors and volume editors under the auspices of a co-branded series of publications including advanced textbooks, Lecture Notes, collections of surveys resulting from international workshops and Summer Schools, SpringerBriefs, monographs as well as contributed volumes and conference proceedings. The works in the series are written in English only, aiming to offer high level research results in the fields of pure and applied mathematics to a global readership of students, researchers, professionals, and policymakers.

Antonio García García

# The Use of Frames in Sampling Theory

Real Sociedad Matemática Española

Springer

Antonio García García
Departments of Mathematics
Universidad Carlos III de Madrid
Madrid, Spain

ISSN 2509-8888 ISSN 2509-8896 (electronic)
RSME Springer Series
ISBN 978-3-031-63241-9 ISBN 978-3-031-63242-6 (eBook)
https://doi.org/10.1007/978-3-031-63242-6



This Springer imprint is published by the registered company Springer Nature Switzerland AG
The registered company address is: Gewerbestrasse 11, 6330 Cham, Switzerland

*For those who enjoy maths...*

# Preface

When introducing a new book on sampling theory, the author should first justify the need for yet another title in an already long list of books and survey papers published on the subject. In spite of the fact that a rich mathematical literature in sampling already exists, the main reason for this book is to build a new narrative from my own experience in sampling research. Since it is almost impossible to be exhaustive, it is necessary to make a choice of contents giving a justification for this selection. Namely, my choice is guided by the idea that entitles the book: *The Use of Frames in Sampling Theory*.

Roughly speaking, sampling theory deals with the reconstruction of functions $f$, belonging, in general, to a Hilbert space $\mathcal{H}$ of continuous functions, through their values (samples) $\{f(t_n)\}$ at an appropriate sequence of points $\{t_n\}$ by means of a sampling expansion $f(t) = \sum_n f(t_n)\, S_n(t)$ involving these values. Frequently, the functions to be recovered belong to a reproducing kernel Hilbert space where the point values can be expressed in terms of the reproducing kernels $\{k_{t_n}\}$ as the inner products $f(t_n) = \langle f, k_{t_n}\rangle$. This leads us to study the properties of the sequence $\{k_{t_n}\}$ in the Hilbert space $\mathcal{H}$.

As we will see through this book, another similar situation is very frequent: the available data is a sequence $\{(\mathcal{L}f)(t_n)\}$, where $\mathcal{L}$ represents the linear device involved in the sampling process. The above samples can be expressed as a sequence of inner products $\{\langle x, x_n\rangle\}$ in an auxiliary Hilbert space $\mathcal{K}$, where $x \in \mathcal{K}$ is an element univocally related to the function $f \in \mathcal{H}$ to be recovered by means of an operator $\mathcal{T} : \mathcal{K} \to \mathcal{H}$—an isomorphism in general–, and $\{x_n\}$ is a fixed sequence in $\mathcal{K}$. This auxiliary space $\mathcal{K}$ will be usually an $\ell^2$ or $L^2$-space.

Thus, the task is to recover $x$ (and consequently $f$) uniquely from the available data $\{\langle x, x_n\rangle\}$ in a stable way, which means that the norm $\|\{\langle x, x_n\rangle\} - \{\langle y, x_n\rangle\}\|_{\ell^2}$ is small if and only if $\|x - y\|_{\mathcal{K}}$ is small.

This leads us to the concept of *frame* in a separable Hilbert space: the sequence $\{x_n\}$ should be a frame for the auxiliary space $\mathcal{K}$. Given a frame $\{x_n\}$ for $\mathcal{K}$, any $x \in \mathcal{K}$ can be expanded as the sum $x = \sum_n \langle x, x_n\rangle\, y_n$ where the sequence $\{y_n\}$ is a dual frame of $\{x_n\}$. Finally, the relationship $f = \mathcal{T}x$ will lead to a sampling formula like $f = \sum_n (\mathcal{L}f)(t_n)\, S_n(t)$, where $S_n = \mathcal{T}y_n$.

Whenever the space $\mathcal{H}$ where the sampling works has a predetermined structure, an appropriate choice, if possible, of the dual frame $\{y_n\}$ will lead us to a sampling formula sharing the same structure as the functions in $\mathcal{H}$. This is the case for shift-invariant spaces.

The main aim of this book is to exhibit how the use of frames, including Riesz and orthonormal bases, in the above sampling schemes comprises a variety of examples: from the classical Shannon sampling theorem up to average sampling in certain subspaces of Hilbert-Schmidt operators. Thus, the book constructs a unified narrative which helps to understand how sampling theory works. Naturally, it does not intend to be exhaustive and, as a consequence, some important topics are not included.

Although the ideas in the book are partially borrowed from some previous research of the author and his collaborators, this does not mean a mere cut and paste from previous work. The author has tried to give a unified and comprehensive treatment on the subject by completing, when necessary, the presentation of previous research.

The book is divided into seven chapters whose contents we now briefly describe:

In Chap. 1, sampling theory is presented in a non-rigorous way including a short historical note on who is considered its father, namely, Claude E. Shannon, without forgetting other important authors, such as J. M. Whittaker, V. Kotel'nikov, K. Ogura or I. Someya. Surely, the vast number of its applications has helped the growth of sampling theory in other related fields, not least in mathematics itself. In the second part of the chapter, an almost self-contained introduction to frame theory is included. Frame theory is the main mathematical axis around which the other chapters are constructed. Of course, frames include, as particular cases, both orthonormal and Riesz bases which probably lead to the more known examples.

Chapter 2 includes a brief introduction to reproducing kernel Hilbert spaces (hereafter RKHSs) leading to the study of an important example of spaces in sampling theory: Paley-Wiener spaces $PW_{\pi\sigma}$. Thus, this chapter may be also useful as a mathematical introductory course in sampling theory, avoiding perhaps some specialized topics. The needed mathematical background just consists of a classical course in harmonic analysis and a basic knowledge of Hilbert spaces, functional analysis and complex analysis.

Chapter 3 is a non-standard collection of results concerning regular sampling in shift-invariant subspaces of $L^2(\mathbb{R}^d)$. The first part is devoted to the search of Riesz bases giving Shannon-type sampling formulas in shift-invariant subspaces of $L^2(\mathbb{R}^d)$. To this end, we follow the main technique in the book. More precisely, in the particular $L^2(\mathbb{R})$ case by searching Riesz bases expansions in an auxiliary space—here $L^2(0, 1)$—having the available (regular) samples as coefficients. Finally, the inverse Fourier transform—the operator $\mathcal{T}$ in this case—gives the desired sampling results.

The second part of Chap. 3 deals with a topic that does not usually appear in the sampling literature. Very often, in interpolation theory, input data are transformed by using differences of the given data: thus we are skipping irrelevant information when changes between neighbors are not very significant. The (regular) Shannon-type sampling formulas obtained are (infinite) interpolation formulas; in intervals where a function does not change too much, the available data can be reduced by using regular samples with a bigger sampling period along with their differences. New sampling formulas can be obtained in a straightforward way by using the same technique as in Sect. 3.1.

Finally, the third part is devoted to generalizing shift-invariant subspaces: the shift operator can be replaced by another unitary operator $U$, leading to the concept of $U$-invariant subspaces. Some results are proven in this new setting that directly apply to the shift-invariant case.

Chapter 4 concerns to the so-called Kramer sampling theorem. The Kramer sampling theorem has been the cornerstone for a significant mathematical literature which has flourished for the past few years on the topic of sampling theorems associated with differential or difference problems. In this chapter, we take a unified and abstract point of view, enlightened with a variety of examples belonging to different areas in mathematics: orthogonal polynomials and moment problems, difference and differential problems, entire functions, reproducing kernel Hilbert spaces, Lagrange and Hermite interpolation, semi-inner-product Banach spaces, or Hilbert spaces of distributions with reproducing distribution. In short, this chapter shows, by means of a unified approach, how sampling theory is transversal to other mathematical fields.

In Chap. 5, a generalized sampling theory is proposed. It is based on dealing with convolution systems on discrete abelian groups as a unifying strategy. On the one hand, working in locally compact abelian groups allows us to gather all the classical groups along with their products covering most of practical situations. On the other hand, we see that most of linear sampling devices can be treated as convolution systems defined on appropriate spaces. Thus a sampling theory is developed by following the master line that guides the proposal of this book. A comprehensive list of examples showcases the richness of the followed approach.

Chapter 6 is devoted to the study of sampling-related frames in finite-dimensional $U$-invariant subspaces of a separable Hilbert space. In applications, frames in finite-dimensional spaces are required and, as in finite dimension a frame is nothing but a spanning set of vectors, finite frames require control of condition numbers and over the spectrum of certain matrices. Thus, frame theory, firstly considered as a part of applied harmonic analysis, meets matrix analysis and numerical linear algebra. As an important example, the theory developed in this chapter encompasses the problem of periodic extensions of finite signals. The chapter includes also a section devoted to finite sampling associated with a non-abelian group defined as the knit product of two groups.

Finally, Chap. 7 is focused on the average sampling problem in shift-invariant-like subspaces of Hilbert-Schmidt operators in $L^2(\mathbb{R}^d)$. This problem is motivated by the recovery of a time-varying system from its diagonal channel samples, a signal processing problem appearing in orthogonal frequency-division multiplexing. Hilbert-Schmidt operators are the simplest examples of time-varying systems whose models are, in general, pseudo-differential operators. Having in mind the properties of certain operators (including the Hilbert-Schmidt ones), we can define shift-invariant-like subspaces of Hilbert-Schmidt operators and average samples as in the classical $L^2(\mathbb{R}^d)$ setting. The diagonal channel samples appear as a particular case. Following the general steps outlined in Chap. 5, we obtain the desired sampling results. These results are intimately related with topics belonging to time-frequency analysis.

A list of references closes each chapter of the book. It does not intend to be exhaustive and the author apologizes in advance for any important omission. The list includes, for obvious reasons, several references due to the author and his collaborators.

I would like to mention my sampling students: P. E. Fernández-Moncada, H. R. Fernández-Morales, M. A. Hernández-Medina, G. Pérez-Villalón and A. Portal, and also my sampling coauthors in papers cited in this book: W. N. Everitt, A. Ibort, K. H. Kwon, L. L. Littlejohn, J. Moro, M. J. Muñoz-Bouzo, A. Ortega, F. H. Szafraniec and A. Zayed. Thanks a lot, it was my pleasure to do mathematics with all of you.

I would like also to express here an anonymous acknowledgment to those authors who, over the years, taught me sampling theory, and the most important thing, to enjoy the mathematics that come with it.

Most of the results stated throughout this book are well-known or have been previously published. The author, at this stage, only claims originality in the way they have been set out. He will be satisfied if this contributes to make sampling theory better known to the overall scientific community.

Finally, I would like to thank Prof. Francisco Marcellán for encouraging me to carry out this nice project. Many thanks go also to the anonymous reviewers for their critical reading of the book; their comments and suggestions helped me to improve both the presentation and the contents of the book.

Leganés-Madrid, Spain
April 2024

Antonio García García

# Contents

# List of Symbols

| | |
|---|---|
| $\mathbb{C}$ | The set of complex numbers |
| $\mathbb{C}^d$ | The $d$-dimensional complex space |
| $\chi_A$ | Characteristic function of the set $A$ |
| $\mathcal{D}(\mathbb{R})$ | Vector space of test functions |
| $\Delta_+^k$ | $k$-th forward difference operator |
| $\Delta_-^k$ | $k$-th backward difference operator |
| $\Delta_0$ | Central difference operator |
| $\delta_{n,m}$ | Kronecker delta |
| $\mathcal{E}(\mathbb{R})$ | Space of functions infinitely differentiable on $\mathbb{R}$ |
| $\operatorname{ess\,inf}_{t\in I}\|f(t)\|$ | The essential infimum of $f$ on the interval $I$ |
| $\ell^2(\mathbb{N})$ | The Hilbert space of square summable sequences on $\mathbb{N}$ |
| $\ell^2_N(G)$ | The product Hilbert space $\ell^2(G) \times \cdots \times \ell^2(G)$ ($N$ times) |
| span $A$ | Linear space spanned by $A$ |
| $\operatorname{ess\,sup}_{t\in I}\|f(t)\|$ | Essential supremum of $f$ on the interval $I$ |
| $\|\|f\|\|_0$ | The essential infimum of $f$ on the interval $I$ |
| $\|\|f\|\|_\infty$ | Essential supremum of $f$ on the interval $I$ |
| $\mathbb{C}^{\mathbb{Z}}$ | Sequences of complex numbers indexed by $\mathbb{Z}$ |
| $\mathbb{C}^{\Omega}$ | The set of functions $f : \Omega \longrightarrow \mathbb{C}$ |
| $\mathbb{R}^d$ | The $d$-dimensional Euclidean space |
| $\mathbb{T}$ | The unidimensional torus |
| $\mathbf{N} \bowtie \mathbf{H}$ | The knit product of the groups **N** and **H** |
| $\mathcal{D}_{2N}$ | Dihedral group of symmetries of the $N$-gon |
| $\mathcal{E}'(\mathbb{R})$ | The space of compact supported distributions |
| $\mathcal{F}(f) = \widehat{f}$ | Fourier transform of $f$ |
| $\mathcal{F}^{-1}(\widehat{f}) = f$ | Inverse Fourier transform of $\widehat{f}$ |
| $\mathcal{F}_s$ | Symplectic Fourier transform |
| $\mathcal{F}_W$ | Fourier-Wigner transform |
| $\mathcal{HS}(\mathbb{R}^d)$ | Hilbert-Schmidt operators on $L^2(\mathbb{R}^d)$ |
| $\mathcal{S}'(\mathbb{R}^d)$ | Tempered distributions in $\mathbb{R}^d$ |
| $\mathcal{S}(\mathbb{R}^d)$ | Schwartz space in $\mathbb{R}^d$ |

| | |
|---|---|
| $\mathcal{S}_0(\mathbb{R}^d)$ | Feichtinger's algebra |
| $\mathcal{T}^1$ | Trace class operators |
| $\mathcal{T}^p$ | Schatten $p$-class |
| $\mathcal{U}(\mathcal{H})$ | Unitary operators on $\mathcal{H}$ |
| $\mathbb{N}$ | The set of natural numbers |
| $\mathbb{N}_0$ | $\mathbb{N} \cup \{0\}$ |
| $\overline{\text{span}} A$ | Closed linear subspace spanned by $A$ |
| $\overline{A}$ | Closure of the set $A$ |
| $\overline{B}(0; r)$ | Closed ball of radius $r$ centered at 0 |
| $\mathbb{R}$ | The set of real numbers |
| $\mathbb{R}^+$ | The set of positive real numbers |
| $\text{supp}\, f$ | Support of the function $f$ |
| $\mathcal{T}_s$ | Symmetric coordinate transformation |
| $\widehat{G}$ | Dual group of the group $G$ |
| $\mathbb{Z}$ | The set of integer numbers |
| $A \otimes B$ | Kronecker product of the matrices $A$ and $B$ |
| $A^\perp$ | Orthogonal subspace of the set $A$ |
| $B(0; r)$ | Open ball of radius $r$ centered at 0 |
| $B\big(L^2(\mathbb{R}^d)\big)$ | The Banach space of bounded operators in $L^2(\mathbb{R}^d)$ |
| $D_N$ | $N$-th Dirichlet kernel |
| $GL(2d, \mathbb{R})$ | General linear group of $2d \times 2d$ real matrices under multiplication |
| $H(\mathbb{C})$ | Set of entire functions |
| $H^2(\mathbb{C}^+)$ | Hardy space in the upper plane |
| $Hf$ | Hilbert transform of $f$ |
| $J_\nu(t)$ | Bessel function of order $\nu$ |
| $K \rtimes_\sigma H$ | Semi-direct product of the groups $K$ and $H$ under the action $\sigma$ |
| $L^2(\mathbb{R})$ | The Hilbert space of square integrable functions on $\mathbb{R}$ |
| $m(E)$ | Lebesgue measure of the measurable set $E$ |
| $M^*$ | Transpose conjugate of the matrix $M$ |
| $M^\dagger$ | Moore-Penrose pseudo-inverse of the matrix $M$ |
| $N_m$ | $B$-spline of order $m - 1$ |
| $O(d)$ | Orthogonal group of order $d$ |
| $PW_\pi$ | Paley-Wiener space |
| $PW_{\pi\sigma}$ | Paley-Wiener spaces with $\sigma > 0$ |
| $T^*$ | Adjoint operator of $T$ |
| $Z\varphi$ | Zak transform of $\varphi$ |
| l.c.m. $(N_1, N_2, )$ | Least common multiple of $N_1$ and $N_2$ |
| LTI | Linear time-invariant system |
| MIMO | Multi-input multi-output system |
| RKBS | Reproducing kernel Banach space |
| RKHS | Reproducing kernel Hilbert space |
| s.i.p. RKBS | semi-inner-product reproducing kernel Banach space |

# Chapter 1
# What Does Sampling Theory Mean?

This introductory chapter is devoted to give a historical flavor to sampling theory. Besides, it serves as a motivation for the main aim in the book: the use of frames in sampling theory. For the sake of completeness it includes the basic frame theory needed throughout the book.

## 1.1 A Historical Note

Roughly speaking, sampling theory deals with the reconstruction of functions through their values (samples) on an appropriate sequence of points by means of sampling expansions involving these values.[1] This is not always possible: for instance, a continuous function $f$ on $\mathbb{R}$ is not completely determined by a sequence $\{f(t_n)\}$ of its values (samples). As a consequence, one needs to impose some additional conditions on the function $f$ to be recovered. The seminal result in this direction is the so-called Shannon's theorem stated in his famous paper *Communication in the presence of noise* (Ref. [41]) as:

> THEOREM 1: If a function $f(t)$ contains no frequencies higher than $W$ cps, it is completely determined by giving its ordinates at a series of points spaced $1/2W$ seconds apart.

In this case, $f$ must be a signal bandlimited to the interval $(-2\pi W, 2\pi W)$—i.e. signals containing no frequencies beyond a critical value, $W$ cycles per second

[1] This should not be confused with *statistical sampling*, which consists in the selection of a subset (statistical sample) of individuals from within a statistical population to estimate characteristics of the whole population.

 
A. García García, *The Use of Frames in Sampling Theory*,
RSME Springer Series 14, https://doi.org/10.1007/978-3-031-63242-6_1

in the above case—from an equidistant sequence of samples whose spacing $1/2W$ depends on $W$. The (sampling) formula allowing this reconstruction is

$$f(t) = \sum_{n=-\infty}^{\infty} f\left(\frac{n}{2W}\right) \frac{\sin \pi(2Wt - n)}{\pi(2Wt - n)}, \quad t \in \mathbb{R}. \tag{1.1}$$

The fact that $f$ is bandlimited to the interval $(-2\pi W, 2\pi W)$ implies that $f$ can be expressed from its Fourier transform $F$ as

$$f(t) = \frac{1}{\sqrt{2\pi}} \int_{-2\pi W}^{2\pi W} F(\omega)\, e^{i\omega t} d\omega, \quad t \in \mathbb{R}.$$

Notice that the function $f$ given by the above series (1.1) is the unique bandlimited function to the interval $(-2\pi W, 2\pi W)$ that interpolates the sequence of values $\{f(\frac{n}{2W})\}_{n=-\infty}^{\infty}$ since the sampling functions used in the reconstruction are

$$S_n(t) := \frac{\sin \pi(2Wt - n)}{\pi(2Wt - n)}, \quad n \in \mathbb{Z},$$

that satisfy the interpolation property $S_n(t_m) = \delta_{n,m}$ at $t_m = \frac{m}{2W}$, $n, m \in \mathbb{Z}$, where $\delta_{n,m}$ equals 1 if $n = m$, and 0 if $n \neq m$ denotes the *Kronecker delta*.

A series as in (1.1) is known as a *cardinal series* because the sampling functions involve the *cardinal sine* function (or *sinc function*) depicted in Fig. 1.1 and defined as

$$\operatorname{sinc}(t) = \begin{cases} \frac{\sin \pi t}{\pi t} & \text{if } t \neq 0, \\ 1 & \text{if } t = 0. \end{cases} \tag{1.2}$$

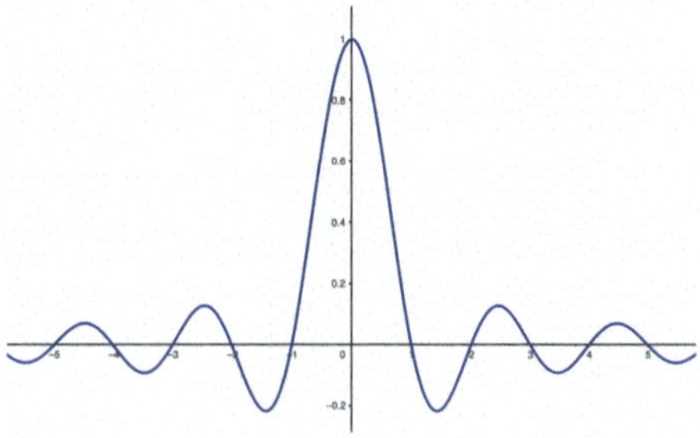

**Fig. 1.1** Sinc function

Thus, the Shannon sampling formula (1.1) generalizes to the infinite setting the finite polynomial interpolation:

Given $N+1$ points $\big(t_n, f(t_n)\big)$, $n = 0, 1, \ldots, N$, with $t_n \neq t_m$ for $n \neq m$, there exists a unique polynomial $p_N$ of degree $N$ that interpolates the data; the *Lagrange interpolation formula* reads

$$p_N(t) = \sum_{n=0}^{N} f(t_n) \frac{P(t)}{P'(t_n)(t - t_n)}, \quad t \in \mathbb{R},$$

where $P(t) = \prod_{n=0}^{N}(t - t_n)$.

The series involved in (1.1) owes its name—cardinal series—to J. M. Whittaker Ref. [48], a reference cited by Shannon in Ref. [41]. To be precise, J. M. Whittaker's work was a refinement of his father's, the eminent British mathematician E. T. Whittaker Ref. [47]. However, it is not clear whether or not they were the first mathematicians to introduce these kinds of expansions. About the fatherhood of the above referred sampling formula there exists some historical controversy, which involves famous mathematicians as E. Borel, A. L. Cauchy, W. L. Ferrar, K. Ogura, among others. The interested reader can find some historical sources in Refs. [7, 26, 27, 32, 50]. See also the master references [6, 10, 17, 36].

The Shannon sampling theorem provides the theoretical foundation for modern pulse code modulation communication systems, which were introduced, independently, by V. Kotel'nikov [31] in 1933 (an English translation from the original Russian manuscript can be found in Ref. [5]) and by Shannon in 1949. It is worth noticing that in the same year, 1949, I. Someya published in Japan a book containing the same theorem along with many extensions and applications Ref. [43].

This sampling theorem is presently known in the mathematical literature as the *Whittaker-Shannon-Kotel'nikov theorem* or *WSK sampling theorem*.

In any case, there is no doubt as to when did modern sampling theory start out: it was in 1949, when Shannon published his famous paper *Communication in the presence of noise*. Although it is almost certain that Shannon was not the discoverer of "his formula", his paper triggered an avalanche of works, which have eventually produced a flourishing body of results on sampling methods and their applications.

What started as a theorem for reconstructing bandlimited signals from uniform samples has now become, from a mathematical point of view, a whole branch of applied mathematics, known as sampling theory. This new field turned out to be very useful in many mathematical areas, such as approximation theory, harmonic analysis, theory of entire functions, theory of distributions, and stochastic processes, among others. The efforts in extending Shannon's fundamental result point in various directions: nonuniform samples, other discrete data taken from the signal, multidimensional signals, and more; see, for instance, Refs. [2, 9, 38]. This, together with the technological impact of sampling in communication theory and signal processing, can provide a clearer idea of the importance this topic has nowadays.

Any interested reader in sampling theory should also consult the books by Eldar [15], Higgins [27, 28], Marks II [34, 35] or Zayed [50], or the surveys in Refs. [20, 21, 26, 30, 46].

Next, we include a brief biography[2] of Claude Elwood Shannon (Petoskey MI, 1916, Medford MA, 2001), the mathematician, computer scientist, communication engineer, and the founder of the field of information theory, whose work has laid the foundation for the telecommunication networks that lace the globe:

- B. S. Engineering Mathematics, University of Michigan, 1936
  Ph. D. Mathematics, MIT, 1940
- Research Mathematician, Bell Labs, 1941–1972; MIT Faculty Member, 1956–1978; Donner Professor of Science 1958
- Major publication: *A Mathematical Theory of Communication*, Bell System Technical Report, 1948
- Honorary Degree, Univ. of Michigan, 1961; The National Medal of Science, 1966; The Audio Engineering Society Gold Medal, 1985; The Kyoto Prize, 1985

### *1.1.1 What's Next?*

Of the previously described, dealing with sampling theory the involved functions $f$ must belong to some suitable spaces. For example, assume that the function $f$ belongs to a Hilbert space $\mathcal{H}$ of functions on $\Omega$ (generally, a subset of $\mathbb{R}$ or $\mathbb{C}$) such that any point evaluation functional $E_t : f \in \mathcal{H} \mapsto f(t) \in \mathbb{C}$ is bounded, i.e., the space $\mathcal{H}$ is a reproducing kernel Hilbert space (RKHS henceforth). Via the Riesz representation theorem, for each $t \in \Omega$ there exists a unique $k_t \in \mathcal{H}$ such that

$$f(t) = \langle f, k_t \rangle_{\mathcal{H}} \quad \text{for every} \quad f \in \mathcal{H}.$$

(See the details in Sect. 2.1.1 below).

In this manner, the stable reconstruction of any $f \in \mathcal{H}$ from the sequence of samples $\{f(t_n)\}$ at $\{t_n\} \subset \Omega$ depends on whether the sequence $\{k_{t_n}\}$ is a frame for $\mathcal{H}$. Following Definition 1.1 below, a sequence $\{x_n\}$ is a *frame* for a separable Hilbert space $\mathcal{H}$ if there exist two constants $A, B > 0$ (frame bounds) such that

$$A\|x\|^2 \le \sum_n |\langle x, x_n \rangle|^2 \le B\|x\|^2 \quad \text{for all } x \in \mathcal{H}.$$

The sequence of frame coefficients $\{\langle x, x_n \rangle\}$ of any $x \in \mathcal{H}$ has the two following properties:

(i) It characterizes the element $x \in \mathcal{H}$ uniquely, and
(ii) the norm $\|\{\langle x, x_n \rangle\} - \{\langle y, x_n \rangle\}\|_{\ell^2}$ is small if and only if $\|x - y\|_{\mathcal{H}}$ is small.

[2] Data partially taken from Wikipedia and IEEE Information Theory Society.

For our sampling problem it means:

1. The function $f \in \mathcal{H}$ is uniquely determined by its samples $\{f(t_n)\}$, and
2. the norm $\|\{f(t_n)\}-\{g(t_n)\}\|_{\ell^2}$ is small if and only if the norm $\|f-g\|_{\mathcal{H}}$ is small.

Given a frame $\{x_n\}$ for $\mathcal{H}$ the representation property of any vector $x \in \mathcal{H}$ as a series $x = \sum_n c_n\, x_n$ is retained, but, unlike the case of Riesz (orthonormal) bases, the uniqueness of this representation is sacrificed. Suitable frame coefficients $c_n$ which depend continuously and linearly on $x$ are obtained by using the dual frames $\{y_n\}$ of $\{x_n\}$, i.e., $\{y_n\}$ is another frame for $\mathcal{H}$ such that

$$x = \sum_n \langle x, y_n \rangle\, x_n = \sum_n \langle x, x_n \rangle\, y_n \quad \text{for each } \; x \in \mathcal{H}. \tag{1.3}$$

In particular, frames in $\mathcal{H}$ include orthonormal and Riesz bases for $\mathcal{H}$ (see Sect. 1.2 below). A Riesz basis in a separable Hilbert space $\mathcal{H}$ is the image of an orthonormal basis by means of a bounded invertible operator. Any Riesz basis $\{x_n\}$ has a unique biorthogonal (dual) Riesz basis $\{y_n\}$, i.e., $\langle x_n, y_m\rangle_{\mathcal{H}} = \delta_{n,m}$, the Kronecker delta, such that the expansions (1.3) hold for every $x \in \mathcal{H}$. An orthonormal basis is a self-dual Riesz basis. An equivalent but different approach to Riesz bases is proposed in Sect. 1.2.

In case the sequence $\{k_{t_n}\}$ forms a frame for the RKHS $\mathcal{H}$, and a dual frame $\{S_n(t)\}$ is available (a difficult problem in general), the sampling formula in $\mathcal{H}$

$$f(t) = \sum_n \langle f, k_t \rangle_{\mathcal{H}}\, S_n(t) = \sum_n f(t_n)\, S_n(t)\,, \quad t \in \Omega$$

holds. Notice that convergence in norm in a RKHS $\mathcal{H}$ of functions defined on $\Omega$ implies pointwise convergence in $\Omega$ (see Proposition 2.5 below).

In this book one can find a variety of examples where these involved dual frames are explicitly obtained; in particular, orthonormal bases and dual Riesz bases. In fact, throughout this work there is a common feature in deducing sampling formulas although the different situations where the corresponding sampling problem are formulated. Indeed:

- The data sampling is a sequence of frame coefficients with respect to a suitable frame in an auxiliary Hilbert space $\mathcal{K}$; this space is, in general, an $\ell^2$-space (or $L^2$-space). Moreover, there exist dual frames having the same structure that the involved frame in the expression of the samples.
- There exists an isomorphism $\mathcal{T}: \mathcal{K} \longrightarrow \mathcal{H}$ between the auxiliary space and the space where the sampled functions are living. This isomorphism preserves the underlying unitary structure in both spaces: the auxiliary $\ell^2$-space (or $L^2$-space) and the initial space where the sampling is carried out.
- The application of the isomorphism $\mathcal{T}$ on an appropriate expansion in the auxiliary space gives the desired sampling formula in $\mathcal{H}$.

## 1.2 A Brief on the Theory of Frames in a Separable Hilbert Space

In this section we include a brief on *frame theory*. It contains the core of the theory needed in this book and it could be followed as an introductory mini-course on the topic. For more details on this nice theory I recommend, for instance, the books Refs. [11, 49] or the seminal paper Ref. [14].

### *1.2.1 The Basic Frame Theory*

**Definition 1.1** A sequence $\{x_n\}_{n=1}^{\infty}$ is a frame for a separable Hilbert space $\mathcal{H}$ if there exist constants $0 < A \le B$, called frame bounds, such that

$$A\|x\|^2 \le \sum_{n=1}^{\infty} |\langle x, x_n\rangle|^2 \le B\|x\|^2 \quad \text{for all } x \in \mathcal{H}.$$

The optimal frame bounds are the largest $A$ and the smallest $B$ constants satisfying the above inequalities.

Recall that a Hilbert space is said to be *separable* if there exists a countable and dense subset in $\mathcal{H}$.

Notice also that a frame is a *complete system* in $\mathcal{H}$ since condition $\langle x, x_n\rangle = 0$ for all $n \in \mathbb{N}$ implies that $x = 0$.

A *Bessel sequence* in $\mathcal{H}$ is a sequence $\{x_n\}_{n=1}^{\infty}$ satisfying at least the right side inequality. That is,

$$\sum_{n=1}^{\infty} |\langle x, x_n\rangle|^2 \le B\|x\|^2 \quad \text{for all } x \in \mathcal{H}.$$

An orthonormal basis for a separable Hilbert space is obviously an example of a frame. At the end of this section we will give a characterization of the set of frames in a separable Hilbert space (see Corollary 1.3).

*(a)* ***Preframe Operator*** The *preframe operator* $T$ (or synthesis operator) associated with $\{x_n\}_{n=1}^{\infty}$ is defined by

$$\begin{aligned} T:\ \ell^2(\mathbb{N}) &\longrightarrow \mathcal{H} \\ \{c_n\}_{n=1}^{\infty} &\longmapsto \sum_{n=1}^{\infty} c_n\, x_n \end{aligned}$$

**Proposition 1.1** *For a Bessel sequence the preframe operator is well-defined and bounded with* $\|T\| \le \sqrt{B}$.

***Proof*** The operator $T$ is well-defined, i.e., the series $\sum_{n=1}^{\infty} c_n x_n$ converges in $\mathcal{H}$ for any $\{c_n\}_{n=1}^{\infty} \in \ell^2(\mathbb{N})$. Indeed, by using that $\|x\| = \sup_{\|y\|=1} |\langle x, y\rangle|$ for every $x$ in a Hilbert space $\mathcal{H}$ and the Cauchy-Schwarz inequality in $\ell^2$, for any $N \geq M$ in $\mathbb{N}$ we have

$$\Big\|\sum_{n=1}^{N} c_n x_n - \sum_{n=1}^{M} c_n x_n\Big\| = \Big\|\sum_{n=M+1}^{N} c_n x_n\Big\| = \sup_{\|y\|=1} \Big|\Big\langle \sum_{n=M+1}^{N} c_n x_n, y\Big\rangle\Big|$$

$$\leq \sup_{\|y\|=1} \sum_{n=M+1}^{N} |c_n \langle x_n, y\rangle| \leq \Big(\sum_{n=M+1}^{N} |c_n|^2\Big)^{1/2} \sup_{\|y\|=1} \Big(\sum_{n=M+1}^{N} |\langle x_n, y\rangle|^2\Big)^{1/2}$$

$$\leq \sqrt{B}\Big(\sum_{n=M+1}^{N} |c_n|^2\Big)^{1/2} \longrightarrow 0 \quad \text{whenever } M, N \to \infty\,.$$

That is, the sequence of partial sums of the series is a Cauchy sequence in $\mathcal{H}$. Hence, since $\mathcal{H}$ is a Hilbert space, the series $\sum_{n=1}^{\infty} c_n x_n$ converges in $\mathcal{H}$.

Analogously, since

$$\|T(\{c_n\})\| = \sup_{\|y\|=1} |\langle T(\{c_n\}), y\rangle| \leq \sqrt{B}\Big(\sum_{n=1}^{\infty} |c_n|^2\Big)^{1/2} \quad \text{for all } \{c_n\} \in \ell^2(\mathbb{N})$$

one proves that the operator $T$ is bounded. □

*(b)* ***Analysis Operator*** The adjoint operator $T^*$ of the preframe operator $T$ associated to a Bessel sequence $\{x_n\}_{n=1}^{\infty}$ is called the *analysis operator* and is given by

$$\begin{aligned} T^* : \mathcal{H} &\longrightarrow \ell^2(\mathbb{N}) \\ x &\longmapsto \{\langle x, x_n\rangle\}_{n=1}^{\infty} \end{aligned}$$

Indeed, for any $x \in \mathcal{H}$ and $\{c_n\} \in \ell^2(\mathbb{N})$ we have

$$\langle T\{c_n\}, x\rangle_{\mathcal{H}} = \Big\langle \sum_{n=1}^{\infty} c_n x_n, x\Big\rangle_{\mathcal{H}} = \sum_{n=1}^{\infty} c_n \overline{\langle x, x_n\rangle_{\mathcal{H}}} = \langle \{c_n\}, T^* x\rangle_{\ell^2(\mathbb{N})}\,.$$

In fact, the preframe operator $T$ associated with a sequence $\{x_n\}_{n=1}^{\infty} \in \ell^2(\mathbb{N})$ is well-defined (and hence bounded) if and only if the sequence$\{x_n\}_{n=1}^{\infty}$ is a Bessel sequence:

**Proposition 1.2** *The sequence $\{x_n\}_{n=1}^{\infty}$ is a Bessel sequence for $\mathcal{H}$ if and only if the series $\sum_{n=1}^{\infty} c_n x_n$ converges for each $\{c_n\}_{n=1}^{\infty}$ in $\ell^2(\mathbb{N})$.*

***Proof*** The necessary condition has been proved above. For the sufficient condition, let us prove that the preframe operator associated with the sequence $\{x_n\}_{n=1}^{\infty}$ is bounded. Indeed, for each $N \in \mathbb{N}$, define the bounded operator $T_N : \ell^2(\mathbb{N}) \longrightarrow \mathcal{H}$ by means of $\{x_n\}_{n=1}^{\infty} \longmapsto \sum_{n=1}^{N} c_n\, x_n$. As to $T_N \to T$ pointwise as $N \to \infty$, the *Banach-Steinhaus theorem* assures that the operator $T$ is bounded. Moreover, its adjoint (bounded) operator $T^*$ is given by $T^* : x \longmapsto \{\langle x, x_n\rangle\}_{n=1}^{\infty}$ (see Lemma 3.2.1 in Ref. [11] for a proof), and the sequence $\{x_n\}_{n=1}^{\infty}$ is a Bessel sequence for $\mathcal{H}$ since

$$\|T^* x\|^2 = \sum_{n=1}^{\infty} |\langle x, x_n\rangle|^2 \le \|T\|^2 \|x\|^2 \quad \text{for all} \quad x \in \mathcal{H}.$$

□

*(c)* ***Frame Operator*** The most important properties of frames come out from the so-called *frame operator* which is defined as $S := TT^*$, that is,

$$\begin{aligned} S : \mathcal{H} &\longrightarrow \mathcal{H} \\ x &\longmapsto Sx = \sum_{n=1}^{\infty} \langle x, x_n\rangle\, x_n \end{aligned}$$

**Proposition 1.3** *Let $\{x_n\}_{n=1}^{\infty}$ be a frame for $\mathcal{H}$ with frame operator $S$. Then $S$ is a bounded, self-adjoint, positive and invertible operator.*

***Proof*** The operator $S = TT^*$ is bounded as it is a composition of bounded operators; besides, $\|S\| \le B$.

The operator $S$ is self-adjoint since

$$S^* = (TT^*)^* = (T^*)^* T^* = TT^* = S.$$

The operator $S$ is a positive operator ($S \ge 0$) since

$$\langle Sx, x\rangle = \sum_{n=1}^{\infty} |\langle x, x_n\rangle|^2 \ge 0 \quad \text{for all } x \in \mathcal{H}.$$

Having in mind the partial order in self-adjoint operators in $\mathcal{H}$, i.e.,

$$S \le R \quad \text{if and only if} \quad R - S \ge 0,$$

the frame definition can be also expressed as $AI \le S \le BI$. Indeed,

$$\begin{aligned} A\|x\|^2 = \langle Ax, x\rangle \le \sum_{n=1}^{\infty} |\langle x, x_n\rangle|^2 &= \langle Sx, x\rangle \\ &\le \langle Bx, x\rangle = B\|x\|^2 \quad \text{for all } x \in \mathcal{H}. \end{aligned}$$

The operator $S$ is invertible since $\|I - B^{-1}S\| < 1$. Indeed, the operator $I - B^{-1}S$ is self-adjoint and its norm can be computed as

$$\|I - B^{-1}S\| = \sup_{\|x\|=1} \left|\langle (I - B^{-1}S)x, x\rangle\right| \le \frac{B - A}{B} < 1\,.$$

*Neumann series* theory ensures that the operator $I - (I - B^{-1}S) = B^{-1}S$ is invertible. Moreover, the inverse operator can be expressed as the series

$$S^{-1} = B^{-1}\sum_{k=0}^{\infty}(I - B^{-1}S)^k\,.$$

□

**Proposition 1.4** *The sequence $\{S^{-1}x_n\}_{n=1}^{\infty}$ is also a frame for $\mathcal{H}$, called the canonical dual frame of $\{x_n\}_{n=1}^{\infty}$, and its frame operator is $S^{-1}$.*

***Proof*** The sequence $\{S^{-1}x_n\}_{n=1}^{\infty}$ is a *frame* for $\mathcal{H}$. Indeed, since the operator $S^{-1}$ is also self-adjoint for each $n \in \mathbb{N}$ we have $\langle x, S^{-1}x_n\rangle = \langle S^{-1}x, x_n\rangle$ and, as a consequence, we obtain that:

$$A\|S^{-1}x\|^2 \le \sum_{n=1}^{\infty} |\langle x, S^{-1}x_n\rangle|^2 = \sum_{n=1}^{\infty} |\langle S^{-1}x, x_n\rangle|^2 \le B\|S^{-1}x\|^2\,, \quad x \in \mathcal{H}.$$

Since $\|S^{-1}x\| \ge \|x\|/\|S\|$ and $\|S^{-1}x\| \le \|S^{-1}\|\|x\|$ for all $x \in \mathcal{H}$, we get the desired result.

Next we compute its frame operator:

$$\begin{aligned}\sum_{n=1}^{\infty}\langle x, S^{-1}x_n\rangle\, S^{-1}x_n &= S^{-1}\Big(\sum_{n=1}^{\infty}\langle x, S^{-1}x_n\rangle\, x_n\Big)\\ &= S^{-1}\Big(\sum_{n=1}^{\infty}\langle S^{-1}x, x_n\rangle\, x_n\Big)\\ &= S^{-1}\big(S(S^{-1}x)\big) = S^{-1}x\,, \qquad x \in \mathcal{H}.\end{aligned}$$

□

If the optimal frame bounds of $\{x_n\}_{n=1}^{\infty}$ are $A$ and $B$, then the optimal frame bounds of $\{S^{-1}x_n\}_{n=1}^{\infty}$ are $1/B$ and $1/A$. Indeed, since $S^{-1}$ is a positive operator which commutes with $S$, the composition with $S^{-1}$ in $AI \le S \le BI$ maintains the inequalities, i.e.,

$$AI \le S \le BI \quad \Rightarrow \quad \frac{1}{B}I \le S^{-1} \le \frac{1}{A}I\,.$$

Notice that the optimal frame bounds of $\{x_n\}_{n=1}^{\infty}$ are $B = \|S\|$ and $A = \|S^{-1}\|^{-1}$.

From the equalities $x = S^{-1}Sx = SS^{-1}x$ for any $x \in \mathcal{H}$, one easily deduces the expansions:

$$x = \sum_{n=1}^{\infty} \langle x, x_n \rangle\, S^{-1}x_n = \sum_{n=1}^{\infty} \langle x, S^{-1}x_n \rangle\, x_n \quad \text{in } \mathcal{H}.$$

A frame $\{y_n\}_{n=1}^{\infty}$ for $\mathcal{H}$ is said to be a *dual frame* of the frame $\{x_n\}_{n=1}^{\infty}$ if we have the expansion

$$x = \sum_{n=1}^{\infty} \langle x, y_n \rangle\, x_n \quad \text{for all } x \in \mathcal{H}.$$

Thus, $\{S^{-1}x_n\}_{n=1}^{\infty}$ is one of the possible dual frames of the frame $\{x_n\}_{n=1}^{\infty}$. It has the property that the sequence $\big\{\langle x, S^{-1}x_n \rangle\big\}_{n=1}^{\infty}$ has minimal $\ell^2$-norm among all $\ell^2$-sequences $\{c_n\}$ for which $x = \sum_{n=1}^{\infty} c_n\, x_n$. It comes out from the following result:

**Proposition 1.5** *For $x \in \mathcal{H}$, let denote $\big\{a_n := a_n(x) = \langle x, S^{-1}x_n \rangle\big\}_{n=1}^{\infty}$. If we have the expansions $x = \sum_{n=1}^{\infty} a_n\, x_n = \sum_{n=1}^{\infty} b_n\, x_n$, then*

$$\sum_{n=1}^{\infty} |b_n|^2 = \sum_{n=1}^{\infty} |a_n|^2 + \sum_{n=1}^{\infty} |a_n - b_n|^2 . \tag{1.4}$$

***Proof*** The proof consists of expressing $\langle x, S^{-1}x \rangle$ in three different ways. Indeed,

$$\langle x, S^{-1}x \rangle = \langle \sum_{n=1}^{\infty} a_n\, x_n, S^{-1}x \rangle = \sum_{n=1}^{\infty} a_n\, \langle x_n, S^{-1}x \rangle = \sum_{n=1}^{\infty} |a_n|^2 ,$$

since $\langle x_n, S^{-1}x \rangle = \langle S^{-1}x_n, x \rangle = \overline{a}_n$ due to the self-adjointness of operator $S^{-1}$.

Analogously,

$$\langle x, S^{-1}x \rangle = \langle \sum_{n=1}^{\infty} b_n\, x_n, S^{-1}x \rangle = \sum_{n=1}^{\infty} b_n\, \overline{a}_n .$$

Finally, as $\langle x, S^{-1}x \rangle = \langle S^{-1}x, x \rangle$, it is also true that $\langle x, S^{-1}x \rangle = \sum_{n=1}^{\infty} a_n\, \overline{b}_n$. Therefore,

$$\sum_{n=1}^{\infty} |a_n|^2 + \sum_{n=1}^{\infty} \big(|a_n|^2 - a_n\, \overline{b}_n - b_n\, \overline{a}_n + |b_n|^2\big) = \sum_{n=1}^{\infty} |b_n|^2 .$$

□

**Proposition 1.6** *Let $\{x_n\}_{n=1}^{\infty}$ be a frame for $\mathcal{H}$ with frame operator $S$. For each $x_m \in \{x_n\}_{n=1}^{\infty}$ the following results hold:*

*(i) If $\langle x_m, S^{-1}x_m\rangle \neq 1$, then $\{x_n\}_{n\neq m}$ is a frame for $\mathcal{H}$.*
*(ii) If $\langle x_m, S^{-1}x_m\rangle = 1$, then $\{x_n\}_{n\neq m}$ is not complete in $\mathcal{H}$ (and consequently it is not a frame).*

***Proof*** The element $x_m$ can be expressed as $x_m = \sum_{n=1}^{\infty} a_n(x_m)\, x_n = \sum_{n=1}^{\infty} \delta_{n,m}\, x_n$. Using Eq. (1.4) we obtain

$$\sum_{n=1}^{\infty} |\delta_{n,m}|^2 = \sum_{n=1}^{\infty} |a_n(x_m)|^2 + \sum_{n=1}^{\infty} |a_n(x_m) - \delta_{n,m}|^2\,,$$

that is,

$$1 = |a_m(x_m)|^2 + |a_m(x_m) - 1|^2 + 2\sum_{n\neq m} |a_n(x_m)|^2\,. \tag{1.5}$$

Whenever $a_m(x_m) = 1$ we obtain $a_n(x_m) = 0$ for all $n \neq m$. Therefore, $\langle x_m, S^{-1}x_n\rangle = \langle S^{-1}x_m, x_n\rangle = 0$ for all $n \neq m$. Since $\langle x_m, S^{-1}x_m\rangle = 1$, the element $S^{-1}x_m \neq 0$. Hence, the system $\{x_n\}_{n\neq m}$ is not complete in $\mathcal{H}$.

Whenever $a_m(x_m) \neq 1$, the element $x_m$ can be expressed in terms of $\{x_n\}_{n\neq m}$ as

$$x_m = \frac{1}{1 - a_m(x_m)} \sum_{n\neq m} a_n(x_m)\, x_n\,.$$

For any $x \in \mathcal{H}$ we get

$$\langle x, x_m\rangle = \frac{1}{1 - a_m(x_m)} \sum_{n\neq m} \overline{a_n(x_m)}\langle x, x_n\rangle\,,$$

and the Cauchy-Schwarz inequality in $\ell^2$ gives

$$|\langle x, x_m\rangle|^2 \leq \Big(\frac{1}{|1 - a_m(x_m)|^2} \sum_{n\neq m} |a_n(x_m)|^2\Big) \sum_{n\neq m} |\langle x, x_n\rangle|^2 = C \sum_{n\neq m} |\langle x, x_n\rangle|^2\,.$$

Hence,

$$A\|x\|^2 \leq \sum_{n=1}^{\infty} |\langle x, x_n\rangle|^2 = |\langle x, x_m\rangle|^2 + \sum_{n\neq m} |\langle x, x_n\rangle|^2 \leq (1 + C) \sum_{n\neq m} |\langle x, x_n\rangle|^2\,,$$

from which

$$\frac{A}{1+C}\|x\|^2 \le \sum_{n\neq m} |\langle x, x_n\rangle|^2 \le \sum_{n=1}^{\infty} |\langle x, x_n\rangle|^2 \le B\|x\|^2 ,$$

that is, the sequence $\{x_n\}_{n\neq m}$ is a frame for $\mathcal{H}$. □

We have seen that for a frame $\{x_n\}_{n=1}^{\infty}$ its preframe operator $T$ is bounded and surjective. In fact, this property characterizes the frames:

**Theorem 1.1** *A sequence $\{x_n\}_{n=1}^{\infty}$ is a frame for $\mathcal{H}$ if and only if its preframe operator $T$ is bounded and surjective.*

***Proof*** The necessary condition has been previously proved. To prove the sufficient condition, as $\{x_n\}_{n=1}^{\infty}$ is a Bessel sequence, we should prove the left-hand inequality in the frame definition.

Since operator $T$ is bounded and surjective, from the *open mapping theorem* there exists a constant $C > 0$ such that for each $x \in \mathcal{H}$ there exists a sequence $\{c_n\} \in \ell^2(\mathbb{N})$ such that

$$T(\{c_n\}) = x \quad \text{and} \quad \|\{c_n\}\| \le C\|x\| .$$

Indeed, since the image of the unit open ball centered at 0, i.e., $T\big(B(0; 1)\big)$, is an open set in $\mathcal{H}$ and $T(0) = 0$, there exists $\delta > 0$ such that $\overline{B}(0; \delta) \subset T\big(B(0; 1)\big)$. For any $x \neq 0$ in $\mathcal{H}$, since $\delta x/\|x\| \in \overline{B}(0; \delta)$, there exists $\{a_n\} \in B(0; 1)$ in $\ell^2(\mathbb{N})$ such that $T(\{a_n\}) = \delta x/\|x\|$.

Hence, choosing the sequence $\big\{c_n := \frac{a_n\|x\|}{\delta}\big\} \in \ell^2(\mathbb{N})$ we obtain $T(\{c_n\}) = x$ and $\|\{c_n\}\| \le \frac{\|x\|}{\delta}$, that is, it is enough to take $C := 1/\delta$.

As a consequence, for any $x \in \mathcal{H}$ we have

$$\langle T^*x, \{c_n\}\rangle = \langle x, T(\{c_n\})\rangle = \|x\|^2 \ge \|x\|\frac{\|\{c_n\}\|}{C} ,$$

from which $\|T^*x\| \ge \|x\|/C$. Hence,

$$\|T^*x\|^2 = \sum_{n=1}^{\infty} |\langle x, x_n\rangle|^2 \ge \frac{\|x\|^2}{C^2} \quad \text{for all } x \in \mathcal{H}.$$

□

**Corollary 1.1** *A sequence $\{x_n\}_{n=1}^{\infty}$ is a frame for $\mathcal{H}$ if and only if its analysis operator $T^*$ is bounded and injective with a closed range.*

***Proof*** It is known that $T$ is bounded if and only if $T^*$ is bounded. Besides, the range of $T$ is closed if and only if the range of $T^*$ is closed. Thus, the result comes out from the relationship between the kernel of $T^*$ and the range of $T$, that is,

$$\left(\ker T^*\right)^\perp = \overline{\text{range}(T)}\,.$$

Thus, $\text{range}(T) = \mathcal{H}$ implies that $\ker T^* = \{0\}$, i.e., $T^*$ is injective with a closed range. Conversely, $T^*$ injective with a closed range implies that

$$\overline{\text{range}(T)} = \text{range}(T) = \mathcal{H}\,,$$

i.e., $T$ is surjective. □

**Corollary 1.2** *A sequence $\{x_n\}_{n=1}^\infty$ is a frame for $\mathcal{H}$ if and only if its preframe operator $T$ is bounded and $TT^*$ is invertible.*

***Proof*** If $\{x_n\}_{n=1}^\infty$ is a frame for $\mathcal{H}$, then $T$ is bounded and $TT^*$ is its frame operator which is invertible. Reciprocally, if $TT^*$ is invertible, then $T$ is surjective and bounded, and the sequence $\{x_n\}_{n=1}^\infty$ is a frame for $\mathcal{H}$. □

### *1.2.2 Two Important Particular Cases*

(A) *The case where the preframe operator $T$ is also injective, that is, $T$ is an isomorphism*

In this case we say that the frame $\{x_n\}_{n=1}^\infty$ is a *Riesz basis* for $\mathcal{H}$. Due to the uniqueness of the expansion $x = \sum_{n=1}^\infty c_n x_n$ for each $x \in \mathcal{H}$ we have that

$$c_n = \langle x, S^{-1}x_n\rangle \quad \text{for any } n \in \mathbb{N}\,.$$

Moreover, from the expansion $x_n = \sum_{m=1}^\infty \langle x_n, S^{-1}x_m\rangle\, x_m$ we deduce that the sequences $\{x_n\}_{n=1}^\infty$ and $\{S^{-1}x_n\}_{n=1}^\infty$ are biorthogonal, i.e., $\langle x_n, S^{-1}x_m\rangle = \delta_{n,m}$, the Kronecker delta.

Besides, the sequence $\{S^{-1}x_n\}_{n=1}^\infty$ is also a Riesz basis since its preframe operator

$$\widetilde{T}\big(\{c_n\}\big) = \sum_{n=1}^\infty c_n S^{-1}x_n = S^{-1}\Big(\sum_{n=1}^\infty c_n x_n\Big) = S^{-1}T\big(\{c_n\}\big)$$

is also an isomorphism. It is the so-called *dual (biorthogonal) Riesz basis* of $\{x_n\}_{n=1}^\infty$.

Since $T$ is an isomorphism, there exist two constants $0 < C \le D$ such that

$$C\|\{c_n\}_{n=1}^{\infty}\|^2 \le \Big\|\sum_{n=1}^{\infty} c_n\, x_n\Big\|^2 \le D\|\{c_n\}_{n=1}^{\infty}\|^2 \quad \text{for all } \{c_n\}_{n=1}^{\infty} \in \ell^2(\mathbb{N})\,, \tag{1.6}$$

and $\overline{\operatorname{span}}\{x_n\}_{n=1}^{\infty} = \mathcal{H}$.

The optimal bounds in (1.6) are called the *Riesz bounds*; they coincide with the optimal frame bounds for $\{x_n\}_{n=1}^{\infty}$ (see Proposition 1.9).

Notice that, in order to be $\{x_n\}_{n=1}^{\infty}$ a Riesz basis for $\mathcal{H}$ it is sufficient that inequalities (1.6) be satisfied for each finite scalar sequence $\{c_n\}$ and $\overline{\operatorname{span}}\{x_n\}_{n=1}^{\infty} = \mathcal{H}$.

Whether, the inequalities (1.6) just hold in the closed subspace $\overline{\operatorname{span}}\{x_n\}_{n=1}^{\infty}$ of $\mathcal{H}$, then $\{x_n\}_{n=1}^{\infty}$ is a *Riesz sequence* in $\mathcal{H}$, i.e., a Riesz basis for the closed span of $\{x_n\}_{n=1}^{\infty}$.

**Theorem 1.2 (Riesz Basis Theorem)** *Any Riesz basis $\{x_n\}_{n=1}^{\infty}$ for a separable Hilbert space $\mathcal{H}$ has a unique biorthogonal (dual) Riesz basis $\{y_n\}_{n=1}^{\infty}$, i.e., $\langle x_n, y_m\rangle_{\mathcal{H}} = \delta_{n,m}$, such that, for every $x \in \mathcal{H}$, the expansions*

$$x = \sum_{n=1}^{\infty} \langle x, y_n\rangle\, x_n = \sum_{n=1}^{\infty} \langle x, x_n\rangle\, y_n \quad \textit{in } \mathcal{H}$$

*hold.*

*(B) The case where the preframe operator T is a unitary operator*

Here $T^* = T^{-1}$ and hence $S = I$. As a consequence, $A = B = 1$ and the frame $\{x_n\}_{n=1}^{\infty}$ is an *orthonormal basis* for $\mathcal{H}$, i.e., $\langle x_n, x_m\rangle = \delta_{n,m}$, and for any $x \in \mathcal{H}$ the expansion

$$x = \sum_{n=1}^{\infty} \langle x, x_n\rangle\, x_n \quad \text{in } \mathcal{H}$$

holds.

Besides, we obtain the Parseval equality $\|x\|^2 = \sum_{n=1}^{\infty} |\langle x, x_n\rangle|^2$ for any $x \in \mathcal{H}$.

In fact, $T$ is a unitary operator if and only if $\{x_n\}_{n=1}^{\infty}$ is an *orthonormal basis* for $\mathcal{H}$. An orthonormal basis is a self-dual Riesz basis.

*At this stage, we have reached the starting point in frame theory: orthonormal bases are the simplest examples of frames.*

Besides, it is easy to prove that every separable Hilbert space $\mathcal{H}$ has an orthonormal basis. Indeed, since $\mathcal{H}$ is separable, there exists a sequence $\{x_n\}_{n=1}^{\infty}$ such that $\overline{\operatorname{span}}\{x_n\}_{n=1}^{\infty} = \mathcal{H}$. By extracting a subsequence if necessary, we can assume that, for each $N \in \mathbb{N}$, $x_{N+1} \notin \operatorname{span}\{x_k\}_{k=1}^N$. Finally, by using the Gram-Schmidt process to $\{x_n\}_{n=1}^\infty$ we get an orthonormal system $\{e_n\}_{n=1}^\infty$ in $\mathcal{H}$ such that

$$\overline{\operatorname{span}}\{e_n\}_{n=1}^\infty = \overline{\operatorname{span}}\{x_n\}_{n=1}^\infty = \mathcal{H}.$$

From orthonormal bases we can describe the whole set of frames in a Hilbert space (see Corollary 1.3 below). As an example, let $\{e_n\}_{n=1}^\infty$ be an orthonormal basis in a separable Hilbert space $\mathcal{H}$. Then, we have the first examples:

1. $\{e_1, e_1, e_2, e_3, e_4, \dots\}$ is a frame for $\mathcal{H}$ with frame bounds $A = 1$ and $B = 2$.
2. $\{e_1, e_1, e_2, e_2, e_3, e_3, \dots\}$ is a frame for $\mathcal{H}$ with frame bounds $A = B = 2$.
3. $\{e_1, \frac{e_2}{\sqrt{2}}, \frac{e_2}{\sqrt{2}}, \frac{e_3}{\sqrt{3}}, \frac{e_3}{\sqrt{3}}, \frac{e_3}{\sqrt{3}}, \dots\}$ is a frame for $\mathcal{H}$ with frame bounds $A = B = 1$.
4. $\{2e_1, e_2, e_3, e_4, \dots\}$ is a Riesz basis with dual basis $\{\frac{e_1}{2}, e_2, e_3, e_4, \dots\}$.
5. The sequence $\{\frac{e_n}{n}\}_{n=1}^\infty$ is a complete orthogonal system which is not a frame since it violates the left-hand inequality in frame definition: Taking $x = e_m$, $m \in \mathbb{N}$, we obtain

$$A\|e_m\|^2 = A \le \sum_{n=1}^\infty |\langle e_m, \frac{e_n}{n}\rangle|^2 = \frac{1}{m^2} \quad \text{for all } m \in \mathbb{N},$$

   that is, $A = 0$.

Next, the orthonormal basis theorem is stated:

**Theorem 1.3 (Orthonormal Basis Theorem)** *Let $\{x_n\}_{n=1}^\infty$ be an orthonormal system in a separable Hilbert space $\mathcal{H}$. The following conditions are equivalent:*

*(a) The system $\{x_n\}_{n=1}^\infty$ is a complete system for $\mathcal{H}$, that is, $\left(\{x_n\}_{n=1}^\infty\right)^\perp = \{0\}$.*
*(b) We have $\overline{\operatorname{span}}\{x_n\}_{n=1}^\infty = \mathcal{H}$, that is, the subspace in $\mathcal{H}$ spanned by $\{x_n\}_{n=1}^\infty$ is dense in $\mathcal{H}$.*
*(c) For each $x \in \mathcal{H}$ the expansion $x = \sum_{n=1}^\infty \langle x, x_n\rangle\, x_n$ in $\mathcal{H}$ holds.*
*(d) For each $x \in \mathcal{H}$ we have*

$$\|x\|^2 = \sum_{n=1}^\infty |\langle x, x_n\rangle|^2 \quad \textit{(Parseval's identity)}.$$

*(e) For each $x, y \in \mathcal{H}$ we have $\langle x, y\rangle = \sum_{n=1}^\infty \langle x, x_n\rangle\, \overline{\langle y, x_n\rangle}$.*

**Proposition 1.7** *In case the sequence $\{x_n\}_{n=1}^\infty$ is a Riesz basis for $\mathcal{H}$ then there exist an orthonormal basis $\{e_n\}_{n=1}^\infty$ for $\mathcal{H}$ and an isomorphism $R : \mathcal{H} \longrightarrow \mathcal{H}$ such that $Re_n = x_n$ for each $n \in \mathbb{N}$. The converse result also holds.*

***Proof*** Indeed, let $\{e_n\}_{n=1}^{\infty}$ be any orthonormal basis in $\mathcal{H}$. Its preframe operator $U : \ell^2(\mathbb{N}) \longrightarrow \mathcal{H}$ is a unitary operator and $U\boldsymbol{\delta}_n = e_n$ for each $n \in \mathbb{N}$, where $\{\boldsymbol{\delta}_n\}_{n=1}^{\infty}$ denotes the sequence of canonical basis vectors, i.e., $\boldsymbol{\delta}_n = \{\delta_{n,m}\}_{m=1}^{\infty}$ which forms an orthonormal basis for $\ell^2(\mathbb{N})$.

Since $\{x_n\}_{n=1}^{\infty}$ is a Riesz basis for $\mathcal{H}$ its preframe operator $T : \ell^2(\mathbb{N}) \longrightarrow \mathcal{H}$ satisfies $T\boldsymbol{\delta}_n = x_n$ for each $n \in \mathbb{N}$. Finally, for the isomorphism $R = TU^{-1}$ we have $Re_n = x_n$ for each $n \in \mathbb{N}$.

The converse result is a consequence of the proposition below. □

**Proposition 1.8** *Suppose that $R : \mathcal{H}' \longrightarrow \mathcal{H}$ is an isomorphism between Hilbert spaces, and let $x_n = Re_n$, $n \in \mathbb{N}$, be a sequence in $\mathcal{H}$ with $\{e_n\}_{n=1}^{\infty}$ an orthonormal basis in $\mathcal{H}'$. Then, the sequence $\{x_n\}_{n=1}^{\infty}$ is a Riesz basis for $\mathcal{H}$ and its dual basis $\{y_n\}_{n=1}^{\infty}$ is given by $y_n = (R^*)^{-1}e_n$, $n \in \mathbb{N}$.*

***Proof*** Let $U$ be the unitary operator such that $U\boldsymbol{\delta}_n = e_n$, $n \in \mathbb{N}$, where $\{\boldsymbol{\delta}_n\}_{n=1}^{\infty}$ denotes the sequence of canonical basis vectors in $\ell^2(\mathbb{N})$. Then, the operator $T = RU$ is an isomorphism and coincides with the preframe operator for $\{x_n\}_{n=1}^{\infty}$. Hence, it is a Riesz basis for $\mathcal{H}$.

Next, we search for its dual Riesz basis $\{y_n\}_{n=1}^{\infty}$. For any $y \in \mathcal{H}$ there exists a unique $x \in \mathcal{H}'$ such that

$$y = Rx = R\Big(\sum_{n=1}^{\infty}\langle x, e_n\rangle\, e_n\Big) = \sum_{n=1}^{\infty}\langle x, e_n\rangle\, x_n\,.$$

Then, due to the uniqueness of the coefficients following a Riesz basis expansion

$$\langle y, y_n\rangle = \langle x, e_n\rangle = \langle R^{-1}y, e_n\rangle = \langle y, (R^{-1})^*e_n\rangle\,, \quad \text{for all } y \in \mathcal{H} \text{ and } n \in \mathbb{N}\,,$$

that is, $y_n = (R^{-1})^*e_n = (R^*)^{-1}e_n$, $n \in \mathbb{N}$.

Observe that the dual Riesz basis of $\{y_n\}_{n=1}^{\infty}$ is, as expected, $\{x_n\}_{n=1}^{\infty}$. Indeed, following the above result the dual Riesz basis of $\{y_n\}_{n\in\mathbb{N}}$ is

$$\big(((R^*)^{-1})^*\big)^{-1}e_n = Re_n = x_n\,, \quad n \in \mathbb{N}\,,$$

since $(R^*)^{-1} = (R^{-1})^*$, $(R^*)^* = R$ and $(R^{-1})^{-1} = R$. □

**Proposition 1.9** *Let $\{x_n\}_{n=1}^{\infty}$ be a Riesz basis for $\mathcal{H}$ such that $Re_n = x_n$, $n \in \mathbb{N}$, for some orthonormal basis $\{e_n\}_{n=1}^{\infty}$ in $\mathcal{H}$ and an isomorphism $R : \mathcal{H} \longrightarrow \mathcal{H}$. Then, the optimal frame bounds coincide with the optimal Riesz bounds, and both are given by*

$$A_{opt} = \|R^{-1}\|^{-2} \quad \textit{and} \quad B_{opt} = \|R\|^2\,.$$

***Proof*** Indeed, concerning the frame bounds, for bound $B$ we have:

$$\sum_{n=1}^{\infty} |\langle x, x_n \rangle|^2 = \sum_{n=1}^{\infty} |\langle x, R\, e_n \rangle|^2 = \sum_{n=1}^{\infty} |\langle R^* x, e_n \rangle|^2 = \|R^* x\|^2 \leq \|R\|^2 \|x\|^2 ,$$

since $\|R^*\| = \|R\|$. For the bound $A$ we get

$$\|x\|^2 = \|(R^*)^{-1} R^* x\|^2 \leq \|(R^*)^{-1}\|^2 \|R^* x\|^2 = \|(R^{-1})^*\|^2 \sum_{n=1}^{\infty} |\langle x, x_n \rangle|^2 .$$

The optimality of these bounds is deduced from the operator norm definition, that is, $A_{opt} = \|R^{-1}\|^{-2}$ and $B_{opt} = \|R\|^2$.

Next, we search for the optimal Riesz bounds in inequalities (1.6). Indeed, concerning the bound $D$ we have

$$\Big\| \sum_{n=1}^{\infty} c_n\, x_n \Big\|^2 = \Big\| R\Big( \sum_{n=1}^{\infty} c_n\, e_n \Big) \Big\|^2 \leq \|R\|^2 \sum_{n=1}^{\infty} |c_n|^2 ,$$

that is, $D_{opt} = \|R\|^2$. Analogously, for bound $C$ we obtain

$$\sum_{n=1}^{\infty} |c_n|^2 = \Big\| \sum_{n=1}^{\infty} c_n\, e_n \Big\|^2 = \Big\| R^{-1} R\Big( \sum_{n=1}^{\infty} c_n\, e_n \Big) \Big\|^2 \leq \|R^{-1}\|^2 \Big\| \sum_{n=1}^{\infty} c_n\, x_n \Big\|^2 ,$$

that is $C_{opt} = \|R^{-1}\|^{-2}$, which concludes the proof. □

Finally, we provide a characterization for frames in a separable Hilbert space:

**Corollary 1.3** *Let* $\{e_n\}_{n=1}^{\infty}$ *be an arbitrary orthonormal basis in a separable Hilbert space* $\mathcal{H}$.

*(a) The frames for* $\mathcal{H}$ *are all the sequences* $\{V e_n\}_{n=1}^{\infty}$ *where* $V : \mathcal{H} \to \mathcal{H}$ *is a bounded and surjective operator.*
*(b) The Riesz bases for* $\mathcal{H}$ *are all the sequences* $\{V e_n\}_{n=1}^{\infty}$ *where* $V : \mathcal{H} \to \mathcal{H}$ *is a bounded and bijective operator (isomorphism).*
*(c) The orthonormal bases for* $\mathcal{H}$ *are all the sequences* $\{V e_n\}_{n=1}^{\infty}$ *where* $V : \mathcal{H} \to \mathcal{H}$ *is a unitary operator.*

A frame is said to be an *exact frame* if it ceases to be a frame when an arbitrary element is removed. Otherwise, the frame is said to be an *overcomplete frame*.

**Proposition 1.10** *A sequence* $\{x_n\}_{n=1}^{\infty}$ *in* $\mathcal{H}$ *is an exact frame if and only if* $\{x_n\}_{n=1}^{\infty}$ *is a Riesz basis for* $\mathcal{H}$.

***Proof*** For the sufficient condition, any Riesz basis is a frame; we deduce that it is an exact frame since any sequence $\{x_n\}_{n\neq m}$ is not complete. Indeed, considering the element $y_m$ of its dual basis we have $\langle y_m, x_n\rangle = 0$ for all $n \neq m$ and $y_m \neq 0$.

For the necessary condition, we prove that the preframe operator $T$ es injective. Assume that $0 = \sum_{n=1}^{\infty} a_n\, x_n$; for each $m \in \mathbb{N}$ we get

$$0 = \langle \sum_{n=1}^{\infty} a_n\, x_n, S^{-1}x_m\rangle = \sum_{n=1}^{\infty} a_n \langle x_n, S^{-1}x_m\rangle = \sum_{n=1}^{\infty} a_n\, \delta_{n,m} = a_m\,,$$

since for an exact frame $\{x_n\}_{n=1}^{\infty}$ the biorthogonal relationship $\langle x_n, S^{-1}x_m\rangle = \delta_{n,m}$ holds.

To prove this, since $\{x_n\}_{n\neq m}$ is not a frame for $\mathcal{H}$, by using Proposition 1.6 we deduce that $\langle x_m, S^{-1}x_m\rangle = 1$. Moreover, having in mind Eq. (1.5) we have

$$0 = 2\sum_{n\neq m} |\langle x_m, S^{-1}x_n\rangle|^2\,.$$

In other words, $\langle x_m, S^{-1}x_n\rangle = \delta_{n,m}$. □

If the frame $\{x_n\}_{n=1}^{\infty}$ is not a Riesz basis (overcomplete frame), then it possesses a infinite number of *dual frames*. A practical criterion to decide whether two sequences are dual frames in $\mathcal{H}$ is the following:

**Proposition 1.11** *Assume that $\{x_n\}_{n=1}^{\infty}$ and $\{y_n\}_{n=1}^{\infty}$ are two Bessel sequences in $\mathcal{H}$. The following conditions are equivalent:*

*(1) For each $x \in \mathcal{H}$ the expansion $x = \sum_{n=1}^{\infty} \langle x, y_n\rangle\, x_n$ holds.*

*(2) For each $x \in \mathcal{H}$ the expansion $x = \sum_{n=1}^{\infty} \langle x, x_n\rangle\, y_n$ holds.*

*(3) For each $x, y \in \mathcal{H}$ we have $\langle x, y\rangle = \sum_{n=1}^{\infty} \langle x, x_n\rangle\, \langle y_n, y\rangle$.*

*If any of the above conditions is satisfied, then the sequences $\{x_n\}_{n=1}^{\infty}$ and $\{y_n\}_{n=1}^{\infty}$ form a pair of* dual frames *for $\mathcal{H}$.*

***Proof*** Consider the preframe operators $T_x$ and $T_y$ of the Bessel sequences $\{x_n\}_{n=1}^{\infty}$ and $\{y_n\}_{n=1}^{\infty}$ respectively. Condition (1) means that $T_x T_y^* = I$, and condition (2) means $T_y T_x^* = I$.

Hence, condition (1) is equivalent to condition (2) since

$$\left(T_x T_y^*\right)^* = \left(T_y^*\right)^* T_x^* = T_y T_x^* = I^* = I\,.$$

Obviously, condition (2) implies condition (3). Finally, condition (3) implies condition (2). Indeed, for each $y \in \mathcal{H}$ we have

$$\Big\langle x - \sum_{n=1}^{\infty} \langle x, x_n \rangle\, y_n, y \Big\rangle = \langle x, y \rangle - \sum_{n=1}^{\infty} \langle x, x_n \rangle\, \langle y_n, y \rangle = 0\,,$$

that is, $x = \sum_{n=1}^{\infty} \langle x, x_n \rangle\, y_n$.

For proving that the sequence $\{x_n\}_{n=1}^{\infty}$ is a frame, we check the left-hand inequality in frame definition. Indeed, taking $x = y$ in (3) and applying the Cauchy-Schwarz inequality in $\ell^2(\mathbb{N})$ one gets

$$\|x\|^4 = \Big(\sum_{n=1}^{\infty} \langle x, x_n \rangle\, \langle y_n, x \rangle\Big)^2 \le \sum_{n=1}^{\infty} |\langle x, x_n \rangle|^2 \sum_{n=1}^{\infty} |\langle x, y_n \rangle|^2\,.$$

Since $\{y_n\}_{n=1}^{\infty}$ is a Bessel sequence, there exists a constant $B > 0$ such that $\sum_{n=1}^{\infty} |\langle x, y_n \rangle|^2 \le B\|x\|^2$ for all $x \in \mathcal{H}$. If $x \neq 0$ we finally obtain

$$\frac{1}{B}\|x\|^2 \le \sum_{n=1}^{\infty} |\langle x, x_n \rangle|^2\,.$$

Analogously we prove that the sequence $\{y_n\}_{n=1}^{\infty}$ is also a frame for $\mathcal{H}$. □

A final comment concerning the unconditional character of a frame expansion is in order: Since any index reordering $\{x_{\sigma(n)}\}_{n\in\mathbb{N}}$ of the elements of a frame $\{x_n\}_{n\in\mathbb{N}}$ is also a frame, any frame expansion is unconditionally convergent in $\mathcal{H}$.

### 1.2.3 The Case of Finite Frames

Next we include some results concerning finite frames which are used in this work.

**Proposition 1.12** *In a finite d-dimensional Hilbert space $\mathcal{H}_d$, a finite set of vectors $\{x_n\}_{n=1}^{M}$ in $\mathcal{H}_d$ is a frame for $\mathcal{H}_d$ if and only if $\{x_n\}_{n=1}^{M}$ is a spanning set for $\mathcal{H}_d$. Then, necessarily, $M \ge d$.*

***Proof*** Assume firstly that $\{x_n\}_{n=1}^{M}$ is a frame for $\mathcal{H}_d$ such that the subspace $V := \text{span}\{x_n\}_{n=1}^{M} \subsetneq \mathcal{H}_d$. Then, there exists a non-null $x \in V^{\perp}$ such that

$$A\|x\|^2 \le \sum_{n=1}^{M} |\langle x, x_n \rangle|^2 = 0\,,$$

which contradicts the frame definition. Therefore, $\operatorname{span}\{x_n\}_{n=1}^M = \mathcal{H}_d$ and necessarily, $M \geq d$.

Next, we prove that any spanning set $\{x_n\}_{n=1}^M$ for $\mathcal{H}_d$ is a frame for $\mathcal{H}_d$. By using the Cauchy-Schwarz inequality we get that it is a Bessel sequence. Namely,

$$\sum_{n=1}^{M} |\langle x, x_n\rangle|^2 \leq \Big(\sum_{n=1}^{M} \|x_n\|^2\Big)\|x\|^2 = B\|x\|^2 \quad \text{for all } x \in \mathcal{H}_d\,.$$

To prove the left-hand inequality in frame definition we consider the continuous function

$$\begin{aligned} \mathcal{H}_d &\longrightarrow \mathbb{R} \\ x &\longmapsto \textstyle\sum_{n=1}^{M} |\langle x, x_n\rangle|^2\,, \end{aligned}$$

which attains its absolute minimum in the compact set $\mathcal{S} := \big\{x \in \mathcal{H}_d \,:\, \|x\| = 1\big\}$. Thus, there exists $y \in \mathcal{S}$ such that

$$A := \min_{\|x\|=1} \Big\{\sum_{n=1}^{M} |\langle x, x_n\rangle|^2\Big\} = \sum_{n=1}^{M} |\langle y, x_n\rangle|^2\,.$$

Besides $A > 0$ since $A = 0$ would imply that $y \perp x_n$ for all $n = 1, 2, \ldots, M$, that is, $y = 0$, which contradicts the fact that $y \in \mathcal{S}$. □

Concerning the optimal frame bounds of a finite frame we have:

**Proposition 1.13** *Let $\{x_n\}_{n=1}^M$ be a frame for a d-dimensional Hilbert space $\mathcal{H}_d$, and let $S : \mathcal{H}_d \to \mathcal{H}_d$ denote its frame operator. Then, the optimal frame bounds of $\{x_n\}_{n=1}^M$ are*

$$A_{opt} = \lambda_1 \quad \text{and} \quad B_{opt} = \lambda_d\,,$$

*where $0 < \lambda_1 \leq \lambda_2 \leq \cdots \leq \lambda_d$ are the eigenvalues of the frame operator $S$.*

***Proof*** Since $S$ is a self-adjoint operator, we can consider an orthonormal basis of eigenvectors $\{e_i\}_{i=1}^d$, that is, $Se_j = \lambda_j e_j$ for each $j = 1, 2, \ldots, d$. For any $x \in \mathcal{H}_d$ we have $x = \sum_{j=1}^d \langle x, e_j\rangle\, e_j$, and hence $Sx = \sum_{j=1}^d \lambda_j \langle x, e_j\rangle\, e_j$. Then,

$$\sum_{n=1}^{M} |\langle x, x_n\rangle|^2 = \langle Sx, x\rangle = \sum_{j=1}^{d} \lambda_j |\langle x, e_j\rangle|^2\,,$$

that is

$$\lambda_1 \sum_{j=1}^{d} |\langle x, e_j\rangle|^2 = \lambda_1\|x\|^2 \leq \sum_{n=1}^{M} |\langle x, x_n\rangle|^2 \leq \lambda_d \sum_{j=1}^{d} |\langle x, e_j\rangle|^2 = \lambda_d\|x\|^2\,.$$

Now, taking $x = e_d$ we obtain $\sum_{n=1}^{M} |\langle e_d, x_n\rangle|^2 = \langle Se_d, e_d\rangle = \lambda_d$ which implies $B_{opt} = \lambda_d$.

Analogously, taking $x = e_1$ we obtain $\sum_{n=1}^{M} |\langle e_1, x_n\rangle|^2 = \langle Se_1, e_1\rangle = \lambda_1$ which implies that $A_{opt} = \lambda_1$. □

In the particular case of $\mathcal{H}_d = \mathbb{C}^d$, a set of vectors $\{\mathbf{x}_1, \mathbf{x}_2, \dots, \mathbf{x}_M\}$ form a frame for $\mathbb{C}^d$ if and only if the $d \times M$ matrix

$$\mathbf{M} := \begin{pmatrix} \vdots & \vdots & & \vdots \\ \mathbf{x}_1 & \mathbf{x}_2 & \dots & \mathbf{x}_M \\ \vdots & \vdots & & \vdots \end{pmatrix}$$

has rank $\mathbf{M} = d$. Besides, the frame operator is $S = \mathbf{M}\mathbf{M}^*$, where $\mathbf{M}^*$ denotes the conjugate transpose of the matrix $\mathbf{M}$. Indeed,

$$S\mathbf{x} = \sum_{i=1}^{M} \langle \mathbf{x}, \mathbf{x}_i\rangle\, \mathbf{x}_i = \mathbf{M} \begin{pmatrix} \langle \mathbf{x}, \mathbf{x}_1\rangle \\ \langle \mathbf{x}, \mathbf{x}_2\rangle \\ \vdots \\ \langle \mathbf{x}, \mathbf{x}_M\rangle \end{pmatrix} = \mathbf{M}\mathbf{M}^*\mathbf{x} \quad \text{for all } \mathbf{x} \in \mathbb{C}^d\,,$$

where $\langle \mathbf{x}, \mathbf{y}\rangle$ denotes the usual inner product in $\mathbb{C}^d$:

$$\langle \mathbf{x}, \mathbf{y}\rangle = \sum_{n=1}^{d} x_n\, \overline{y}_n \ \text{ for } \mathbf{x} = (x_1, x_2, \dots, x_d)^\top \text{ and } \mathbf{y} = (y_1, y_2, \dots, y_d)^\top \text{ in } \mathbb{C}^d\,.$$

The optimal frame bounds are $A_{opt} = \sigma_1^2$ and $B_{opt} = \sigma_N^2$, where $\sigma_1$ (respectively $\sigma_N$) denotes the lowest (respectively the largest) singular value of the matrix $\mathbf{M}$. Recall that the square root of the eigenvalues of $S = \mathbf{M}\mathbf{M}^*$ are the singular values of the matrix $\mathbf{M}$.

## 1.3 A Note Concerning the Needed Mathematical Background

As it was claimed in the preface, the intention of the author is to gather in a book form some sampling theory obtained, as the title suggests, by using frame theory. The book comprises topics of different level of difficulty: some parts can be used by graduate students (Chaps. 1–3), while others are more specific and they are addressed to researchers interested in sampling theory. As a consequence, the needed mathematical background depends on this. An advice for the potential reader: choose what you want according to your own sampling interests.

First of all, a background on frame theory is necessary. For the sake of completeness we have included a non-exhaustive introduction to frame theory. This material can be completed by using, for instance, the classical Christensen's book Ref. [11].

Other mathematical background is also necessary, but is not included in detail here. Let me cite, among others, a basic knowledge in Hilbert spaces, including orthogonal projections, orthonormal bases and the basic theory of bounded operators in a Hilbert space, which can be found in any classical book. As particular examples of Hilbert spaces a knowledge of $L^2$ and $\ell^2$ spaces is also required. We will use some basic results in functional analysis such as the Banach-Steinhaus theorem, the open mapping theorem, the closed graph theorem or the isomorphism theorem. Some basic results of complex analysis are also used. See, for instance, the references included in the book Refs. [12, 23, 25, 37, 40, 49]. The required measure theory can be found in the classical Ref. [39].

As mentioned in this introduction, sampling theory heavily depends on the Fourier transform concept. As a consequence, a background in classical harmonic analysis in $L^2(\mathbb{R}^d)$ or $L^2(\mathbb{R}^d)$ will be necessary mainly in Chaps. 2 and 3:

**The Fourier Transform in $L^2(\mathbb{R})$**
The *Fourier transform* is defined in $L^1(\mathbb{R}) \cap L^2(\mathbb{R})$ as

$$\hat{f}(\omega) := \frac{1}{\sqrt{2\pi}} \int_{-\infty}^{\infty} f(t)\, \mathrm{e}^{-i\omega t}\, dt\,, \quad \omega \in \mathbb{R}\,,$$

and extended to $L^2(\mathbb{R})$ in the usual way. Briefly, we remind the followed steps:

1. We consider a dense subset of $L^2(\mathbb{R})$ where the above defined Fourier transforms works well. For instance we can take the subspace $C^2_{00}(\mathbb{R}) \subset L^1(\mathbb{R}) \cap L^2(\mathbb{R})$ of functions having two continuous derivatives and vanishing outside a bounded interval of $\mathbb{R}$.
2. By using the properties of the Fourier transform in $L^1(\mathbb{R})$ it is easy to prove that

$$\langle f, g\rangle = \langle \widehat{f}, \widehat{g}\rangle \quad \text{for all } f, g \in C^2_{00}(\mathbb{R})$$

Indeed,

$$\begin{aligned}\langle f, g\rangle &= \int_{-\infty}^{\infty} f(t)\, \overline{g(t)} dt = \int_{-\infty}^{\infty} f(t) \Big(\frac{1}{\sqrt{2\pi}} \int_{-\infty}^{\infty} \overline{\widehat{g}(w)}\, \mathrm{e}^{-itw} dw\Big) dt \\ &= \int_{-\infty}^{\infty} \overline{\widehat{g}(w)} \Big(\frac{1}{\sqrt{2\pi}} \int_{-\infty}^{\infty} f(t)\, \mathrm{e}^{-itw} dt\Big) dw \\ &= \int_{-\infty}^{\infty} \widehat{f}(w)\, \overline{\widehat{g}(w)} dw = \langle \widehat{f}, \widehat{g}\rangle\,,\end{aligned}$$

where we have used the Fourier inversion formula in $L^1(\mathbb{R})$ for $g(t)$ and we have interchanged the integrals since the function $f(t)\overline{\widehat{g}(w)}e^{-itw}$ belongs to $L^1(\mathbb{R}^2)$.

3. Next, given $f \in L^2(\mathbb{R})$ there exists a sequence $\{f_n\}_{n=1}^{\infty}$ in $C_{00}^2(\mathbb{R})$ such that $f = \lim_{n\to\infty} f_n$ in $L^2(\mathbb{R})$-norm. Since $\|f_n - f_m\| = \|\widehat{f_n} - \widehat{f_m}\|$, we get a Cauchy sequence $\{\widehat{f_n}\}_{n=1}^{\infty}$ in $L^2(\mathbb{R})$. Thus, defining

$$\widehat{f} := \lim_{n\to\infty} \widehat{f_n}$$

we extend the Fourier transform to $L^2(\mathbb{R})$. The above limit is independent of the choice of the sequence $\{f_n\}_{n=1}^{\infty}$ in $C_{00}^2(\mathbb{R})$ such that $f = \lim_{n\to\infty} f_n$.

4. Furthermore, for each $f, g \in L^2(\mathbb{R})$ we have

$$\langle f, g\rangle = \lim_{n\to\infty} \langle f_n, g_n\rangle = \lim_{n\to\infty} \langle \widehat{f_n}, \widehat{g_n}\rangle = \langle \widehat{f}, \widehat{g}\rangle .$$

5. The Fourier transform in $L^2(\mathbb{R})$ can be understood as

$$\widehat{f}(\omega) = \frac{1}{\sqrt{2\pi}} \int_{-\infty}^{\infty} f(t)\, e^{-i\omega t}\, dt = \lim_{N\to\infty} \frac{1}{\sqrt{2\pi}} \int_{-N}^{N} f(t)\, e^{-i\omega t}\, dt$$

in $L^2(\mathbb{R})$-sense.

Summarizing, the *Fourier transform* $\mathcal{F} : f \mapsto \widehat{f}$ defines a unitary operator from $L^2(\mathbb{R})$ onto $L^2(\mathbb{R})$, i.e., $\mathcal{F}$ is a bijective operator satisfying the *Plancherel-Parseval formula,*

$$\langle \mathcal{F}(f), \mathcal{F}(g)\rangle = \langle f, g\rangle \quad \text{for all } f, g \in L^2(\mathbb{R}) . \tag{1.7}$$

In particular, we have the *Parseval identity*

$$\|\mathcal{F}(f)\| = \|f\| \quad \text{for each } f \in L^2(\mathbb{R}) . \tag{1.8}$$

In this case, the inverse Fourier transform $\mathcal{F}^{-1} : \widehat{f} \mapsto f$ from $L^2(\mathbb{R})$ onto $L^2(\mathbb{R})$ reads

$$f(t) = \frac{1}{\sqrt{2\pi}} \int_{-\infty}^{\infty} \widehat{f}(\omega)\, e^{i\omega t}\, d\omega = \lim_{N\to\infty} \frac{1}{\sqrt{2\pi}} \int_{-N}^{N} \widehat{f}(\omega)\, e^{i\omega t}\, d\omega \quad \text{in } L^2(\mathbb{R})\text{-sense} .$$

Also, analogously, in the multidimensional case, the Fourier transform is defined in $L^1(\mathbb{R}^d) \cap L^2(\mathbb{R}^d)$ as

$$\widehat{f}(\omega) := \frac{1}{(2\pi)^{d/2}} \int_{\mathbb{R}^d} f(t)\, e^{-i\omega\cdot t}\, dt\,, \quad \omega \in \mathbb{R}^d\,, \tag{1.9}$$

and extended to $L^2(\mathbb{R}^d)$ in the usual way.

Here $t = (t_1, t_2, \dots, t_d)$ and $\omega = (\omega_1, \omega_2, \dots, \omega_d)$ belong to $\mathbb{R}^d$ and, as usual, $\omega \cdot t = \sum_{j=1}^{d} \omega_j t_j$.

Thus, the Fourier transform $\mathcal{F}$ defines a unitary operator from $L^2(\mathbb{R}^d)$ onto $L^2(\mathbb{R}^d)$. The inverse Fourier transform reads

$$f(t) = \frac{1}{(2\pi)^{d/2}} \int_{\mathbb{R}^d} \widehat{f}(\omega)\, \mathrm{e}^{i\omega\cdot t}\, d\omega\,, \quad t \in \mathbb{R}^d\,. \tag{1.10}$$

Concerning the Fourier transform in $L^2(\mathbb{R})$ just to say that other equivalent definitions are used in the literature. For instance:

$$\widehat{f}(\omega) = \int_{-\infty}^{\infty} f(t)\, \mathrm{e}^{-i\omega t}\, dt \quad \text{or} \quad \widehat{f}(\omega) = \int_{-\infty}^{\infty} f(t)\, \mathrm{e}^{-2\pi i\omega t}\, dt\,,$$

with associated inverse Fourier transforms:

$$f(t) = \frac{1}{2\pi} \int_{-\infty}^{\infty} \widehat{f}(\omega)\, \mathrm{e}^{i\omega t}\, d\omega \quad \text{or} \quad f(t) = \int_{-\infty}^{\infty} \widehat{f}(\omega)\, \mathrm{e}^{2\pi i\omega t}\, d\omega\,.$$

In Chap. 5, while dealing with generalized sampling associated with countable discrete groups, some basic concepts in harmonic analysis for these particular groups will be needed. A summary, accompanied by references, has been included there.

In any case, the reader can be find the harmonic analysis needed in this book in Refs. [3, 11, 13, 18, 19, 22, 27, 37].

In Chaps. 2 and 4 some touches of distributional sampling are included. For an introduction to the distribution theory including the distributional Fourier transform, we refer the interested reader to Ref. [22].

In Chap. 4 an extended Kramer sampling theory is developed. Here, the sampling examples taken from different areas in mathematics shows the transversality of sampling theory with other mathematical fields. Thus, mathematical concepts appear which are not standard in the sampling literature such as orthogonal polynomials, moment problems, Bessel functions, difference and differential problems, entire functions, semi-inner products in Banach spaces or distributions theory. In each particular case, suitable references will be provided for the interested and unfamiliar reader along with a short ad hoc explanation.

In Chaps. 2 and 4 is included all the necessary theory on reproducing kernel Hilbert spaces (RKHSs) needed throughout the book.

In Chap. 6 finite frames are used. Since in finite dimension a frame is nothing but a spanning set of vectors, finite frame theory meets matrix analysis and numerical linear algebra. See the references [29, 33] included in the book.

Surely, Chap. 7 is the more specialized chapter in the book. It uses a wide range of concepts used very often in time-frequency analysis. For our sampling purposes, the most important is to exhibit the parallelism with classical shift-invariant spaces by using the generalized sampling results included in Chap. 5. We have tried to make it as self-contained as possible. However, a basic knowledge on Hilbert-Schmidt operators can help the unfamiliar reader to follow it. Most of the necessary background to follow its content is provided in Gröchenig's book [24], and in Ref. [42].

Finally to say that the reader should be forewarned of the omission of important sampling topics such as irregular sampling, local average sampling, compressed sensing, dynamical sampling, stochastic sampling or the relationship between Shannon sampling with other mathematical areas, error theory in sampling, among others. In any case, the interested reader might consult some seminal or recent works; we refer to the works [1, 2, 4, 8, 9, 16, 38, 44, 45] and references therein.

## References

1. Aldroubi, A., Gröchenig, K.: Nonuniform sampling and reconstruction in shift-invariant spaces. SIAM Rev. **43**, 585–620 (2001)
2. Aldroubi, A., Cabrelli, C., Molter, U., Tang, S.: Dynamical sampling. Appl. Comput. Harmon. Anal. **42**(3), 378–401 (2017)
3. Bachman, G., Narici, L., Beckenstein, E.: Fourier and Wavelet Analysis. Springer, New York (2000)
4. Benedetto, J.J.: Irregular frames and sampling. In: Chui, C.K. (ed.) Wavelets-A Tutorial in Theory and Applications, pp. 445–507. Academic Press, San Diego, CA (1992)
5. Benedetto, J.J., Ferreira, P.J.S.G. (eds.): Modern Sampling Theory: Mathematics and Applications. Birkhäuser, Boston (2001)
6. Borel, E.: Sur l'interpolation. C. R. Acad. Sci. Paris **124**, 673–676 (1897)
7. Butzer, P.L., Stens, R.L.: Sampling theory for not necessarily band–limited functions: A historical overview. SIAM Rev. **34**, 40–53 (1992)
8. Butzer, P.L., et al.: The sampling theorem, Poisson's summation formula, general Parseval formula, reproducing kernel formula and the Paley-Wiener theorem for bandlimited signals – their interconnections. Appl. Anal. **90**(3–4), 431–5461 (2011)
9. Casey, S., et al. (eds.): Sampling: Theory and Applications. Birkhäuser, Boston (2020)
10. Cauchy, A.L.: Méthode pour développer des fonctions d'une ou plusieurs variables en séries composées de fonctions de même espèce. Oeuvres de Cauchy, série II, tome VII, pp. 366–392 (1827)
11. Christensen, O.: An Introduction To Frames and Riesz Bases, 2nd edn. Birkhäuser, Boston (2016)
12. Conway, J.B.: A Course in Operator Theory. AMS, Providence RI (2000)
13. Deitmar, A., Echterhoff, S.: Principles of Harmonic Analysis, 2nd edn. Universitext. Springer, Berlin (2014)
14. Duffin, R.J., Shaeffer, A.C.: A class of nonharmonic Fourier series. Trans. Amer. Math. Soc. **72**, 341–366 (1952)
15. Eldar, Y.: Sampling Theory: Beyond Bandlimited Systems. Cambridge University Press, Cambridge UK (2015)
16. Feichtinger, H.G., Gröchenig, K.: Theory and practice of irregular sampling. In: Benedetto, J., Frazier, M. (eds.) Wavelets: Mathematics and Applications, pp. 305–363. CRC Press, Boca Raton, FL (1994)
17. Ferrar, W.L.: On the cardinal function of interpolation theory. Proc. Roy. Soc. Edinburgh **46**, 323–333 (1926)
18. Folland, G.B.: A Course in Abstract Harmonic Analysis, CRC Press, Boca Raton, FL (1995)

19. Führ, H.: Abstract Harmonic Analysis of Continuous Wavelet Transform. Springer, Berlin-Heidelberg (2005)
20. García, A.G.: Orthogonal sampling formulas: a unified approach. SIAM Rev. **42**, 499–512 (2000)
21. García, A.G.: A brief walk through sampling theory. In: Hawkes, P.W. (ed.) Advances in Imaging and Electron Physics, vol. 124, pp. 63–137. Academic Press, San Diego (2002)
22. Gasquet, C., Witomski, P.: Fourier Analysis and Applications. Springer, New York (1998)
23. Goffman, C., Pedrick, G.: First Course in Functional Analysis. American Mathematical Society, Providence RI (2017)
24. Gröchenig, K.: Foundations of Time-Frequency Analysis. Birkhäuser, Basel (2001)
25. Heuser, H.: Functional Analysis. John Wiley, New York (1982)
26. Higgins, J.R.: Five short stories about cardinal series. Bull. Amer. Math. Soc. **12**, 45–89 (1985)
27. Higgins, J.R.: Sampling Theory in Fourier and Signal Analysis: Foundations. Oxford University Press, Oxford (1996)
28. Higgins, J.R., Stens, R.L. (eds.): Sampling Theory in Fourier and Signal Analysis: Advanced Topics. Oxford University Press, Oxford (1999)
29. Horn, R.A., Johnson, C.R.: Matrix Analysis. Cambridge University Press, Cambridge UK (1999)
30. Jerri, A.J.: The Shannon sampling theorem–its various extensions and applications: A tutorial review. Proc. IEEE **65**, 1565–1598 (1977)
31. Kotel'nikov, V.: On the carrying capacity of the "ether" and wire in telecommunications, Material for the first All–Union Conference on Questions of Communications. Izd. Red. Upr. Svyazy RKKA (1933)
32. Lacaze, B.: La formule d'echantillonnage et A. L. Cauchy. Traitement du Signal **15**(4), 289–295 (1998)
33. Laub, A.J.: Matrix Analysis. SIAM, Philadelphia (2005)
34. Marks II, R.J.: Introduction to Shannon Sampling and Interpolation Theory. Springer, New York (1991)
35. Marks II, R.J. (ed.): Advanced Topics in Shannon Sampling and Interpolation Theory. Springer, New York (1992)
36. Ogura, K.: On a certain transcendental integral function in the theory of interpolation. Tôhoku Math. J. **17**, 64–72 (1920)
37. Partington, J.R.: Interpolation, Identification and Sampling. Clarendon Press, Oxford (1997)
38. Pfander, G.E. (ed.): Sampling Theory, a Renaissance. Birkhäuser, Boston (2015)
39. Royden, H.L.: Real Analysis, 3rd edn. Macmillan Publishing Company, New York (1988)
40. Saitoh, S., Sawano, Y.: Theory of Reproducing Kernels and Applications. Developments in Mathematics, vol. 44. Springer, Singapore (2016)
41. Shannon, C.E.: Communication in the presence of noise. Proc. IRE **137**, 10–21 (1949)
42. Skrettingland, E.: Quantum harmonic analysis on lattices and Gabor multipliers. J. Fourier Anal. Appl. **26**, 48 (2020)
43. Someya, I.: Waveform Transmission. Shukyo, Tokyo (1949)
44. Sun, W.: Reconstruction of band-limited signals from local averages. IEEE Trans. Inform. Theory **48**(11), 2955–2963 (2002)
45. Sun, Q.: Reconstruction of band-limited signals from local averages. SIAM J. Math. Anal. **38**(5), 1389–1422 (2007)
46. Unser, M.: Sampling 50 Years After Shannon. Proc. IEEE **88**(4), 569–587 (2000)
47. Whittaker, E.T.: On the functions which are represented by the expansion on the interpolation theory. Proc. Roy. Soc. Edinburgh Sect. A **35**, 181–194 (1915)
48. Whittaker, J.M.: Interpolatory Function Theory. Cambridge University Press, Cambridge, UK (1935)
49. Young, R.: An Introduction to Nonharmonic Fourier Series. Academic Press, San Diego (2001)
50. Zayed, A.I.: Advances in Shannon's Sampling Theory. CRC Press, Boca Raton (1993)

# Chapter 2
# Basic Sampling Theory

As announced in Chap. 1, reproducing kernel Hilbert spaces (RKHSs) are suitable for sampling purposes. In this chapter we briefly study these spaces where, under an appropriate hypothesis, a sampling theorem can be established (see Theorem 2.1 below). In Chap. 4 we will add some needed complementary theory for this important class of Hilbert spaces.

Concerning sampling theory, the most important example of RKHSs is given by the $L^2$-subspaces of bandlimited functions, that is, the so-called Paley-Wiener spaces. This chapter introduces the Paley-Wiener spaces together with their most important properties, most of which can be rephrased in terms of signal processing theory. Thus, some familiar concepts for the engineering community such as the Shannon sampling formula, the equality between the energy of a bandlimited signal and the energy of the sampled signal, the oversampling/undersampling setting or the oversampling technique, will be shown to have a mathematical counterpart by using concepts from harmonic analysis such as orthonormal bases expansions and its associated Parseval identities, the Poisson summation formula or a regularization process. Not only regular samples of a bandlimited function, as in the classical Shannon sampling theorem, are considered: nonuniform sampling or sampling using samples of a related function are also taken into account.

As we will see throughout this chapter, a sampling expansion in a Paley-Wiener space is nothing but an orthonormal (or a Riesz) expansion in a Hilbert space. The additional pointwise convergence is obtained because Paley-Wiener spaces are RKHSs. This chapter follows, in general terms, the scheme proposed in Ref. [14].



A. García García, *The Use of Frames in Sampling Theory*,
RSME Springer Series 14, https://doi.org/10.1007/978-3-031-63242-6_2

## 2.1 RKHSs Are Suitable Spaces for Sampling

Firstly, we introduce RKHSs and their basic properties:

### 2.1.1 *The Basic RKHSs Theory*

Let $\mathcal{H}$ denote a Hilbert space of functions $f : \Omega \longrightarrow \mathbb{C}$ defined on the same domain $\Omega$ (in particular, a subset of $\mathbb{R}$ or $\mathbb{C}$), endowed with the inner product $\langle \cdot\,, \cdot \rangle_{\mathcal{H}}$ (for simplicity, we denote it as $\langle \cdot\,, \cdot \rangle$). For each $t \in \Omega$, the map

$$\begin{aligned} E_t : \mathcal{H} &\longrightarrow \mathbb{C} \\ f &\longmapsto E_t(f) := f(t)\,, \end{aligned}$$

denotes the *point evaluation functional* at $t \in \Omega$.

**Definition 2.1** A Hilbert space $\mathcal{H}$ of functions defined on $\Omega$ is a reproducing kernel Hilbert space (RKHS) if each point evaluation functional is bounded on $\mathcal{H}$. In other words, for each $t \in \Omega$ there exists a positive constant $M_t$ such that $|f(t)| \le M_t\, \|f\|$ for all $f \in \mathcal{H}$.

By using the *Riesz representation theorem*, for each $t \in \Omega$, there exists a unique element $k_t \in \mathcal{H}$ such that $f(t) = \langle f, k_t \rangle$ for all $f \in \mathcal{H}$. This leads to the reproducing kernel in $\mathcal{H}$

**Definition 2.2** The function $k : \Omega \times \Omega \longrightarrow \mathbb{C}$ defined by

$$k(t, s) := \langle k_s, k_t \rangle = k_s(t)\,, \quad (t, s) \in \Omega \times \Omega\,,$$

is the reproducing kernel of $\mathcal{H}$.

From the definition of reproducing kernel $k$ we have that:

1. For each fixed $s \in \Omega$, the function $k(\cdot, s) = k_s(\cdot)$ belongs to $\mathcal{H}$.
2. The *reproducing property* in $\mathcal{H}$ holds:

$$f(s) = \big\langle f, k(\cdot, s)\big\rangle = \big\langle f, k_s(\cdot)\big\rangle \quad \text{for all} \quad f \in \mathcal{H} \text{ and } s \in \Omega\,. \tag{2.1}$$

**Proposition 2.1** *Let $\mathcal{H}$ be a Hilbert space of functions defined on $\Omega$ such that there exists a function $k : \Omega \times \Omega \longrightarrow \mathbb{C}$ satisfying the properties 1–2 above. Then, the space $\mathcal{H}$ is a RKHS.*

***Proof*** Applying the Cauchy-Schwarz inequality in the reproducing formula (2.1) one proves that each point evaluation functional $E_t$ at $t \in \Omega$ is bounded. □

**Proposition 2.2** *The reproducing kernel $k$ in a RKHS is unique.*

***Proof*** Assume that $k'(t,s) = k'_s(t)$ is another reproducing kernel. For $t, s \in \Omega$ we have

$$k'_s(t) = \langle k'_s, k_t \rangle = \overline{\langle k_t, k'_s \rangle} = \overline{k_t(s)} = \langle k_s, k_t \rangle = k_s(t)\,,$$

from which we deduce that $k = k'$. □

From an orthonormal basis $\{e_n(t)\}_{n=1}^{\infty}$ for $\mathcal{H}$ we obtain an expression for $k$:

**Proposition 2.3** *Let $\{e_n(t)\}_{n=1}^{\infty}$ be an orthonormal basis for a RKHS $\mathcal{H}$. The reproducing kernel k in $\mathcal{H}$ can be expressed as:*

$$k(t,s) = \sum_{n=1}^{\infty} e_n(t)\,\overline{e_n(s)}\,, \quad t, s \in \Omega\,.$$

***Proof*** Expanding $k_t$ and $k_s$ following the orthonormal basis $\{e_n(t)\}_{n=1}^{\infty}$ we obtain

$$k_t = \sum_{n=1}^{\infty} \langle k_t, e_n \rangle\, e_n \quad \text{and} \quad k_s = \sum_{n=1}^{\infty} \langle k_s, e_n \rangle\, e_n\,,$$

from which we obtain

$$k(t,s) = \langle k_s, k_t \rangle = \sum_{n=1}^{\infty} \langle k_s, e_n \rangle\, \overline{\langle k_t, e_n \rangle} = \sum_{n=1}^{\infty} \overline{e_n(s)}\, e_n(t)\,, \quad t, s \in \Omega\,.$$

□

**Proposition 2.4** *Let $\{e_n(t)\}_{n=1}^{\infty}$ be a Riesz basis for a RKHS $\mathcal{H}$ with dual basis $\{\widetilde{e}_n(t)\}_{n=1}^{\infty}$. The following expressions for the reproducing kernel k hold:*

$$k(t,s) = \sum_{n=1}^{\infty} e_n(t)\,\overline{\widetilde{e}_n(s)} = \sum_{n=1}^{\infty} \widetilde{e}_n(t)\,\overline{e_n(s)}\,, \quad t, s \in \Omega\,.$$

***Proof*** Expanding $k_s$ with respect to the dual basis $\{\widetilde{e}_n\}_{n=1}^{\infty}$ and $k_t$ with respect to $\{e_n\}_{n=1}^{\infty}$ we obtain

$$k_s = \sum_{n=1}^{\infty} \langle k_s, e_n \rangle\, \widetilde{e}_n = \sum_{n=1}^{\infty} \overline{e_n(s)}\, \widetilde{e}_n\,; \quad k_t = \sum_{m=1}^{\infty} \langle k_t, \widetilde{e}_m \rangle\, e_m = \sum_{m=1}^{\infty} \overline{\widetilde{e}_m(t)}\, e_m\,.$$

As a consequence,

$$k(t,s) = \langle k_s, k_t \rangle = \sum_{n,m} \overline{e_n(s)}\, \widetilde{e}_m(t) \langle \widetilde{e}_n, e_m \rangle = \sum_{n=1}^{\infty} \overline{e_n(s)}\, \widetilde{e}_n(t)\,, \quad t, s \in \Omega\,,$$

since $\langle \widetilde{e}_n, e_m \rangle = \delta_{n,m}$. Interchanging the bases, the other expression follows. □

The next property relates convergence in $\mathcal{H}$-norm with pointwise convergence in the set $\Omega$:

**Proposition 2.5** *In a RKHS $\mathcal{H}$ the convergence in $\mathcal{H}$-norm implies pointwise convergence in $\Omega$, which is uniform in those subsets of $\Omega$ where the function $t \mapsto k(t,t)$ is bounded.*

***Proof*** Let $\{f_n\}_{n=1}^{\infty}$ be a sequence in $\mathcal{H}$ such that $f_n \to f$ in $\mathcal{H}$ as $n \to \infty$. The reproducing property (2.1) applied to $f_n - f \in \mathcal{H}$ gives

$$f_n(t) - f(t) = \big\langle f_n - f, k(\cdot, t)\big\rangle, \quad t \in \Omega,$$

and by using the Cauchy-Schwarz inequality we get

$$|f_n(t) - f(t)| \le \|f_n - f\| \, \|k(\cdot, t)\| = \sqrt{k(t,t)}\, \|f_n - f\| \to 0, \text{ when } n \to \infty.$$

Moreover, the pointwise convergence will be uniform on subsets of $\Omega$ where the function $t \mapsto \sqrt{k(t,t)} = \|k_t\|$ is bounded. □

Assume now that our RKHS $\mathcal{H}$ is a closed subspace of a larger Hilbert space $\widetilde{\mathcal{H}}$ (not necessarily with a reproducing kernel). In this case we have the following result:

**Proposition 2.6** *If a RKHS $\mathcal{H}$ is a closed subspace of a Hilbert space $\widetilde{\mathcal{H}}$ then, for each function $f \in \widetilde{\mathcal{H}}$, we have*

$$\langle f, k(\cdot, s)\rangle = P_{\mathcal{H}} f(s), \quad s \in \Omega,$$

*where $P_{\mathcal{H}}$ denotes the orthogonal projection onto $\mathcal{H}$.*

***Proof*** Every $f \in \widetilde{\mathcal{H}}$ can be decomposed as $f = f_1 + f_2$ with $f_1 \in \mathcal{H}$ and $f_2 \in \mathcal{H}^{\perp}$. As a consequence,

$$\langle f, k(\cdot, s)\rangle = \langle f_1 + f_2, k(\cdot, s)\rangle = \langle f_1, k(\cdot, s)\rangle + \langle f_2, k(\cdot, s)\rangle = f_1(s) = P_{\mathcal{H}} f(s),$$

since $f_2 \perp k(\cdot, s)$ and $f_1 \in \mathcal{H}$. □

A straightforward sampling result, involving orthonormal bases, is the following:

**Theorem 2.1 (Sampling Theorem in a RKHS)** *Let $\mathcal{H}$ be a RKHS of functions defined on a subset $\Omega$ with reproducing kernel k. Assume that there exists a sequence $\{t_n\}_{n=1}^{\infty} \subset \Omega$ such that $\big\{k(\cdot, t_n)\big\}_{n=1}^{\infty}$ is an orthogonal basis for $\mathcal{H}$. Then, any $f \in \mathcal{H}$ can be expanded as*

$$f(t) = \sum_{n=1}^{\infty} f(t_n) \frac{k(t, t_n)}{k(t_n, t_n)}, \quad t \in \Omega, \tag{2.2}$$

*with pointwise convergence which is absolute and uniform on subsets of $\Omega$ where the function $t \mapsto k(t,t)$ is bounded.*

***Proof*** This result follows from the expansion of $f \in \mathcal{H}$ in the orthonormal basis $\{k(\cdot,t_n)/\sqrt{k(t_n,t_n)}\}_{n=1}^{\infty}$. Indeed, for each $f \in \mathcal{H}$ we obtain

$$\begin{aligned} f &= \sum_{n=1}^{\infty} \Big\langle f, \frac{k(\cdot,t_n)}{\sqrt{k(t_n,t_n)}} \Big\rangle_{\mathcal{H}} \frac{k(\cdot,t_n)}{\sqrt{k(t_n,t_n)}} \\ &= \sum_{n=1}^{\infty} f(t_n) \frac{k(\cdot,t_n)}{k(t_n,t_n)} \quad \text{in } \mathcal{H}. \end{aligned}$$

Now, the convergence in norm in a RKHS $\mathcal{H}$ implies pointwise convergence in $\Omega$ which is uniform on subsets of $\Omega$ where the function $t \mapsto k(t,t)$ is bounded. Moreover, since an orthonormal basis is an unconditional basis, the above sampling series is pointwise unconditionally convergent for each $t \in \Omega$ and hence absolutely convergent. □

The standard Hilbert space $\ell^2(\mathbb{N})$ is a RKHS whose reproducing kernel $k$ is the Kronecker delta, i.e., $k(m,n) = \delta_{m,n}$, $m,n \in \mathbb{N}$. In this case, for any $\{x(m)\}_{m=1}^{\infty}$ in $\ell^2(\mathbb{N})$ formula (2.2) trivially reads:

$$x(m) = \sum_{n=1}^{\infty} x(n)\,\delta_{m,n}\,, \quad m \in \mathbb{N}\,.$$

### 2.1.2 *Some Examples in the Finite Dimensional Setting*

Since in finite dimension every linear functional is bounded, then any finite dimensional Hilbert space of functions defined on $\Omega$ is a RKHS. Next we give two interesting examples in the finite dimensional setting:

**Example 2.1 (Trigonometric Polynomials)** Consider the space $\mathcal{H}_N$ of $2\pi$-periodic trigonometric polynomials of degree $\le N$. The space $\mathcal{H}_N$ is a closed subspace of $L^2[-\pi,\pi]$ endowed with its usual inner product and $\dim \mathcal{H}_N = 2N+1$. An orthonormal basis for $\mathcal{H}_N$ is given by the set of complex exponentials $\{e^{ikt}/\sqrt{2\pi}\}_{k=-N}^{N}$. Therefore, the reproducing kernel for $\mathcal{H}_N$ is

$$k_N(t,s) = \frac{1}{2\pi} \sum_{k=-N}^{N} e^{ik(t-s)} = \frac{1}{2\pi} D_N(t-s)\,,$$

where $D_N$ denotes the *Nth Dirichlet kernel* (see Ref. [27, p. 9])

$$D_N(t) := \sum_{k=-N}^{N} e^{ikt} = \begin{cases} \frac{\sin(N+\frac{1}{2})t}{\sin\frac{t}{2}} & \text{if } t \in \mathbb{R}\setminus 2\pi\mathbb{Z}\,, \\ 2N+1 & \text{if } t \in 2\pi\mathbb{Z}\,. \end{cases}$$

At the points $s_n = \frac{2\pi n}{2N+1} \in [-\pi, \pi]$, $-N \le n \le N$, the sequence $\{k_N(\cdot, s_n)\}_{n=-N}^{N}$ is an orthogonal basis for $\mathcal{H}_N$ since

$$\begin{aligned}\langle k_N(\cdot, s_n), k_N(\cdot, s_m)\rangle_{L^2[-\pi,\pi]} &= k_N(s_m, s_n) \\ &= \frac{1}{2\pi}\frac{\sin\pi(m-n)}{\sin\frac{\pi(m-n)}{2N+1}} = \frac{2N+1}{2\pi}\,\delta_{m,n}\,.\end{aligned}$$

A direct application of the sampling formula (2.2) gives, for every trigonometric polynomial $p(t) = \sum_{k=-N}^{N} c_k e^{ikt}$ in $\mathcal{H}_N$, the trigonometric interpolation formula:

$$p(t) = \frac{1}{2N+1}\sum_{n=-N}^{N} p\Big(\frac{2\pi n}{2N+1}\Big)\frac{\sin\big(\frac{2N+1}{2}\big)\big(t-\frac{2\pi n}{2N+1}\big)}{\sin\frac{1}{2}\big(t-\frac{2\pi n}{2N+1}\big)}\,, \quad t \in [-\pi, \pi)\,.$$

This trigonometric interpolation formula goes back to Cauchy, see Ref. [8].

**Example 2.2 (Orthogonal Polynomials)** Another important class of examples is given by finite families of orthogonal polynomials on an interval of the real line. Consider, as an example, the particular case of the *Legendre polynomials* $\{P_n\}_{n=0}^{\infty}$ defined, for instance, by means of the *Rodrigues formula*

$$P_n(t) = \frac{1}{2^n n!}\frac{d^n}{dt^n}\big[(t^2-1)^n\big]\,, \quad n = 0, 1, \dots$$

They form an orthogonal basis for $L^2[-1, 1]$ with $\|P_n\|^2 = (n+\frac{1}{2})^{-1}$.

The subspace $\mathcal{H}_N$ of $L^2[-1, 1]$ spanned by $\{P_0, P_1, \dots, P_N\}$ has $\dim\mathcal{H}_N = N+1$.

The *Christoffel-Darboux formula* for Legendre polynomials gives its reproducing kernel

$$\begin{aligned}k_N(t, s) &= \sum_{n=0}^{N}(n+\frac{1}{2})\,P_n(t)P_n(s) \\ &= \frac{2N+1}{2}\,\frac{P_{N+1}(t)P_N(s) - P_N(t)P_{N+1}(s)}{t-s}\,.\end{aligned}$$

Note that

$$k_N(t,t) = \frac{2N+1}{2}\left[P'_{N+1}(t)P_N(t) - P'_N(t)P_{N+1}(t)\right].$$

We seek points $\{s_n\}_{n=0}^N$ in $[-1,1]$ such that $k_N(s_m,s_n) = 0$ for $m \neq n$, i.e.,

$$\frac{P_{N+1}(s_m)}{P_N(s_m)} = \frac{P_{N+1}(s_n)}{P_N(s_n)}.$$

In particular we can take for $\{s_n\}_{n=0}^N$ the $N+1$ simple roots of $P_{N+1}$ in $(-1,1)$. Thus, for every $f(t) = \sum_{k=0}^{N} c_k \sqrt{k+\frac{1}{2}}\, P_k(t)$ in $\mathcal{H}_N$ we have the finite sampling formula

$$f(t) = \sum_{n=0}^{N} f(s_n) \frac{P_{N+1}(t)}{(t-s_n)P'_{N+1}(s_n)}, \quad t \in \mathbb{R}.$$

This formula is nothing but the Lagrange interpolation formula for the samples $\{f(s_n)\}_{n=0}^N$. In general, one can take as sampling points $\{s_n\}_{n=0}^N$ the $N+1$ simple roots of the polynomial $P_{N+1}(t) - cP_N(t)$ in $(-1,1)$, where $c \in \mathbb{R}$. The details on orthogonal polynomials can be found in Refs. [29, 30].

## 2.2 The Paradigmatic Example: Paley-Wiener Spaces $PW_{\pi\sigma}$

Throughout this section, the *Fourier transform* is defined in $L^1(\mathbb{R}) \cap L^2(\mathbb{R})$ as

$$\widehat{f}(\omega) := \frac{1}{\sqrt{2\pi}} \int_{-\infty}^{\infty} f(t)\, \mathrm{e}^{-i\omega t}\, dt, \quad \omega \in \mathbb{R},$$

and extended to $L^2(\mathbb{R})$ in the usual way as announced in Sect. 1.3.

**Definition 2.3** A function $f \in L^2(\mathbb{R})$ is said to be bandlimited to the interval $[-\pi,\pi]$ if its Fourier transform $\widehat{f}$ vanishes outside $[-\pi,\pi]$, i.e., the function $\widehat{f}$ is supported in $[-\pi,\pi]$.

For notational ease we consider the interval $[-\pi,\pi]$ instead of a general interval $[-\pi\sigma,\pi\sigma]$ with $\sigma > 0$; its study is completely similar.

The space of bandlimited functions to $[-\pi,\pi]$ is known in the mathematical literature as the *Paley-Wiener space* and denoted by $PW_\pi$. That is,

$$PW_\pi := \left\{ f \in L^2(\mathbb{R}) \,:\, \operatorname{supp} \widehat{f} \subseteq [-\pi,\pi] \right\}.$$

For each $f \in PW_\pi$, having in mind the inverse Fourier transform, we consider its integral representation

$$f(t) = \frac{1}{\sqrt{2\pi}} \int_{-\pi}^{\pi} \widehat{f}(w)\, e^{iwt}\, dw\,, \quad t \in \mathbb{R}\,.$$

This parametric integral defines a $C^\infty(\mathbb{R})$ function $f$ and $\widehat{f^{(p)}}(w) = (iw)^p \widehat{f}(w)$ for any $p \in \mathbb{N}$. As a consequence, any derivative of $f$ also belongs to $PW_\pi$. Since, in particular, the function $\widehat{f}$ belongs to $L^1[-\pi, \pi]$, from the *Riemann-Lebesgue lemma* we get that $f(t) \to 0$ as $|t| \to \infty$.

### 2.2.1 Properties of the Paley-Wiener Space $PW_\pi$

Next, we list some properties of Paley-Wienes spaces:

1. Since the Fourier transform $\mathcal{F}: L^2(\mathbb{R}) \longrightarrow L^2(\mathbb{R})$ is a unitary operator and

$$PW_\pi = \mathcal{F}^{-1}\big(L^2[-\pi, \pi]\big)\,,$$

the space $PW_\pi$ is a closed subspace of $L^2(\mathbb{R})$. The space $L^2[-\pi, \pi]$ is identified to a closed subspace of $L^2(\mathbb{R})$ by extending to 0 on $\mathbb{R}$ the functions of $L^2[-\pi, \pi]$.

2. Any $f \in PW_\pi$ can be expressed as

$$f(t) = \frac{1}{\sqrt{2\pi}} \int_{-\pi}^{\pi} \widehat{f}(w) e^{iwt}\, dw = \Big\langle \widehat{f}, \frac{e^{-iwt}}{\sqrt{2\pi}} \Big\rangle_{L^2[-\pi,\pi]}\,, \quad t \in \mathbb{R}\,. \tag{2.3}$$

Cauchy-Schwarz's inequality and Parseval equality, $\|f\|_{L^2(\mathbb{R})} = \|\widehat{f}\|_{L^2[-\pi,\pi]}$, give, for every $t \in \mathbb{R}$,

$$|f(t)| \le \|\widehat{f}\|_{L^2[-\pi,\pi]} \Big\| \frac{e^{-iwt}}{\sqrt{2\pi}} \Big\|_{L^2[-\pi,\pi]} = \|f\|_{L^2(\mathbb{R})}\,, \quad f \in PW_\pi\,.$$

In other words, evaluation functionals are bounded on $PW_\pi$, which consequently is a RKHS. Its reproducing kernel is

$$k_\pi(t, s) = \frac{\sin \pi(t-s)}{\pi(t-s)}\,, \quad t, s \in \mathbb{R}\,.$$

Indeed, using Plancherel-Parseval formula (1.7), for each $f \in PW_\pi$ we have

$$f(s) = \Big\langle \widehat{f}, \frac{e^{-iws}}{\sqrt{2\pi}} \Big\rangle_{L^2[-\pi,\pi]} = \Big\langle f, \frac{\sin \pi(\cdot - s)}{\pi(\cdot - s)} \Big\rangle_{L^2(\mathbb{R})}\,, \quad s \in \mathbb{R}\,.$$

Notice that

$$\mathcal{F}^{-1}\Big(\frac{e^{-iws}}{\sqrt{2\pi}}\chi_{[-\pi,\pi]}(w)\Big)(t) = \frac{\sin\pi(t-s)}{\pi(t-s)}, \quad t \in \mathbb{R}.$$

3. Since the sequence $\left\{\frac{e^{-inw}}{\sqrt{2\pi}}\right\}_{n=-\infty}^{\infty}$ is an orthonormal basis for $L^2[-\pi,\pi]$ and $\mathcal{F}^{-1}$ is a unitary operator we obtain that the sequence

$$\left\{\frac{\sin\pi(t-n)}{\pi(t-n)}\right\}_{n=-\infty}^{\infty} \tag{2.4}$$

is an orthonormal basis for the Paley-Wiener space $PW_\pi$.
4. Moreover, having in mind that $k_\pi(t,t) = 1$ for all $t \in \mathbb{R}$, the formula (2.2) yields the famous Shannon's sampling theorem:

**Theorem 2.2 (Shannon's Sampling Theorem)** *Any function $f \in PW_\pi$, i.e., bandlimited to $[-\pi,\pi]$, can be recovered from the sequence of its samples $\{f(n)\}_{n=-\infty}^{\infty}$ by means of the formula*

$$f(t) = \sum_{n=-\infty}^{\infty} f(n)\frac{\sin\pi(t-n)}{\pi(t-n)}, \quad t \in \mathbb{R}. \tag{2.5}$$

*The convergence of the series is in $L^2$-norm, absolute and uniform on $\mathbb{R}$.*

Another proof of the above theorem is the following due to Hardy (see Ref. [20]): Given $f \in PW_\pi$, the expansion of its Fourier transform $\widehat{f} \in L^2[-\pi,\pi]$ with respect to the orthonormal basis $\left\{e^{-inw}/\sqrt{2\pi}\right\}_{n=-\infty}^{\infty}$ for $L^2[-\pi,\pi]$ gives

$$\widehat{f} = \sum_{n=-\infty}^{\infty}\langle\widehat{f}, \frac{e^{-inw}}{\sqrt{2\pi}}\rangle\frac{e^{-inw}}{\sqrt{2\pi}} = \sum_{n=-\infty}^{\infty} f(n)\frac{e^{-inw}}{\sqrt{2\pi}} \quad \text{in } L^2[-\pi,\pi]. \tag{2.6}$$

The inverse Fourier transform $\mathcal{F}^{-1}$ in (2.6) gives

$$\begin{aligned} f &= \sum_{n=-\infty}^{\infty} f(n)\mathcal{F}^{-1}\Big(\frac{e^{-inw}}{\sqrt{2\pi}}\chi_{[-\pi,\pi]}(w)\Big) \\ &= \sum_{n=-\infty}^{\infty} f(n)\frac{\sin\pi(t-n)}{\pi(t-n)} \quad \text{in } L^2(\mathbb{R}). \end{aligned}$$

The convergence properties come again since $PW_\pi$ is a RKHS.

5. Since Shannon's sampling formula is an orthonormal expansion in $PW_\pi$, Parseval's identity says that

$$\|f\|^2 = \sum_{n=-\infty}^{\infty} |f(n)|^2 \quad \text{for all } f \in PW_\pi .$$

In other words, the energy $E_f := \|f\|^2$ of the bandlimited function $f \in PW_\pi$ is contained in the sequence of its samples $\{f(n)\}_{n=-\infty}^{\infty}$.
The following commutative diagram goes into the meaning of the sampling formula in $PW_\pi$:

$$\begin{array}{ccc} f \in PW_\pi & \xrightarrow{\mathcal{F}} & \widehat{f} \in L^2[-\pi,\pi] \\ S\big\downarrow & & \big\downarrow\mathcal{P} \\ \{f(n)\}_{n\in\mathbb{Z}} \in \ell^2(\mathbb{Z}) & \xrightarrow{\mathcal{F}} & \widehat{f}_p \in L^2_p[-\pi,\pi] . \end{array}$$

All mappings included in this diagram are unitary operators:

(a) $S$ denotes the *sampling mapping* with sampling period $T_s = 1$.
(b) $\mathcal{P}$ is the $2\pi$-*periodization mapping* which extends a function $\widehat{f}$ in $[-\pi,\pi]$ to the whole $\mathbb{R}$ with period $2\pi$.
(c) The other two mappings are, respectively, the Fourier transform in $L^2(\mathbb{R})$ and the Fourier transform in $\ell^2(\mathbb{Z})$, defined here as the *Fourier series*

$$\mathcal{F}(\{a_n\})(w) := \sum_{n=-\infty}^{\infty} a_n \frac{e^{-inw}}{\sqrt{2\pi}} .$$

The situation described by the diagram is depicted in Fig. 2.1:

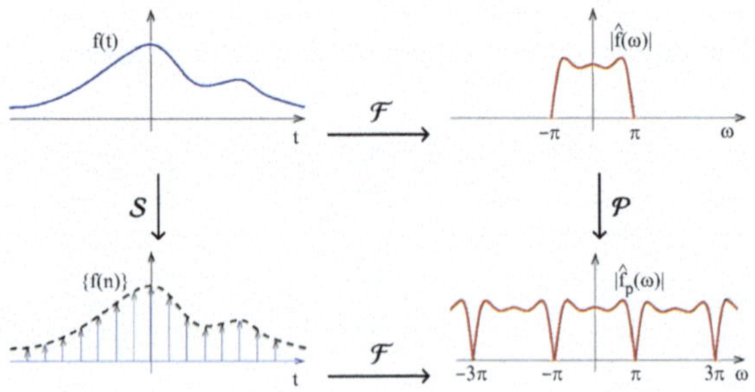


**Fig. 2.1** Time-frequency interpretation of Shannon's sampling theorem

6. As the space $PW_\pi$ is a RKHS contained in the Hilbert space $L^2(\mathbb{R})$, the reproducing formula applied to any $f \in L^2(\mathbb{R})$ gives its orthogonal projection onto $PW_\pi$

$$P_{PW_\pi} f(s) = \Big\langle f, \frac{\sin \pi(\cdot - s)}{\pi(\cdot - s)} \Big\rangle_{L^2(\mathbb{R})} = \big(f * \operatorname{sinc}\big)(s), \quad s \in \mathbb{R},$$

where $*$ means the convolution operator and sinc denotes the cardinal sine function defined in (1.2).

7. The crucial feature in $PW_\pi$ is that the sampling period is $T_s = 1$. Any function in $PW_\pi$ can be recovered by sampling at any sequence $\{a+n\}_{n=-\infty}^{\infty}$ of real numbers.

   In fact, any function $f \in PW_\pi$ can be recovered from the sequence of samples $\big\{f(n+a)\big\}_{n=-\infty}^{\infty}$, where $a$ is any fixed real number. Observe that the sequence $\Big\{e^{-i(n+a)w}/\sqrt{2\pi}\Big\}_{n=-\infty}^{\infty}$ is also an orthonormal basis for $L^2[-\pi,\pi]$ which goes, via $\mathcal{F}^{-1}$, onto the orthonormal basis $\Big\{\dfrac{\sin \pi(t-n-a)}{\pi(t-n-a)}\Big\}_{n=-\infty}^{\infty}$ for $PW_\pi$. The expansion of any $f \in PW_\pi$ with respect to this basis yields the new sampling formula:

$$f(t) = \sum_{n=-\infty}^{\infty} f(n+a)\frac{\sin \pi(t-n-a)}{\pi(t-n-a)}, \qquad t \in \mathbb{R}.$$

8. The Shannon sampling formula (2.5) is nothing but a *Lagrange-type interpolation series*. Indeed, formula (2.5) can be rewritten as

$$f(t) = \sum_{n=-\infty}^{\infty} f(n)\frac{(-1)^n \sin \pi t}{\pi(t-n)} = \sum_{n=-\infty}^{\infty} f(n)\frac{P(t)}{P'(n)(t-n)}, \qquad t \in \mathbb{R}.$$

   where $P(t) := \sin \pi t,\ t \in \mathbb{R}$.

9. In general, one can consider the Paley-Wiener space $PW_{\pi\sigma}$, $\sigma > 0$, of $L^2$-bandlimited functions to the interval $[-\pi\sigma, \pi\sigma]$ defined as

$$PW_{\pi\sigma} := \Big\{f \in L^2(\mathbb{R}) \,:\, \operatorname{supp} \widehat{f} \subseteq [-\pi\sigma, \pi\sigma]\Big\}.$$

   In this case the associated sampling period is $T_s = 1/\sigma$. Indeed, for $f \in PW_{\pi\sigma}$ define $g(t) := f(t/\sigma)$. Since $\widehat{g}(w) = \sigma \widehat{f}(\sigma w)$, the function $g \in PW_\pi$. Therefore

$$g(t) = f(t/\sigma) = \sum_{n=-\infty}^{\infty} f\Big(\frac{n}{\sigma}\Big)\frac{\sin \pi(t-n)}{\pi(t-n)}, \qquad t \in \mathbb{R}.$$

The change of variable $t/\sigma = s$ gives, for any $f \in PW_{\pi\sigma}$, the sampling formula:

$$f(s) = \sum_{n=-\infty}^{\infty} f\left(\frac{n}{\sigma}\right) \frac{\sin \pi(\sigma s - n)}{\pi(\sigma s - n)}, \qquad s \in \mathbb{R}. \tag{2.7}$$

The reproducing kernel for $PW_{\pi\sigma}$ is $k_{\pi\sigma}(t,s) = \sigma \operatorname{sinc} \sigma(t-s)$, $t, s \in \mathbb{R}$.

10. Usually, the band of frequencies is centered at 0 since this is the case for real-valued bandlimited functions. Indeed, for a real-valued function $f$ one has

$$|\widehat{f}(w)|^2 = \widehat{f}(w)\overline{\widehat{f}(w)} = \widehat{f}(w)\widehat{f}(-w) \quad \text{a.e. } w\,,$$

i.e., it is an even function.
Let $f$ be a function in $L^2(\mathbb{R})$ bandlimited to the interval $[w_0 - \pi, w_0 + \pi]$. The Fourier transform of the function $g(t) := e^{-iw_0 t} f(t)$ is $\widehat{g}(w) = \widehat{f}(w + w_0)$. Then $g$ is bandlimited to the interval $[-\pi, \pi]$ and, as a consequence,

$$g(t) = e^{-iw_0 t} f(t) = \sum_{n=-\infty}^{\infty} e^{-iw_0 n} f(n) \frac{\sin \pi(t-n)}{\pi(t-n)}, \qquad t \in \mathbb{R},$$

from which the sampling formula for $f$, an $L^2$-bandlimited function to the interval $[w_0 - \pi, w_0 + \pi]$, reads:

$$f(t) = \sum_{n=-\infty}^{\infty} f(n)\, e^{iw_0(t-n)} \frac{\sin \pi(t-n)}{\pi(t-n)}, \qquad t \in \mathbb{R}. \tag{2.8}$$

11. In the Paley-Wiener space $PW_\pi$, for each $p \in \mathbb{N}$, the following *Bernstein-type inequality* holds:

$$\|f^{(p)}\| \leq \pi^p \|f\| \quad \text{for all } f \in PW_\pi\,. \tag{2.9}$$

It suffices to prove the inequality for $p = 1$. Let $f$ be in $PW_\pi$; since the Fourier transform of its derivative is $\widehat{f'}(\omega) = i\omega \widehat{f}(\omega)$ one gets

$$\|f'\|^2 = \|\widehat{f'}\|^2 = \|\omega \widehat{f}(\omega)\|^2 \leq \pi^2 \|\widehat{f}\|^2 = \pi^2 \|f\|^2,$$

where Parseval's equality (1.8) has been used twice.

12. The Shannon sampling formula also holds for any function $f$ whose Fourier transform $\widehat{f}$ belongs to $L^1[-\pi, \pi] \setminus L^2[-\pi, \pi]$.
Indeed, the function $\widehat{f} \in L^1[-\pi, \pi]$ has the formal Fourier series

$$\widehat{f}(w) \sim \frac{1}{\sqrt{2\pi}} \sum_{n=-\infty}^{\infty} f(n)\, e^{-inw}\,.$$

It is known that this formal Fourier series can be multiplied by $e^{itw}/\sqrt{2\pi}$ and integrated over $[-\pi, \pi]$ (see, for instance, Ref. [3, p. 214]). On the left hand one obtains

$$f(t) = \frac{1}{\sqrt{2\pi}} \int_{-\pi}^{\pi} \widehat{f}(w)\, e^{itw} dw\,,$$

meanwhile on the right hand one gets

$$\frac{1}{2\pi} \sum_{n=-\infty}^{\infty} f(n) \int_{-\pi}^{\pi} e^{i(t-n)w}\, dw = \sum_{n=-\infty}^{\infty} f(n)\, \mathrm{sinc}(t-n)\,, \qquad t \in \mathbb{R}\,,$$

that is, the cardinal sampling formula for $f$. The convergence of the series is uniform on compact sets but can fail to be absolute. See the details in Ref. [21, Theorem 6.16].

**Example 2.3 (Undersampling and Oversampling)** If one samples a function $f$ in $PW_\pi$ with a general sampling period $T_s > 0$, the question arises whether it is possible to reconstruct it from the sequence of its samples $\{f(nT_s)\}_{n=-\infty}^{\infty}$.

It is indeed possible in the case where $0 < T_s \le 1$, i.e., sampling the signal at a frequency higher than that given by its bandwidth $[-\pi, \pi]$. For sampling periods $T_s > 1$, we cannot reconstruct the signal due to the *aliasing phenomenon*, which will be explained below.

Firstly, we study the relationship between the Fourier transform (series) of the sequence of samples $\{f(nT_s)\}_{n=-\infty}^{\infty}$ taken from a function $f \in PW_\pi$ with a sampling period $T_s > 0$ and the $\frac{2\pi}{T_s}$-periodized version $\widehat{f}_P$ of $\widehat{f}$ defined as

$$\widehat{f}_P(\omega) := \sum_{n=-\infty}^{\infty} \widehat{f}\big(\omega + \frac{2\pi}{T_s} n\big)\,.$$

Obviously, $\widehat{f}_P$ is a $\frac{2\pi}{T_s}$-periodic function which belongs to $L^2[0, \frac{2\pi}{T_s}]$. Its Fourier expansion with respect to the orthonormal basis $\big\{\sqrt{\frac{T_s}{2\pi}}\, e^{-imT_s\omega}\big\}_{m=-\infty}^{\infty}$ of $L^2[0, \frac{2\pi}{T_s}]$ has Fourier coefficients

$$\begin{aligned} c_m &= \sqrt{\frac{T_s}{2\pi}} \int_0^{\frac{2\pi}{T_s}} \widehat{f}_P(\omega) e^{imT_s\omega} d\omega = \sqrt{\frac{T_s}{2\pi}} \int_0^{\frac{2\pi}{T_s}} \sum_{n=-\infty}^{\infty} \widehat{f}\big(\omega + \frac{2\pi}{T_s} n\big) e^{imT_s\omega} d\omega \\ &= \sqrt{\frac{T_s}{2\pi}} \sum_{n=-\infty}^{\infty} \int_0^{\frac{2\pi}{T_s}} \widehat{f}\big(\omega + \frac{2\pi}{T_s} n\big) e^{imT_s\omega} d\omega\,, \qquad m \in \mathbb{Z}\,. \end{aligned}$$

The change of variable $\omega + \frac{2\pi}{T_s}n = x$ allows us to obtain

$$\begin{aligned} c_m &= \sqrt{\frac{T_s}{2\pi}} \sum_{n=-\infty}^{\infty} \int_{\frac{2\pi}{T_s}n}^{\frac{2\pi}{T_s}(n+1)} \widehat{f}(x) e^{imT_s x} dx \\ &= \sqrt{\frac{T_s}{2\pi}} \int_{-\pi}^{\pi} \widehat{f}(x) e^{imT_s x} dx \\ &= \sqrt{T_s} f(mT_s), \qquad m \in \mathbb{Z}. \end{aligned}$$

Thus, the Fourier expansion for $\widehat{f_p}$ is

$$\widehat{f_p}(\omega) = \sum_{n=-\infty}^{\infty} \widehat{f}\Big(\omega + \frac{2\pi}{T_s}n\Big) = T_s \sum_{m=-\infty}^{\infty} f(mT_s) \frac{e^{-imT_s\omega}}{\sqrt{2\pi}}. \tag{2.10}$$

Formula (2.10) express a well-known fact for the signal processing community:

*The Fourier transform of the sequence* $\{f(mT_s)\}_{m\in\mathbb{Z}}$, *i.e., the sampled function with period* $T_s$, *is precisely (up to a scale factor) the* $\frac{2\pi}{T_s}$*-periodized version of the Fourier transform* $\widehat{f}$ *of* $f$.

Formula (2.10) is nothing but the so-called *Poisson summation formula* applied to $\widehat{f}$ with period $2\pi/T_s$. For the Poisson summation formula see, for instance, Refs. [18, 21, 27].

Two cases can occur:

(*a*) In the *oversampling case*, where $0 < T_s \leq 1$, the Fourier transform $\widehat{f}$ of $f$ can be recovered from the Fourier transform of the sampled function. Hence, the function $f$ can be also recovered.

This situation is depicted in Fig. 2.2:

In terms of the Shannon sampling theorem, the explanation is easy: if a function is bandlimited to the interval $[-\pi, \pi]$, is also bandlimited to any interval $[-\pi\sigma, \pi\sigma]$ with $\sigma \geq 1$.

(*b*) In the *undersampling case*, where $T_s > 1$, we cannot obtain the Fourier transform of $f$ from the Fourier transform of the sampled function because the copies of $\widehat{f}$ overlap in $\widehat{f_p}$. Hence, it is impossible to recover the function from its samples.

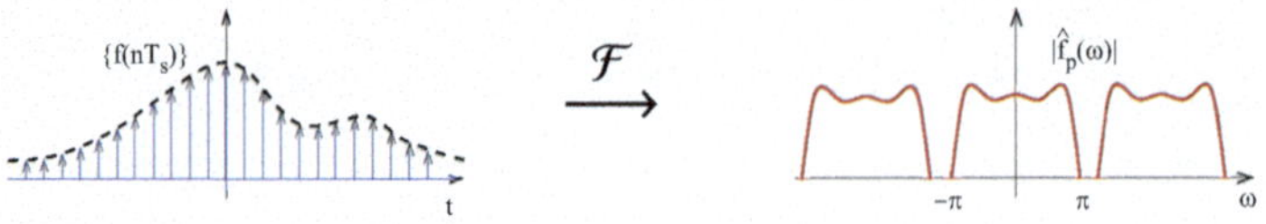


**Fig. 2.2** Oversampling case

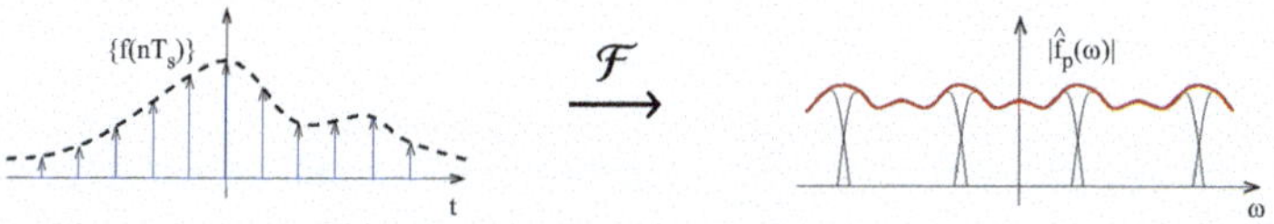


**Fig. 2.3** Undersampling case

The alluded overlap produces the *aliasing phenomenon*, i.e., some frequencies go under the name of another ones. As pointed out in Ref. [19], this is a familiar phenomenon to the watchers of TV and western movies. As the stage coach starts up, the wheels start going faster and faster, but then they gradually slow down, stop, go backwards, slow down, stop, go forward, etc. This effect is due solely to the sampling the picture makes of the real scene.

The undersampling situation is depicted in Fig. 2.3:

This undersampling/oversampling discussion clarifies the crucial role of the critical *Nyquist period* which is given by $T_s = 1/\sigma$ whenever $\operatorname{supp} \widehat{f} \subseteq [-\pi\sigma, \pi\sigma]$.

**Example 2.4 (Robust Reconstruction: The Oversampling Technique)** The actual computation of the cardinal series presents some numerical difficulties since the cardinal sine function behaves like $1/t$ as $|t| \to \infty$.

An easy example is given by the numerical computation of $f(1/2)$, for a function $f$ in $PW_\pi$, from a noisy sequence of samples $\{f(n) + \delta_n\}_{n=-\infty}^{\infty}$. The error in this case $\left| \sum_n \frac{(-1)^n \delta_n}{\pi(n-\frac{1}{2})} \right|$, even when all $|\delta_n| \leq \delta$, could be infinity.

One way to overcome this difficulty is the *oversampling technique*, i.e., sampling the signal at a frequency higher than that given by its bandwidth. In this way we obtain sampling functions converging to zero at infinity faster than the cardinal sine functions.

Indeed, consider the bandlimited function

$$f(t) = \frac{1}{\sqrt{2\pi}} \int_{-\pi\sigma}^{\pi\sigma} F(\omega)\, \mathrm{e}^{i\omega t}\, d\omega\,, \ t \in \mathbb{R}, \quad \text{with } F \in L^2[-\pi\sigma, \pi\sigma] \ \text{ and } \sigma < 1\,.$$

Extending $F$ to be zero in $[-\pi, \pi] \setminus [-\pi\sigma, \pi\sigma]$, we have

$$F(\omega) = \sum_{n=-\infty}^{\infty} f(n) \frac{\mathrm{e}^{-in\omega}}{\sqrt{2\pi}} \quad \text{in } L^2[-\pi, \pi]\,.$$

Let $\theta(\omega)$ be a smooth function such that

$$\theta(\omega) = \begin{cases} 1 & \text{if } \omega \in [-\pi\sigma, \pi\sigma], \\ 0 & \text{if } \omega \notin [-\pi, \pi]. \end{cases}$$

As a consequence,

$$F(\omega) = \theta(\omega)F(\omega) = \sum_{n=-\infty}^{\infty} f(n)\theta(\omega)\frac{e^{-in\omega}}{\sqrt{2\pi}} \quad \text{in } L^2[-\pi,\pi],$$

and the sampling expansion

$$f(t) = \sum_{n=-\infty}^{\infty} f(n)S_\theta(t-n), \quad t \in \mathbb{R},$$

holds, where $S_\theta(t)$ is the inverse Fourier transform $\mathcal{F}^{-1}$ of the function $\theta(w)/\sqrt{2\pi}$. Consequently,

$$S_\theta(t-n) = \mathcal{F}^{-1}[\theta(\omega)\,e^{-in\omega}/\sqrt{2\pi}](t), \quad t \in \mathbb{R}.$$

Furthermore, using the properties of the Fourier transform, as smoother $\theta$ is, faster is the decay of $S_\theta$ as $|t| \to \infty$. However, the new sampling functions $\{S_\theta(t-n)\}_{n=-\infty}^{\infty}$ are no longer orthogonal and they do not belong to $PW_{\pi\sigma}$.

Next, let us consider an illustrative example. Take $\sigma = 1-\epsilon$ with $0 < \epsilon < 1$, and consider for $\theta(w)$ the trapezoidal function

$$\theta(\omega) = \begin{cases} 1 & \text{if } |\omega| \le \pi(1-\epsilon), \\ \dfrac{1}{\epsilon}\Big(1 - \dfrac{|\omega|}{\pi}\Big) & \text{if } \pi(1-\epsilon) \le |\omega| \le \pi \\ 0 & \text{if } |\omega| \ge \pi. \end{cases}$$

One can easily obtain that $S_\theta(t) = \dfrac{\sin \epsilon\pi t}{\epsilon\pi t}\dfrac{\sin \pi t}{\pi t}$, $t \in \mathbb{R}$, which behaves like $1/t^2$ as $|t| \to \infty$.

Thus, any $f \in PW_{\pi\sigma}$, where $\sigma = 1-\epsilon$, can be recovered from its samples at the integers by means of the sampling expansion

$$f(t) = \sum_{n=-\infty}^{\infty} f(n)\frac{\sin \epsilon\pi(t-n)}{\epsilon\pi(t-n)}\frac{\sin \pi(t-n)}{\pi(t-n)}, \quad t \in \mathbb{R}.$$

In this example, if each sample $f(n)$ is subject to an error $\delta_n$ such that $|\delta_n| \le \delta$, then the total error in the above calculated $f(t)$ is bounded by a constant depending only on the constants $\delta$ and $\epsilon$ (see Ref. [27, p. 211]).

*The Paley-Wiener space $PW_\pi$ as a RKHS of entire functions*

Any function $f \in PW_\pi$ can be extended to the complex plane $\mathbb{C}$ as

$$f(z) = \frac{1}{\sqrt{2\pi}}\int_{-\pi}^{\pi} \widehat{f}(\omega)e^{iz\omega}d\omega, \quad z \in \mathbb{C}. \tag{2.11}$$

This extended function $f$ is proved to be a continuous function on $\mathbb{C}$ by using a standard argument allowing interchange the limit with the integral. Now, taking $\gamma : [a, b] \longrightarrow \mathbb{C}$ a closed curve in $\mathbb{C}$, the integral

$$\int_\gamma f(z)\,dz = \frac{1}{\sqrt{2\pi}} \int_a^b \left( \int_{-\pi}^{\pi} \widehat{f}(\omega) e^{i\gamma(t)\omega} d\omega \right) \gamma'(t) dt$$

is shown to be zero by interchanging the order of the integrals. Hence, *Morera's theorem* says that $f$ is an entire function.

Moreover, $f$ is a function of *exponential type at most* $\pi$, i.e., $f$ satisfies an inequality $|f(z)| \le A e^{\pi|z|}$ for all $z \in \mathbb{C}$ for some positive constant $A$. It follows from (2.11) by using the Cauchy-Schwarz inequality. Indeed, for $z = x + iy \in \mathbb{C}$ one has

$$|f(x+iy)| \le \frac{1}{\sqrt{2\pi}} \int_{-\pi}^{\pi} |\widehat{f}(\omega)|\, e^{-y\omega}\, d\omega \le \frac{e^{\pi|y|}}{\sqrt{2\pi}} \int_{-\pi}^{\pi} |\widehat{f}(\omega)|\, d\omega \le e^{\pi|z|} \|f\|_{PW_\pi}\,.$$

In particular, for any $t \in \mathbb{R}$ we have $|f(t)| \le \|f\|_{PW_\pi}$.

Conversely, the *Paley-Wiener theorem*, whose proof can be found, for instance, in Ref. [35, p. 101], tells us that these properties characterize the space $PW_\pi$:

**Theorem 2.3 (Paley-Wiener Theorem)** *Let $f$ be an entire function such that $|f(z)| \le C e^{\pi|z|}$, for any $z \in \mathbb{C}$, and $f|_{\mathbb{R}}$ belongs to $L^2(\mathbb{R})$. Then there exists a function $F \in L^2[-\pi, \pi]$ such that*

$$f(z) = \frac{1}{\sqrt{2\pi}} \int_{-\pi}^{\pi} F(w)\, e^{-izw} dw\,, \quad z \in \mathbb{C}\,.$$

Consequently, $PW_\pi \equiv \Big\{ f \in H(\mathbb{C}) \,:\, |f(z)| \le A e^{\pi|z|}, \quad f|_{\mathbb{R}} \in L^2(\mathbb{R}) \Big\}$.

Considering the space $PW_\pi$ as a RKHS of entire functions, its reproducing kernel is given by

$$k_\pi(z, w) = \operatorname{sinc}(z - \overline{w})\,, \quad z, w \in \mathbb{C}\,.$$

Indeed, for any $f \in PW_\pi$ we have

$$f(w) = \langle \widehat{f}, \frac{e^{-i\overline{w}\cdot}}{\sqrt{2\pi}} \rangle_{L^2[-\pi,\pi]} = \Big\langle f, \frac{\sin \pi(\cdot - \overline{w})}{\pi(\cdot - \overline{w})} \Big\rangle_{L^2(\mathbb{R})} \quad \text{for any } w \in \mathbb{C}\,.$$

### 2.2.2 Irregular Sampling: Paley-Wiener-Levinson Theorem

Let $\{t_n\}_{n=-\infty}^{\infty}$ be a sequence of real numbers such that

$$D := \sup_{n\in\mathbb{Z}} |t_n - n| < 1/4\,.$$

The so-called $\frac{1}{4}$-*Kadec's theorem* (see Ref. [35, p. 42]) assures us that the sequence of complex exponentials $\{\mathrm{e}^{-it_n w}/\sqrt{2\pi}\}_{n=-\infty}^{\infty}$ forms a Riesz basis for $L^2[-\pi,\pi]$. The proof of the $\frac{1}{4}$-Kadec theorem uses a Paley-Wiener theorem on perturbation of orthonormal bases (see Ref. [35]).

Denote as $\{h_n\}_{n=-\infty}^{\infty}$ the dual Riesz basis of $\{\mathrm{e}^{-it_n w}/\sqrt{2\pi}\}_{n=-\infty}^{\infty}$ in $L^2[-\pi,\pi]$. Given $f \in PW_\pi$, expand its Fourier transform $\widehat{f} \in L^2[-\pi,\pi]$ with respect to the dual basis $\{h_n\}_{n=-\infty}^{\infty}$ obtaining

$$\widehat{f} = \sum_{n=-\infty}^{\infty} \Big\langle \widehat{f}, \frac{\mathrm{e}^{-it_n w}}{\sqrt{2\pi}} \Big\rangle h_n = \sum_{n=-\infty}^{\infty} f(t_n)\, h_n \quad \text{in } L^2[-\pi,\pi]\,.$$

The inverse Fourier transform $\mathcal{F}^{-1}$ gives in $PW_\pi$ the sampling formula

$$f(t) = \sum_{n=-\infty}^{\infty} f(t_n)\,(\mathcal{F}^{-1}h_n)(t)\,, \quad t \in \mathbb{R}\,. \tag{2.12}$$

The problem consists of identifying the sampling functions $\mathcal{F}^{-1}h_n,\ n \in \mathbb{Z}$. By using entire functions techniques, Paley-Wiener-Levinson (see Ref. [23]) proved that

$$(\mathcal{F}^{-1}h_n)(t) = \frac{G(t)}{(t-t_n)G'(t_n)}\,, \quad t \in \mathbb{R}\,,$$

where $G$ denotes the convergent infinite product

$$G(t) = (t - t_0)\prod_{n=1}^{\infty}\Big(1 - \frac{t}{t_n}\Big)\Big(1 - \frac{t}{t_{-n}}\Big)\,.$$

In other words, sampling formula (2.12) is again a Lagrange-type interpolation series. As a consequence, the sequences

$$\Big\{\frac{\sin\pi(t-t_n)}{\pi(t-t_n)}\Big\}_{n=-\infty}^{\infty} \quad \text{and} \quad \Big\{\frac{G(t)}{(t-t_n)G'(t_n)}\Big\}_{n=-\infty}^{\infty}$$

form a pair of dual Riesz bases for $PW_\pi$.

The irregular sampling alluded here corresponds to that associated with the *time-jitter error* where the sampling points are taken at the integers affected with an error $\delta_n$, i.e., $t_n = n + \delta_n,\ n \in \mathbb{Z}$.

The general case concerns with real sequences $\{t_n\}_{n=-\infty}^{\infty}$ for which there exist constants $0 < A \le B < \infty$ such that

$$A\|f\|^2 \le \sum_{n=-\infty}^{\infty} |f(t_n)|^2 \le B\|f\|^2 \quad \text{for all } f \in PW_\pi\,.$$

This means that the sequence $\{e^{-it_n w}/\sqrt{2\pi}\}_{n=-\infty}^{\infty}$ is a frame for $L^2[-\pi,\pi]$.

This case will not be considered here; see, for instance, Refs. [4, 10, 12]. Below we will deal with the general case of Riesz bases.

### Zeros of Bandlimited Functions

The problem of signal recovery can also be considered from a different point of view. As we know, the signals in the space $PW_\pi$ are entire functions of exponential type at most $\pi$ whose restriction to $\mathbb{R}$ belongs to $L^2(\mathbb{R})$. Although entire functions are not completely determined by the location of their zeros, as can be seen from the *Hadamard factorization theorem* (see, for instance, Ref. [35, p. 74]), bandlimited functions are, as can be deduced from a *Titchmarsh's theorem* [31]; this result has been improved in Ref. [2].

As a consequence, any bandlimited function is uniquely determined by its zeros up to an exponential factor depending on its spectral interval $[a, b]$. If the spectral interval is of the form $[-a, a]$, this exponential factor reduces to a constant. Recall that it is the case for real-valued bandlimited signals as already pointed out.

For our sampling purposes we state a corollary from Ref. [2, Theorem 5] providing the needed mathematical foundation:

**Theorem 2.4** *Let $f \in L^2(\mathbb{R})$ be a bandlimited function, and let $[a, b]$ be the smallest interval that contains the support of the Fourier transform of $f$. If $a < b$, then $f$ has infinitely many zeros $z_n$, $z_n \ne 0$, ordered with non-decreasing absolute values, such that*

$$f(z) = Az^m\, e^{i\frac{(a+b)z}{2}} \prod_{n=1}^{\infty}\Big(1 - \frac{z}{z_n}\Big),$$

*where $m$ is a non-negative integer. The infinite product converges conditionally, and the convergence is uniform on compact subsets of $\mathbb{C}$.*

If $f$ is bandlimited to $[-a, a]$, then

$$f(z) = f(0) \prod_{n=1}^{\infty} \left(1 - \frac{z}{z_n}\right),$$

provided $f(0) \neq 0$, or

$$f(z) = Az^m \prod_{n=1}^{\infty} \left(1 - \frac{z}{z_n}\right),$$

if $z = 0$ is a zero of $f$ of order $m$.

Notice that the zeros $z_n$ may be complex. This poses a difficulty from a practical point of view, as complex zeros are harder to detect than real zeros. Whenever they are real, this theorem provides a useful tool for signal recovery, usually referred to as *real-zero interpolation* (see Refs. [6, 25]). One way to deal only with real zeros is by using the so-called *sine wave crossings* technique (see Ref. [25]) involved in the following result from Duffin and Schaeffer [9]:

> *Let f be an entire function of exponential type at most $\gamma$ such that $|f(x)| \le 1$ on the real axis. Then for every real $\alpha$ the function $\cos(\gamma z + \alpha) - f(z)$ has only real zeros, or vanishes identically. Moreover all the zeros are simple, excepts perhaps at points on the real axis where $f(x) = \pm 1$.*

For a deeper study of the oscillatory properties of Paley-Wiener functions see Refs. [21, 26, 33].

### The Paley-Wiener-Levinson Theorem Revisited

In this section we will prove that the Paley-Wiener-Levinson theorem is true for any sequence $\{t_n\}_{n=-\infty}^{\infty} \subset \mathbb{C}$ such that the sequence $\{e^{-it_n w}/\sqrt{2\pi}\}_{n=-\infty}^{\infty}$ is a Riesz basis for $L^2[-\pi, \pi]$. For the sake of simplicity we confine ourselves to real sequences $\{t_n\}_{n=-\infty}^{\infty}$. See Ref. [2] for the complex case. First, by using frame theory, we characterize the Riesz basis $\{e^{-it_n w}/\sqrt{2\pi}\}_{n=-\infty}^{\infty}$ from the sequence $\{t_n\}_{n=-\infty}^{\infty}$:

**Proposition 2.7** *The sequence $\{e^{-it_n w}/\sqrt{2\pi}\}_{n=-\infty}^{\infty}$ is a Riesz basis for $L^2[-\pi, \pi]$ if and only if the following two conditions hold:*

*(i) For any sequence $\{c_n\}_{n=-\infty}^{\infty} \in \ell^2(\mathbb{Z})$ the interpolation problem $f(t_n) = c_n$, $n \in \mathbb{Z}$, has a (unique) solution $f \in PW_\pi$.*
*(ii) There exist two constants $0 < A \le B$ such that*

$$A\|f\|^2 \le \sum_{n=-\infty}^{\infty} |f(t_n)|^2 \le B\|f\|^2 \quad \textit{for all } f \in PW_\pi\,. \tag{2.13}$$

***Proof*** Since the Fourier transform is a unitary operator between $PW_\pi$ and $L^2[-\pi,\pi]$, inequalities in $(ii)$ can be written as

$$A\|\widehat{f}\|^2 \leq \sum_{n=-\infty}^{\infty} |\langle \widehat{f}, \frac{e^{-it_n w}}{\sqrt{2\pi}}\rangle|^2 \leq B\|\widehat{f}\|^2 \quad \text{for all } \widehat{f} \in L^2[-\pi,\pi],$$

which implies that the sequence $\{\frac{e^{-it_n w}}{\sqrt{2\pi}}\}_{n=-\infty}^{\infty}$ is a frame for $L^2[-\pi,\pi]$.

Besides, the interpolation problem can be written as

$$\langle \widehat{f}, \frac{e^{-it_n w}}{\sqrt{2\pi}}\rangle = c_n\,, \quad n \in \mathbb{Z},$$

which implies that the analysis operator $T^*$ associated with the frame $\{\frac{e^{-it_n w}}{\sqrt{2\pi}}\}_{n=-\infty}^{\infty}$ is onto and, as a consequence, $T^*$ (and $T$) is an isomorphism from $L^2[-\pi,\pi]$ onto $\ell^2(\mathbb{Z})$. Hence, the sequence $\{e^{-it_n w}/\sqrt{2\pi}\}_{n=-\infty}^{\infty}$ is a Riesz basis for $L^2[-\pi,\pi]$.

Notice that the uniqueness in $(i)$ comes out from the fact that $T^*$ is always injective for a frame. □

A sequence $\{t_n\}_{n=-\infty}^{\infty}$ satisfying $(i)$ and $(ii)$ is said to be a *complete interpolation sequence (CIS)* for $PW_\pi$.

**Proposition 2.8** *Any complete interpolating sequence* $\{t_n\}_{n=-\infty}^{\infty}$ *is a* uniformly discrete set, *that is, there exists* $\delta > 0$ *such that*

$$\inf_{n\neq m} |t_n - t_m| = \delta > 0\,.$$

*Thus, unless otherwise specified the* $t_n$ *are enumerated as* $t_n \leq t_{n+1}$, *and such that* $t_n \leq 0$ *for* $n \leq 0$, *and* $t_n \geq 0$ *for* $n \geq 0$. *If* $0$ *belongs to the sequence we set* $t_0 = 0$.

***Proof*** The solutions $f_n \in PW_\pi$, $n \in \mathbb{Z}$, of the interpolation problems $f_n(t_m) = \delta_{n,m}$ are uniformly bounded. In fact, they satisfy that $\|f_n\|^2 \leq \frac{1}{A}$.

Indeed, using $(ii)$ we obtain

$$A\|f_n\|^2 \leq \sum_{m=-\infty}^{\infty} |f_n(t_m)|^2 = 1\,.$$

On the other hand, for $n \neq m$ we have

$$\begin{aligned}
1 = f_n(t_n) - f_n(t_m) &= \int_{t_m}^{t_n} f_n'(t)\,dt \leq |t_n - t_m| \sup_{t\in\mathbb{R}} |f_n'(t)| \\
&\leq |t_n - t_m|\,\|f_n'\| \leq |t_n - t_m|\,\pi\,\|f_n\| \\
&\leq \frac{\pi}{\sqrt{A}}|t_n - t_m|\,,
\end{aligned}$$

where we have used the Bernstein inequality (2.9) for $p = 1$. Hence, for $n \neq m$ we have $|t_n - t_m| \geq \frac{\sqrt{A}}{\pi}$, and the sequence $\{t_n\}_{n=-\infty}^{\infty}$ is a uniformly discrete set. □

Next, we include a generalization of the Paley-Wiener-Levinson theorem borrowed from Ref. [2]. As it was said above, the real character of the sampling sequence $\{t_n\}_{n=-\infty}^{\infty}$ can be removed.

**Theorem 2.5** *Let $\{t_n\}_{n=-\infty}^{\infty}$ be a (real) CIS for $PW_\pi$. Then, the following interpolation formula holds in $PW_\pi$*

$$f(z) = \sum_{n=-\infty}^{\infty} f(t_n) \frac{G(z)}{G'(t_n)(z - t_n)}, \qquad z \in \mathbb{C},$$

*where $G$ is the entire function*

$$G(z) = (z - t_0) \prod_{n=1}^{\infty} \Big(1 - \frac{z}{t_n}\Big)\Big(1 - \frac{z}{t_{-n}}\Big).$$

*The series converges absolutely, in $L^2(\mathbb{R})$-norm, and uniformly on any horizontal strip in $\mathbb{C}$ of finite width.*

***Proof*** Let $G_0$ be the unique solution in $PW_\pi$ of the interpolation problem:

$$f(t_n) = \delta_{0,n}, \quad n \in \mathbb{Z}. \tag{2.14}$$

We prove some facts about this function:

- The function $G_0$ does not have any other zeros than $\{t_n\}_{n\neq 0}$. If $\alpha$ is another zero, then the function $g(z) = \frac{t_0-\alpha}{z-\alpha} G_0(z)$, which belongs to $PW_\pi$ by using Theorem 2.3, would be another solution of the interpolation problem (2.14). If 0 belongs to $\{t_n\}_{n=-\infty}^{\infty}$ we set $t_0 = 0$. In any case we obtain that $G_0(0) \neq 0$.
- Any $t_n$, $n \neq 0$, must be a simple zero of $G_0$. Otherwise, the function $g(z) = \frac{t_0-t_n}{z-t_n} G_0(z)$ in $PW_\pi$ would be another solution of (2.14).
- The interval $[-\pi, \pi]$ is the smallest interval containing the support of $\widehat{G}_0$. Indeed, if the support is contained in a proper interval of $[-\pi, \pi]$, there exists $\tau \neq 0$ in $\mathbb{R}$ such that the function $e^{i\tau z}\, G_0(z)$ belongs to $PW_\pi$; then, the function $g(z) = e^{i\tau(z-t_0)}\, G_0(z)$ would be another solution of (2.14).

From Theorem 2.4, we have

$$G_0(z) = G_0(0) \lim_{N\to\infty} \prod_{1\leq|n|\leq N} \Big(1 - \frac{z}{t_n}\Big),$$

and

$$G(z) = (z - t_0)\,\frac{G_0(z)}{G_0(0)} = (z - t_0)\,\lim_{N\to\infty} \prod_{1\le|n|\le N} \Big(1 - \frac{z}{t_n}\Big)$$

is an entire function such that the functions $G_n(z) := \frac{G(z)}{G'(t_n)(z-t_n)}$, $n \in \mathbb{Z}$, are the unique solutions in $PW_\pi$ of the interpolation problems

$$f_n(t_m) = \delta_{n,m}\,,\ m \in \mathbb{Z}\,, \quad \text{for } n \in \mathbb{Z}\,.$$

Next we prove that $\{\widehat{G}_n\}_{n=-\infty}^{\infty}$ is the biorthogonal Riesz basis of $\big\{e^{-it_n w}/\sqrt{2\pi}\big\}_{n=-\infty}^{\infty}$ in $L^2[-\pi,\pi]$. Indeed, for $n, m \in \mathbb{Z}$ we have

$$\begin{aligned}\delta_{n,m} = G_n(t_m) &= \frac{1}{\sqrt{2\pi}} \int_{-\pi}^{\pi} \widehat{G}_n(\omega)\, e^{it_m\omega}\, d\omega \\ &= \Big\langle \widehat{G}_n, \frac{e^{-it_m w}}{\sqrt{2\pi}} \Big\rangle .\end{aligned}$$

Since the Fourier transform is a unitary operator between $PW_\pi$ and $L^2[-\pi,\pi]$, the sequences $\big\{G_n\big\}_{n=-\infty}^{\infty}$ and $\Big\{\frac{\sin \pi(z-t_n)}{\pi(z-t_n)}\Big\}_{n=-\infty}^{\infty}$ form a pair of dual Riesz bases for $PW_\pi$. Therefore, for $f \in PW_\pi$ we have

$$f = \sum_{n=-\infty}^{\infty} \Big\langle f, \frac{\sin \pi(t-t_n)}{\pi(t-t_n)} \Big\rangle G_n = \sum_{n=-\infty}^{\infty} f(t_n)\, G_n \quad \text{in } PW_\pi\,,$$

that is,

$$f(z) = \sum_{n=-\infty}^{\infty} f(t_n)\, \frac{G(z)}{G'(t_n)(z-t_n)}\,, \qquad z \in \mathbb{C}\,.$$

The convergence of the series is absolute because $\big\{G_n\big\}_{n=-\infty}^{\infty}$ is a Riesz basis; the rest of properties comes out since $PW_\pi$ is a RKHS. □

Finally, notice that the characterization of the Riesz bases $\big\{e^{-it_n w}/\sqrt{2\pi}\big\}_{n=-\infty}^{\infty}$ for $L^2[-\pi,\pi]$, without the explicit hypothesis on the frame condition (2.13), is a very much deeper problem solved by B. S. Pavlov in 1979. For a brief discussion on this problem see, for instance, Ref. [21, p. 105].

### 2.2.3 Sampling Through Another Type of Samples

It is possible to recover any function in a Paley-Wiener space by using samples of another related function such as the derivative, the Hilbert transform or, in general, some filtered (obtained by means of a convolution with a fixed function) versions of the function itself.

#### Sampling in $PW_\pi$ by using Samples of the Derivative

Any function $f \in PW_\pi$ can be recovered by using its samples $\{f(2n)\}_{n=-\infty}^{\infty}$ taken at half the due sampling rate, along with the samples $\{f'(2n)\}_{n=-\infty}^{\infty}$ taken from its first derivative.

**Proposition 2.9** *Any function $f \in PW_\pi$ can be recovered from the two sets of samples $\{f(2n)\}_{n=-\infty}^{\infty}$ and $\{f'(2n)\}_{n=-\infty}^{\infty}$ by means of the formula:*

$$f(t) = \sum_{n=-\infty}^{\infty} \{f(2n)+(t-2n)f'(2n)\}\left[\frac{\sin\frac{\pi}{2}(t-2n)}{\frac{\pi}{2}(t-2n)}\right]^2, \qquad t \in \mathbb{R}. \qquad (2.15)$$

***Proof*** For the proof, let $\widehat{f} \in L^2[-\pi,\pi]$ be the Fourier transform of $f$; having in mind its $2\pi$-periodic extension, the following Fourier expansions in $L^2[-\pi,\pi]$ hold

$$\widehat{f}(\omega) = \sum_{n=-\infty}^{\infty} f(n)\frac{e^{-in\omega}}{\sqrt{2\pi}} \quad \text{and} \quad \widehat{f}(\omega-\pi) = \sum_{n=-\infty}^{\infty} (-1)^n f(n)\frac{e^{-in\omega}}{\sqrt{2\pi}}\,.$$

As a consequence, the $\pi$-periodic extension of the function

$$S(\omega) = \frac{1}{2}[\widehat{f}(\omega)+\widehat{f}(\omega-\pi))]$$

admits the Fourier expansion

$$S(\omega) = \sum_{n=-\infty}^{\infty} f(2n)\frac{e^{-i2n\omega}}{\sqrt{2\pi}} \quad \text{in } L^2[0,\pi]\,.$$

In a similar way, since

$$f'(t) = \frac{1}{\sqrt{2\pi}}\int_{-\pi}^{\pi} i\omega\widehat{f}(\omega)e^{it\omega}d\omega\,, \quad t \in \mathbb{R}\,,$$

the following expansions in $L^2[-\pi, \pi]$ hold:

$$i\omega \widehat{f}(\omega) = \sum_{n=-\infty}^{\infty} f'(n) \frac{e^{-in\omega}}{\sqrt{2\pi}},$$

and

$$i(\omega - \pi)\widehat{f}(\omega - \pi) = \sum_{n=-\infty}^{\infty} (-1)^n f'(n) \frac{e^{-in\omega}}{\sqrt{2\pi}}.$$

Hence, the $\pi$-periodic extension of the function

$$R(\omega) = \frac{i}{2}[\omega \widehat{f}(\omega) + (\omega - \pi)\widehat{f}(\omega - \pi)]$$

admits the Fourier expansion

$$R(\omega) = \sum_{n=-\infty}^{\infty} f'(2n) \frac{e^{-i2n\omega}}{\sqrt{2\pi}} \quad \text{in } L^2[0, \pi].$$

Grouping both expansions, for $\omega \in [0, \pi]$, we have

$$\begin{pmatrix} S(\omega) \\ R(\omega) \end{pmatrix} = \frac{1}{2} \begin{pmatrix} 1 & 1 \\ i\omega & i(\omega - \pi) \end{pmatrix} \begin{pmatrix} \widehat{f}(\omega) \\ \widehat{f}(\omega - \pi) \end{pmatrix},$$

and, inverting the matrix

$$\begin{pmatrix} \widehat{f}(\omega) \\ \widehat{f}(\omega - \pi) \end{pmatrix} = \frac{2i}{\pi} \begin{pmatrix} i(\omega - \pi) & -1 \\ -i\omega & 1 \end{pmatrix} \begin{pmatrix} S(\omega) \\ R(\omega) \end{pmatrix}. \tag{2.16}$$

Introducing this splitting of $\widehat{f}$ into (2.3), after some calculations, yields

$$\begin{aligned}
f(t) &= \frac{1}{\sqrt{2\pi}} \int_{-\pi}^{\pi} \widehat{f}(\omega) e^{it\omega} d\omega \\
&= \frac{1}{\sqrt{2\pi}} \int_{-\pi}^{0} \sum_{n=-\infty}^{\infty} [\frac{2}{\pi}(\omega + \pi) f(2n) + \frac{2i}{\pi} f'(2n)] \frac{e^{-i2n\omega}}{\sqrt{2\pi}} e^{it\omega} d\omega \\
&+ \frac{1}{\sqrt{2\pi}} \int_{0}^{\pi} \sum_{n=-\infty}^{\infty} [\frac{2}{\pi}(\pi - \omega) f(2n) - \frac{2i}{\pi} f'(2n)] \frac{e^{-i2n\omega}}{\sqrt{2\pi}} e^{it\omega} d\omega \\
&= \frac{1}{\sqrt{2\pi}} \sum_{n=-\infty}^{\infty} \Big\{ \int_{-\pi}^{\pi} \sqrt{\frac{2}{\pi}} (1 - \frac{|\omega|}{\pi}) f(2n) e^{i(t-2n)\omega} d\omega \\
&+ \frac{2}{\pi} \int_{-\pi}^{\pi} \frac{(-i \operatorname{sgn} \omega)}{\sqrt{2\pi}} f'(2n) e^{i(t-2n)\omega} d\omega \Big\}, \quad t \in \mathbb{R}.
\end{aligned}$$

The desired result comes by using the Fourier duals

$$\operatorname{sinc}\left(\frac{t}{2}\right)\sin\left(\frac{\pi t}{2}\right)=\frac{1}{\sqrt{2\pi}}\int_{-\pi}^{\pi}\frac{(-i\,\operatorname{sgn}\omega)}{\sqrt{2\pi}}\mathrm{e}^{it\omega}d\omega$$

and

$$\operatorname{sinc}^2\left(\frac{t}{2}\right)=\frac{1}{\sqrt{2\pi}}\int_{-\pi}^{\pi}\sqrt{\frac{2}{\pi}}\left(1-\frac{|\omega|}{\pi}\right)\mathrm{e}^{it\omega}d\omega\,.$$

□

It is worth noticing that formula (2.15) is nothing but a Riesz basis expansion in $PW_\pi$ following a pair of dual Riesz bases for $PW_\pi$, having, via Fourier transform, its counterpart in $L^2[-\pi,\pi]$. In the present case, the involved dual Riesz bases for $L^2[-\pi,\pi]$ are:

$$\left\{\frac{1}{\sqrt{\pi}}\,\mathrm{e}^{-2in\omega}\right\}_{n=-\infty}^{\infty}\cup\left\{\frac{1}{\sqrt{\pi}}\,i\omega\,\mathrm{e}^{-2in\omega}\right\}_{n=-\infty}^{\infty} \tag{2.17}$$

whose biorthogonal basis is given by

$$\left\{\left(1-\frac{|\omega|}{\pi}\right)\frac{\mathrm{e}^{-2in\omega}}{\sqrt{\pi}}\right\}_{n=-\infty}^{\infty}\cup\left\{\frac{-i\,\operatorname{sgn}\omega}{\pi}\,\frac{\mathrm{e}^{-2in\omega}}{\sqrt{\pi}}\right\}_{n=-\infty}^{\infty} \tag{2.18}$$

To prove this, observe that we can write

$$\begin{aligned}f(t)=\frac{1}{\sqrt{2\pi}}\int_{-\pi}^{\pi}\sum_{n=-\infty}^{\infty}\Big[&f(2n)\sqrt{\frac{2}{\pi}}\big(1-\frac{|\omega|}{\pi}\big)\,\mathrm{e}^{-i2n\omega}\\&+f'(2n)\frac{2}{\pi}\frac{(-i\,\operatorname{sgn}\omega)}{\sqrt{2\pi}}\,\mathrm{e}^{-i2n\omega}\Big]\mathrm{e}^{it\omega}d\omega\,,\end{aligned}$$

where we have used the continuity of the inner product in $L^2[-\pi,\pi]$ with respect to the $L^2$-norm convergence for interchanging the sum and the integral.

Hence, we obtain the expansion in $L^2[-\pi,\pi]$:

$$\begin{aligned}\widehat{f}(\omega)&=\sum_{n=-\infty}^{\infty}\Big[f(2n)\sqrt{\frac{2}{\pi}}\,\big(1-\frac{|\omega|}{\pi}\big)\mathrm{e}^{-i2n\omega}+f'(2n)\frac{2}{\pi}\frac{(-i\,\operatorname{sgn}\omega)}{\sqrt{2\pi}}\mathrm{e}^{-i2n\omega}\Big]\\&=\sum_{n=-\infty}^{\infty}\Big[\big\langle\widehat{f},\frac{\mathrm{e}^{-2in\omega}}{\sqrt{\pi}}\big\rangle\big(1-\frac{|\omega|}{\pi}\big)\frac{\mathrm{e}^{-i2n\omega}}{\sqrt{\pi}}+\big\langle\widehat{f},\frac{i\omega\mathrm{e}^{-2in\omega}}{\sqrt{\pi}}\big\rangle\frac{(-i\,\operatorname{sgn}\omega)}{\pi}\frac{\mathrm{e}^{-i2n\omega}}{\sqrt{\pi}}\Big].\end{aligned}$$

Having in mind the above expansion, to obtain the desired result it is enough to prove that (2.17) is a Riesz basis for $L^2[-\pi,\pi]$, and (2.18) is a Bessel sequence in $L^2[-\pi,\pi]$ (or vice-versa).

To this end, we define the linear operator $\mathbf{\Phi} : L^2[0,\pi] \times L^2[0,\pi] \longrightarrow L^2[-\pi,\pi]$ as

$$\mathbf{\Phi}(f,g)(\omega) = f(\omega) + i\omega g(\omega)\,, \qquad \text{a.e. } \omega \in [-\pi,\pi] \tag{2.19}$$

where we are considering the $\pi$-periodic extensions of $f$ and $g$ and the Hilbert product space $L^2[0,\pi] \times L^2[0,\pi]$ with its usual inner product

$$\langle (f_1,g_1),(f_2,g_2)\rangle = \langle f_1,f_2\rangle + \langle g_1,g_2\rangle\,.$$

Since the sequence

$$\Big\{\Big(\frac{\mathrm{e}^{-2in\omega}}{\sqrt{\pi}},0\Big)\Big\}_{n=-\infty}^{\infty} \cup \Big\{\Big(0,\frac{\mathrm{e}^{-2in\omega}}{\sqrt{\pi}}\Big)\Big\}_{n=-\infty}^{\infty}$$

is an orthonormal basis for $L^2[0,\pi] \times L^2[0,\pi]$, and

$$\mathbf{\Phi}\Big(\frac{\mathrm{e}^{-2in\omega}}{\sqrt{\pi}},0\Big) = \frac{\mathrm{e}^{-2in\omega}}{\sqrt{\pi}}\,; \quad \mathbf{\Phi}\Big(0,\frac{\mathrm{e}^{-2in\omega}}{\sqrt{\pi}}\Big) = \frac{i\omega \mathrm{e}^{-2in\omega}}{\sqrt{\pi}}\,, \qquad n \in \mathbb{Z}\,,$$

we have to prove that $\mathbf{\Phi}$ is a bounded bijective operator, that is an isomorphism between Hilbert spaces.

Having in mind the inequality $|a+b|^2 \le 2(|a|^2+|b|^2)$ for $a,b \in \mathbb{C}$, it is easy to get (see next section for a complete proof)

$$\|\mathbf{\Phi}(f,g)\|^2_{L^2[-\pi,\pi]} \le 4\pi^2 \|(f,g)\|^2_{L^2[0,\pi]\times L^2[0,\pi]}\,,$$

for all $(f,g) \in L^2[0,\pi] \times L^2[0,\pi]$, i.e., the operator $\mathbf{\Phi}$ is bounded.

To prove that $\mathbf{\Phi}$ is bijective, let $F$ be in $L^2[-\pi,\pi]$ and we are looking for $f,g \in L^2[0,\pi]$ such that $\mathbf{\Phi}(f,g) = F$. In other words,

$$F(\omega) = f(\omega) + i\omega g(\omega)\,, \qquad \text{a.e. } \omega \in [0,\pi]$$

and

$$\begin{aligned} F(\omega-\pi) &= f(\omega-\pi) + i(\omega-\pi)g(\omega-\pi) \\ &= f(\omega) + i(\omega-\pi)g(\omega)\,, \quad \text{a.e. } \omega \in [0,\pi]\,, \end{aligned}$$

or, in matrix notation

$$\begin{pmatrix} F(\omega) \\ F(\omega-\pi) \end{pmatrix} = \begin{pmatrix} 1 & i\omega \\ 1 & i(\omega-\pi) \end{pmatrix} \begin{pmatrix} f(\omega) \\ g(\omega) \end{pmatrix}, \qquad \text{a.e. } \omega \in [0,\pi]$$

Inverting the matrix

$$\begin{pmatrix} f(\omega) \\ g(\omega) \end{pmatrix} = \frac{i}{\pi} \begin{pmatrix} i(\omega-\pi) & -i\omega \\ -1 & 1 \end{pmatrix} \begin{pmatrix} F(\omega) \\ F(\omega-\pi) \end{pmatrix},$$

we obtain $f, g$ in $L^2[0,\pi]$ such that $\mathbf{\Phi}(f,g) = F$; obviously, $\mathbf{\Phi}$ is also injective.

Finally, to prove that (2.18) is a Bessel sequence, it suffices to prove that the operator

$$\mathbf{\Psi}(f,g)(\omega) = \Big(1 - \frac{|\omega|}{\pi}\Big) f(\omega) + \frac{-i\,\operatorname{sgn}\omega}{\pi} g(\omega), \qquad \text{a.e. } \omega \in [-\pi,\pi]$$

is also bounded; the proof goes in the same manner.

By using the inverse Fourier transform, the counterpart in $PW_\pi$ of the pair of dual Riesz bases (2.17) and (2.18) for $L^2[-\pi,\pi]$ are, respectively, the pair of dual Riesz bases

$$\big\{\sqrt{2}\operatorname{sinc}(t-2n)\big\}_{n=-\infty}^{\infty} \cup \big\{\sqrt{2}\operatorname{sinc}'(t-2n)\big\}_{n=-\infty}^{\infty}$$

and

$$\Big\{\operatorname{sinc}^2\Big(\frac{t-2n}{2}\Big)\Big\}_{n=-\infty}^{\infty} \cup \Big\{\frac{2}{\pi}\operatorname{sinc}\Big(\frac{t-2n}{2}\Big)\sin\Big(\frac{\pi(t-2n)}{2}\Big)\Big\}_{n=-\infty}^{\infty}$$

See Ref. [22, pp. 56–77] for an account of derivative sampling in $PW_\pi$, including historical notes and different mathematical techniques.

### Two-Channel Sampling in the Paley-Wiener Space $PW_\pi$

In general, we can replace the particular *multipliers* $m_1(\omega) = 1$ and $m_2(\omega) = i\omega$, $\omega \in [-\pi,\pi]$ in (2.19) by two measurable functions $m_1$ and $m_2$ defined in $[-\pi,\pi]$ and consider the new linear operator $\mathbf{\Phi} : L^2[0,\pi] \times L^2[0,\pi] \longrightarrow L^2[-\pi,\pi]$ defined as

$$\mathbf{\Phi}(f,g)(\omega) = m_1(\omega)\, f(\omega) + m_2(\omega)\, g(\omega), \qquad \text{a.e. } \omega \in [-\pi,\pi],$$

where we take $\pi$-periodic extensions of $f$ and $g$.

(*a*) In the same manner, under the following hypotheses on $m_1$ and $m_2$:

(i) The functions $m_1, m_2 \in L^\infty[-\pi,\pi]$, and

(ii) $|\det M(\omega)| > 0$ a.e. $\omega \in [0,\pi]$ where $M(\omega) = \begin{pmatrix} m_1(\omega) & m_2(\omega) \\ m_1(\omega-\pi) & m_2(\omega-\pi) \end{pmatrix}$, we deduce that the sequence

$$\left\{m_1(\omega)\frac{\mathrm{e}^{-2in\omega}}{\sqrt{\pi}}\right\}_{n=-\infty}^{\infty} \cup \left\{m_2(\omega)\frac{\mathrm{e}^{-2in\omega}}{\sqrt{\pi}}\right\}_{n=-\infty}^{\infty} \tag{2.20}$$

is a Riesz basis for $L^2[-\pi, \pi]$. As before, the operator $\mathbf{\Phi}$ should be an isomorphism. Having in mind the open mapping theorem, it is sufficient to prove that $\mathbf{\Phi}$ is bounded and bijective.

For the boundedness, for each $f, g \in L^2[0, \pi]$ we have

$$\begin{aligned}
\|\mathbf{\Phi}(f, g)\|^2_{L^2[-\pi,\pi]} &= \int_{-\pi}^{\pi} |m_1(\omega)\, f(\omega) + m_2(\omega)\, g(\omega)|^2 d\omega \\
&\le \max\left(\|m_1\|^2_\infty, \|m_2\|^2_\infty\right) \int_{-\pi}^{\pi} \left(|f(\omega)| + |g(\omega)|\right)^2 d\omega \\
&\le 2K \int_{-\pi}^{\pi} \left(|f(\omega)|^2 + |g(\omega)|^2\right) d\omega = 4K \int_{0}^{\pi} \left(|f(\omega)|^2 + |g(\omega)|^2\right) d\omega \\
&= 4K\left(\|f\|^2_{L^2[0,\pi]} + \|g\|^2_{L^2[0,\pi]}\right) = 4K\, \|(f, g)\|^2_{L^2[0,\pi]\times L^2[0,\pi]},
\end{aligned}$$

where $K = \max\left(\|m_1\|^2_\infty, \|m_2\|^2_\infty\right)$, i.e., the operator $\mathbf{\Phi}$ is bounded.

To prove that $\mathbf{\Phi}$ is bijective, observe that for each $F \in L^2[-\pi, \pi]$ we can write

$$\begin{aligned}
F(\omega) &= m_1(\omega) f(\omega) + m_2(\omega) g(\omega)\,, \\
F(\omega - \pi) &= m_1(\omega - \pi) f(\omega) + m_2(\omega - \pi) g(\omega)\,, \qquad \text{a.e. } \omega \in [0, \pi]\,,
\end{aligned}$$

or, in matrix notation, for a.e. $\omega \in [0, \pi]$, we get

$$\begin{pmatrix} F(\omega) \\ F(\omega - \pi) \end{pmatrix} = \begin{pmatrix} m_1(\omega) & m_2(\omega) \\ m_1(\omega - \pi) & m_2(\omega - \pi) \end{pmatrix} \begin{pmatrix} f(\omega) \\ g(\omega) \end{pmatrix} = M(\omega) \begin{pmatrix} f(\omega) \\ g(\omega) \end{pmatrix}.$$

At this point, we use that $M(\omega)$ is an invertible matrix with $|\det M(\omega)| > 0$ a.e. $\omega \in [0, \pi]$ in order to obtain $f, g \in L^2[0, \pi]$ such that $\Phi(f, g) = F$.

Observe that we have proved that the operator $\mathbf{\Phi}$ can be represented by the matrix-valued function $M(\omega)$ and vice-versa, any matrix $M(\omega)$ defines an operator $\mathbf{\Phi}$.

Notice that, for bounded multipliers $m_1$ and $m_2$, in case the sequence (2.20) is a Riesz basis for $L^2[-\pi, \pi]$ the sufficient condition $|\det M(\omega)| > 0$ a.e. $\omega$ in $[0, \pi]$ is also necessary.

(*b*) According to the Riesz bases theory (see Proposition 1.8), the dual Riesz basis of the sequence (2.20) is given by

$$\left\{(\mathbf{\Phi}^*)^{-1}\Big(\frac{\mathrm{e}^{-2in\omega}}{\sqrt{\pi}}, 0\Big)\right\}_{n=-\infty}^{\infty} \cup \left\{(\mathbf{\Phi}^*)^{-1}\Big(0, \frac{\mathrm{e}^{-2in\omega}}{\sqrt{\pi}}\Big)\right\}_{n=-\infty}^{\infty}.$$

Since the operator $\mathbf{\Phi}$ is associated with the matrix-valued function $M(\omega)$, one can easily deduce that the operator $(\mathbf{\Phi}^*)^{-1}$ is associated with the matrix-valued function $(M^*)^{-1}(\omega)$. A straightforward calculation gives

$$(M^*)^{-1}(\omega) = \frac{1}{\det M(\omega)} \begin{pmatrix} \overline{m_2(\omega-\pi)} & -\overline{m_1(\omega-\pi)} \\ -\overline{m_2(\omega)} & \overline{m_1(\omega)} \end{pmatrix}, \qquad \text{a.e. } \omega \in [0,\pi]$$

Hence,

$$(\mathbf{\Phi}^*)^{-1}(f,g)(\omega) = n_1(\omega)\, f(\omega) + n_2(\omega)\, g(\omega), \qquad \text{a.e. } \omega \in [-\pi,\pi]$$

where

$$n_1(\omega) = \frac{1}{\det M(\omega)} \begin{cases} \overline{m_2(\omega-\pi)} & \text{if } \omega \in [0,\pi] \\ -\overline{m_2(\omega+\pi)} & \text{if } \omega \in [-\pi,0] \end{cases},$$

and

$$n_2(\omega) = \frac{-1}{\det M(\omega)} \begin{cases} \overline{m_1(\omega-\pi)} & \text{if } \omega \in [0,\pi] \\ -\overline{m_1(\omega+\pi)} & \text{if } \omega \in [-\pi,0] \end{cases}.$$

Finally, the dual Riesz basis of (2.20) is

$$\Big\{n_1(\omega)\frac{e^{-2in\omega}}{\sqrt{\pi}}\Big\}_{n=-\infty}^{\infty} \cup \Big\{n_2(\omega)\frac{e^{-2in\omega}}{\sqrt{\pi}}\Big\}_{n=-\infty}^{\infty}.$$

(*c*) Next we deduce the associated sampling formula in $PW_\pi$. To this end, for each $f \in PW_\pi$ we expand $\widehat{f} \in L^2[-\pi,\pi]$ following the above dual Riesz bases obtaining

$$\begin{aligned} \widehat{f} &= \sum_{n=-\infty}^{\infty} \Big[\Big\langle \widehat{f}, n_1(\omega)\frac{e^{-2in\omega}}{\sqrt{\pi}}\Big\rangle m_1(\omega)\frac{e^{-2in\omega}}{\sqrt{\pi}} + \Big\langle \widehat{f}, n_2(\omega)\frac{e^{-2in\omega}}{\sqrt{\pi}}\Big\rangle m_2(\omega)\frac{e^{-2in\omega}}{\sqrt{\pi}}\Big] \\ &= \sum_{n=-\infty}^{\infty} \Big[\sqrt{2} f_1(2n)\, m_1(\omega)\frac{e^{-2in\omega}}{\sqrt{\pi}} + \sqrt{2} f_2(2n)\, m_2(\omega)\frac{e^{-2in\omega}}{\sqrt{\pi}}\Big], \end{aligned} \tag{2.21}$$

where

$$f_1(t) = \frac{1}{\sqrt{2\pi}} \int_{-\pi}^{\pi} \widehat{f}(\omega)\, \overline{n_1(\omega)}\, e^{it\omega} d\omega = \frac{1}{\sqrt{2\pi}}(f * h_1)(t) \quad \text{with } h_1 = \mathcal{F}^{-1}(\overline{n}_1)$$

and

$$f_2(t)=\frac{1}{\sqrt{2\pi}}\int_{-\pi}^{\pi}\widehat{f}(\omega)\,\overline{n_2(\omega)}\,e^{it\omega}d\omega=\frac{1}{\sqrt{2\pi}}\big(f*h_2\big)(t)\quad\text{with } h_2=\mathcal{F}^{-1}(\overline{n}_2)\,.$$

In other words, we are taking samples $\{f_1(2n)\}_{n=-\infty}^{\infty}\cup\{f_2(2n)\}_{n=-\infty}^{\infty}$ of two filtered versions $f_1$ and $f_2$ of the function $f$ to be recovered. The *transfer functions (multipliers)* of the involved filters, whose *impulse responses* are $h_1$ and $h_2$, are $\overline{n}_1$ and $\overline{n}_2$ respectively.

Notice that, in this case, for $f, g \in L^2(\mathbb{R})$ the convolution $f*g$ belongs to $L^2(\mathbb{R})$ since $\widehat{f}\,\widehat{g}\in L^2(\mathbb{R})$. Besides, we have $\widehat{f*g}=\sqrt{2\pi}\,\widehat{f}\,\widehat{g}$ (see, [13, Theorem 2.62]).

Now, the inverse Fourier transform $\mathcal{F}^{-1}$ in (2.21) gives, in the Paley-Wiener space $PW_\pi$, the sampling formula

$$f(t)=\sum_{n=-\infty}^{\infty}\Big[f_1(2n)\,S_1(t-2n)+f_2(2n)\,S_2(t-2n)\Big],\qquad t\in\mathbb{R},$$

where

$$S_1(t)=\sqrt{\frac{2}{\pi}}\mathcal{F}^{-1}(m_1)(t)\quad\text{and}\quad S_2(t)=\sqrt{\frac{2}{\pi}}\mathcal{F}^{-1}(m_2)(t)\,,\quad t\in\mathbb{R}\,.$$

The series converges in $L^2(\mathbb{R})$-norm, absolutely and uniformly on $\mathbb{R}$.

It is worth mentioning here that Chap. 5 in the present book is devoted to generalize the average sampling to different general settings. For multichannel sampling, see also Refs. [17, 21].

### Sampling by using Samples of the Hilbert Transform

The aim here is to obtain a sampling formula for each real-valued *passband function*, that is, a function $f\in L^2(\mathbb{R})$ whose Fourier transform $\widehat{f}$ is zero outside a set $[-w_0-\pi,-w_0]\cup[w_0,w_0+\pi]$ by using the sequences of samples $\{f(2n)\}_{n=-\infty}^{\infty}$ of $f$ and $\{Hf(2n)\}_{n=-\infty}^{\infty}$ of its Hilbert transform $Hf$.

**Definition 2.4** The Hilbert transform $H$ in $L^2(\mathbb{R})$ is defined as:

$$H: L^2(\mathbb{R})\longrightarrow L^2(\mathbb{R})$$
$$f\longmapsto Hf:=\mathcal{F}^{-1}[(-i\,\mathrm{sgn}\,w)\mathcal{F}f(w)]\,.$$

**Proposition 2.10** *The Hilbert transform $H$ defined above is a unitary operator on $L^2(\mathbb{R})$.*

***Proof*** Since the Fourier transform and its inverse are linear, then $H$ is also linear. The operator $H$ is an isometry in $L^2(\mathbb{R})$. Indeed, for $f \in L^2(\mathbb{R})$ we get

$$\|Hf\|_2 = \|\mathcal{F}(Hf)\|_2 = \|(-i\,\text{sgn})(\mathcal{F}f)\|_2 = \|\mathcal{F}f\|_2 = \|f\|_2\,.$$

The Hilbert transform $H$ is bijective and its inverse is $H^{-1} = -H$. Indeed,

$$\mathcal{F}[H(Hf)] = (-i\,\text{sgn})\mathcal{F}(Hf) = (-i\,\text{sgn})(-i\,\text{sgn})\mathcal{F}f = -\mathcal{F}f = \mathcal{F}(-f)\,,$$

that is, $H^2 = -I_{L^2(\mathbb{R})}$. □

Besides, for a real-valued function $f \in L^2(\mathbb{R})$, its Hilbert transform $Hf$ is also a real-valued function. Indeed, since

$$Hf(t) = \lim_{N\to\infty} \frac{1}{\sqrt{2\pi}} \int_{-N}^{N} (-i\,\text{sgn}\,\omega)\widehat{f}(\omega)\,\mathrm{e}^{i\omega t}\,d\omega \quad \text{in } L^2(\mathbb{R})\,,$$

and $\overline{\widehat{f}(w)} = \widehat{f}(-w)$ as $f$ is real-valued, we get

$$\overline{Hf(t)} = \lim_{N\to\infty} \frac{1}{\sqrt{2\pi}} \int_{-N}^{N} i\,\text{sgn}\,\omega\,\widehat{f}(-\omega)\,\mathrm{e}^{-i\omega t}\,d\omega\,.$$

The change of variable $\xi = -\omega$ allows us to conclude that $\overline{Hf(t)} = Hf(t)$ a.e. in $\mathbb{R}$.

For a real-valued function $f \in L^2(\mathbb{R})$, its associated *analytic signal* $f_a$ is defined by

$$f_a := f + iHf\,. \tag{2.22}$$

Its Fourier transform satisfies $\mathcal{F}f_a = \mathcal{F}f + i(-i\,\text{sgn})\mathcal{F}f = 2(\mathcal{F}f)\,u$, where $u$ denotes the *Heaviside function*

$$u(t) = \begin{cases} 1 & \text{if } t \geq 0 \\ 0 & \text{if } t < 0\,. \end{cases}$$

In order to deduce a sampling formula for a real-valued passband function $f$ to $[-w_0 - \pi, -w_0] \cup [w_0, w_0 + \pi]$ we proceed as follows: The analytic signal $f_a$ associated with $f$ is bandlimited to the interval $[w_0, w_0 + \pi]$ because

$$f_a(t) = \frac{1}{\sqrt{2\pi}} \int_{w_0}^{w_0+\pi} 2\widehat{f}(w)\,\mathrm{e}^{iwt}\,dw\,, \quad t \in \mathbb{R}\,.$$

From formulas (2.7) and (2.8) we obtain for $f_a$ the sampling formula

$$f_a(t) = \sum_{n=-\infty}^{\infty} f_a(2n)\, e^{i w_1 (t-2n)} \frac{\sin\pi(\frac{t}{2}-n)}{\pi(\frac{t}{2}-n)}, \qquad t \in \mathbb{R},$$

where $w_1 = w_0 + \frac{\pi}{2}$. As $f = \operatorname{Re} f_a$, we deduce for $f$ the sampling formula

$$f(t) = \sum_{n=-\infty}^{\infty} \big\{f(2n)\cos w_1(t-2n) - Hf(2n)\sin w_1(t-2n)\big\} \frac{\sin\pi(\frac{t}{2}-n)}{\pi(\frac{t}{2}-n)}, \qquad t \in \mathbb{R}.$$

In particular, for a real-valued $f \in PW_\pi$, as $w_0 = 0$, we obtain the sampling formula

$$f(t) = \sum_{n=-\infty}^{\infty} \Big\{f(2n)\cos\frac{\pi}{2}(t-2n) - Hf(2n)\sin\frac{\pi}{2}(t-2n)\Big\} \frac{\sin\pi(\frac{t}{2}-n)}{\pi(\frac{t}{2}-n)}, \qquad t \in \mathbb{R}.$$

We close this section with a couple of comments on the analytic signal associated with a real-valued function:

(*i*) The analytic signal is used in signal processing, for example, to define the *instantaneous frequency* of a real-valued signal $f$. We can write its analytic signal as $f_a(t) = A(t)\, e^{i\varphi(t)}$ in order to define its instantaneous frequency as $\omega := \varphi'(t)$.

Thus, for a fixed time $u$, the *Wigner-Ville time-frequency distribution* of $f_a$, given by

$$Wf_a(u,\xi) := \int_{-\infty}^{\infty} f_a\Big(u+\frac{\tau}{2}\Big)\,\overline{f_a\Big(u-\frac{\tau}{2}\Big)}\, e^{-i\tau\xi}\, d\tau,$$

is typically concentrated in a neighborhood of the instantaneous frequency $\xi = \varphi'(u)$ since

$$\varphi'(u) = \frac{\int_{-\infty}^{\infty} \xi\, Wf_a(u,\xi)\, d\xi}{\int_{-\infty}^{\infty} Wf_a(u,\xi)\, d\xi}$$

(see Ref. [24, p. 108]).

(*ii*) A mathematical comment concerning the term *analytic signal* used in this section is pertinent. The analytic signal $f_a$ defined in (2.22), is obviously not an analytic function in the mathematical context of complex analysis. However, if we define the function

$$F(z) = \frac{1}{\sqrt{2\pi}} \int_0^{\infty} 2\widehat{f}(\omega) e^{iz\omega} d\omega, \qquad z = t + iy \in \mathbb{C},$$

then we obtain an analytic function on the upper half-plane $\mathbb{C}^+ = \{z = t + iy \in \mathbb{C},\ y > 0\}$. Besides, its boundary function $\lim_{y\to 0} F(t+iy)$ coincides almost everywhere with $f_a \in L^2(\mathbb{R})$, the analytic signal associated with $f \in L^2(\mathbb{R})$. The mathematical details, involving the *Hardy space* $H^2(\mathbb{C}^+)$, can be found in the classical Ref. [11].

### 2.2.4 Multidimensional Shannon Sampling Formula

In this section the Fourier transform is defined as in (1.9) with inverse Fourier transform given in (1.10). For notational ease we consider here the bidimensional case; the $d$-dimensional case goes in the same manner.

Let $PW_{\mathcal{R}_{\sigma,\tau}}$ denote the set of functions in $L^2(\mathbb{R}^2)$ bandlimited to the rectangle $\mathcal{R}_{\sigma,\tau} := [-\sigma\pi, \sigma\pi] \times [-\tau\pi, \tau\pi]$.

Any $f \in PW_{\mathcal{R}_{\sigma,\tau}}$ admits the representation

$$f(t,s) = \left\langle \widehat{f}(\omega,\xi), \frac{1}{2\pi} e^{-it\omega} e^{-is\xi} \right\rangle_{L^2(\mathcal{R}_{\sigma,\tau})}, \quad (t,s) \in \mathbb{R}^2, \tag{2.23}$$

where $\widehat{f}(\omega,\xi)$ denotes the Fourier transform of $f$.

Having in mind that the sequence $\left\{\frac{1}{2\pi\sqrt{\sigma\tau}} e^{-in\omega/\sigma} e^{-im\xi/\tau}\right\}_{(n,m)\in\mathbb{Z}^2}$ is an orthonormal basis for $L^2(\mathcal{R}_{\sigma,\tau})$, we have the expansion

$$\widehat{f}(\omega,\xi) = \sum_{n,m} \frac{1}{\sqrt{\sigma\tau}} f\Big(\frac{n}{\sigma}, \frac{m}{\tau}\Big) \frac{e^{-in\omega/\sigma} e^{-im\xi/\tau}}{2\pi\sqrt{\sigma\tau}} \quad \text{in } L^2(\mathcal{R}_{\sigma,\tau}).$$

Introducing the above expansion in (2.23) one gets

$$\begin{aligned} f(t,s) &= \sum_{n,m} \frac{1}{(2\pi)^2\sigma\tau} f\Big(\frac{n}{\sigma}, \frac{m}{\tau}\Big) \int_{-\sigma\pi}^{\sigma\pi} e^{i(t-\frac{n}{\sigma})\omega}\, d\omega \int_{-\tau\pi}^{\tau\pi} e^{i(t-\frac{m}{\tau})\xi}\, d\xi \\ &= \sum_{n,m} \frac{1}{(2\pi)^2\sigma\tau} f\Big(\frac{n}{\sigma}, \frac{m}{\tau}\Big) 2\pi\sigma \frac{\sin\pi(\sigma t - n)}{\pi(\sigma t - n)} 2\pi\tau \frac{\sin\pi(\tau s - m)}{\pi(\tau s - m)}, \end{aligned}$$

that is, any function $f \in PW_{\mathcal{R}_{\sigma,\tau}}$ can be expanded as

$$f(t,s) = \sum_{(n,m)\in\mathbb{Z}^2} f\Big(\frac{n}{\sigma}, \frac{m}{\tau}\Big) \frac{\sin\pi(\sigma t - n)}{\pi(\sigma t - n)} \frac{\sin\pi(\tau s - m)}{\pi(\tau s - m)}, \quad (t,s) \in \mathbb{R}^2.$$

As in the unidimensional case, the convergence of the series is in $L^2$-norm, absolute and uniformly on $\mathbb{R}^2$ since the Paley-Wiener space

$$PW_{\mathcal{R}_{\sigma,\tau}} := \mathcal{F}^{-1}\big(L^2(\mathcal{R}_{\sigma,\tau})\big)$$

is also a RKHS. It comes out by applying Cauchy-Schwarz's inequality for (2.23). Indeed,

$$|f(t,s)| \leq \|\widehat{f}\|\,\sqrt{\sigma\tau} = \sqrt{\sigma\tau}\,\|f\|\,, \quad (t,s)\in\mathbb{R}^2\,,$$

i.e., each evaluation functional is bounded in $PW_{\mathcal{R}_{\sigma,\tau}}$, and the $L^2$-convergence implies pointwise convergence which is uniform on $\mathbb{R}^2$.

If the function $f$ belongs to the Paley-Wiener space $PW_{\mathcal{R}_\sigma}$, where $\mathcal{R}_\sigma$ denotes the $d$-dimensional rectangle

$$\mathcal{R}_\sigma = [-\pi\sigma_1, \pi\sigma_1]\times\cdots\times[-\pi\sigma_d, \pi\sigma_d]\,,$$

then the $d$-dimensional Shannon sampling formula reads:

$$f(t) = \sum_{n\in\mathbb{Z}^d} f\Big(\frac{n_1}{\sigma_1}, \frac{n_2}{\sigma_2}, \cdots, \frac{n_d}{\sigma_d}\Big)\operatorname{sinc}(\sigma_1 t_1 - n_1)\operatorname{sinc}(\sigma_2 t_2 - n_2)\cdots\operatorname{sinc}(\sigma_d t_d - n_d)\,,$$

where $t = (t_1, t_2, \cdots, t_d)\in\mathbb{R}^d$ and $n = (n_1, n_2, \cdots, n_d)\in\mathbb{Z}^d$.

In case the band-region $\Omega$ is any bounded set in $\mathbb{R}^d$, it is clear that $\Omega$ might be enclosed by a $d$-dimensional rectangle and to use the above Shannon sampling formula. But usually by paying an oversampling price. It is proved in Ref. [21, Chapter 14] that virtually all of the nice features of the one-dimensional regular case can be retained, provided that the band-region $\Omega$ has the special property that it, and copies of it obtained from a group of translations, form a partition of $\mathbb{R}^d$. See in this reference all the details. In some way, they are related to sampling at lattices, concept that will appear in Chaps. 5 and 7.

For sampling in a shift-invariant subspace of $L^2(\mathbb{R}^d)$ at a lattice $\Lambda = M\mathbb{Z}^d$ of $\mathbb{R}^d$, where $M$ is a $d\times d$ matrix with integer entries and $\det M > 0$, see, for instance, Ref. [15] and references therein; see also the examples in Chap. 5.

### 2.2.5 *Recovering Bandlimited Functions in the Distributional Sense*

Let $\mathcal{E}'(\mathbb{R})$ be the space of distributions with compact support. We define the Paley-Wiener space $PW$ of bandlimited functions in the distributional sense as

$$PW := \mathcal{F}^{-1}\big(\mathcal{E}'(\mathbb{R})\big)\,.$$

Thus, any $f \in PW$ can be expressed as

$$f(z) = \frac{1}{\sqrt{2\pi}}\langle T, e^{i\omega z}\rangle, \quad z \in \mathbb{C}.$$

As usually, we denote $\langle T, \varphi\rangle := T(\varphi)$, the duality between a distribution $T$ and a corresponding test function $\varphi$.

In general, the Shannon sampling formula does not work for any function in $PW$. Indeed, the derivative $\delta'_a$ of the *Dirac delta* at a point $a \in (-\pi, \pi)$ belongs to $\mathcal{E}'(\mathbb{R})$. However, it has inverse Fourier transform

$$f(z) = \frac{1}{\sqrt{2\pi}}\langle \delta'_a, e^{i\omega z}\rangle = \frac{-1}{\sqrt{2\pi}}\langle \delta_a, iz e^{i\omega z}\rangle = \frac{-1}{\sqrt{2\pi}} iz\, e^{-iaz}, \quad z \in \mathbb{C},$$

whose samples $f(n)$ are $O(|n|)$ as $|n|$ goes to $\infty$. As a consequence, the corresponding expansion $f(z) = \sum_{n=-\infty}^{\infty} f(n)\,\operatorname{sinc}(z-n)$ diverges.

According to the *Paley-Wiener-Schwartz theorem* (see Refs. [18, 28]) the space $PW$ can be described as

$$PW = \Big\{ f \in \mathcal{H}(\mathbb{C}) \,:\, \exists A, B \geq 0,\ N \in \mathbb{N}_0,\ |f(z)| \leq A(1+|z|)^N\, e^{B|\operatorname{Im} z|}\ \forall z \in \mathbb{C} \Big\}.$$

Thus, in particular, any Paley-Wiener space $PW_{\pi\sigma}$ is contained in $PW$.

We describe briefly the topology on $PW$ which makes the distributional Fourier transform

$$\begin{aligned} \mathcal{F} \colon PW &\longrightarrow \mathcal{E}'(\mathbb{R}) \\ f &\longmapsto T = \mathcal{F}(f), \end{aligned}$$

a topological isomorphism. For this, we first consider, for each $n \in \mathbb{N}_0$, the space $PW^{(n)}$ of entire functions such that

$$\|f\|_n = \sup_{z\in\mathbb{C}} \big(|f(z)| e^{-n|\operatorname{Im} z|}(1+|z|)^{-n}\big) < \infty.$$

Each $PW^{(n)}$ is a Banach space endowed with the norm $\|\cdot\|_n$. Furthermore, we have $PW^{(n)} \subset PW^{(n+1)}$ for each $n$ with continuous inclusion and $PW = \bigcup_{n=0}^{\infty} PW^{(n)}$.

The *inductive limit topology* on $PW$ is defined as the finest locally convex topology under which all the inclusions $i_n : PW^{(n)} \hookrightarrow PW$ are continuous. Considering this topology in $PW$, the Paley-Wiener-Schwartz-Ehrenpreis theorem asserts that the Fourier transform is a topological isomorphism between $PW$ and $\mathcal{E}'(\mathbb{R})$ equipped with the strong topology $\beta(\mathcal{E}'(\mathbb{R}), \mathcal{E}(\mathbb{R}))$, i.e., the topology of uniform convergence on weakly bounded sets of $\mathcal{E}(\mathbb{R}) = C^\infty(\mathbb{R})$ (see Ref. [5, Theorem 1.4.16]).

One can easily see that convergence in a particular $PW^{(n)}$ implies uniform convergence on compact sets of $\mathbb{C}$. This implies the continuity of every inclusion $i_n : PW^{(n)} \hookrightarrow PW$ where $PW$ is equipped with the topology of uniform convergence on compact sets. Now, since the inductive limit topology is by definition the finest one making the $i_n$ continuous, we obtain the following result:

**Lemma 2.1** *Any sequence in $PW$ converging in the inductive limit topology is also uniformly convergent on compact subsets of $\mathbb{C}$.*

As a consequence, any expansion converging in $\mathcal{E}'(\mathbb{R})$ is automatically transformed by $\mathcal{F}^{-1}$ in an expansion converging uniformly on compact sets of $\mathbb{C}$.

We conclude with a remark on the convergence in $\mathcal{E}'(\mathbb{R})$. Since $C^\infty(\mathbb{R})$ is a *Montel space* (in short, every closed bounded subset is compact), every weakly converging sequence in its strong dual space $\mathcal{E}'(\mathbb{R})$ is also strongly converging. Hence, we have that

$$T_n \to T \text{ in } \mathcal{E}'(\mathbb{R}) \iff \langle T_n, \varphi\rangle \to \langle T, \varphi\rangle \quad \text{for every } \varphi \in \mathcal{E}(\mathbb{R})\,.$$

See Ref. [32, Corollary 2 in §34.4]. For a nice introduction to the theory of distributions, see Ref. [18].

## The Campbell Sampling Theorem

Let $T \in \mathcal{E}'(\mathbb{R})$ be a distribution with support $\operatorname{supp} T$ contained, say, in the open interval $(-\pi, \pi)$, and consider its $2\pi$-periodic extension

$$T_{per} := T * \Delta_{2\pi} = \sum_{n=-\infty}^{\infty} T * \delta_{2\pi n} = \sum_{n=-\infty}^{\infty} \tau_{2\pi n} T\,,$$

where $\Delta_{2\pi} = \sum_{n=-\infty}^{\infty} \delta_{2n\pi}$ is the *Dirac comb* with period $2\pi$, and $\tau_{2\pi n}$ denotes the translation by $2\pi n$. Then, it is well-known that $T_{per}$ is in $\mathcal{S}'(\mathbb{R})$, the space of tempered distributions, and that it admits a Fourier expansion

$$T_{per} = \sum_{n=-\infty}^{\infty} c_n\, \mathrm{e}^{-in\omega}$$

converging in the topology of $\mathcal{S}'(\mathbb{R})$ (see, for instance, Ref. [34]). Furthermore, the coefficients above are $c_n = \frac{1}{2\pi}\langle T, \mathrm{e}^{in\omega}\rangle = \frac{f(n)}{\sqrt{2\pi}}$ for $f = \mathcal{F}^{-1}(T)$, and $c_n = O(|n|^p)$ for some $p \in \mathbb{Z}$.

To apply our duality argument, however, we need convergence to $T$ in the space $\mathcal{E}'(\mathbb{R})$. To do this we introduce a convergence factor, i.e., a smooth function which is identically one on a neighborhood of $\operatorname{supp} T$ (note that we may always choose this function to be even, so that its Fourier transform is a real function). Thus, one can prove:

**Lemma 2.2** *Let* $T \in \mathcal{E}'(\mathbb{R})$ *with support contained in the open interval* $(-\pi, \pi)$, $f = \mathcal{F}^{-1}(T) \in PW$ *and* $\widehat{\theta} \in \mathcal{D}(\mathbb{R})$ *with* $\operatorname{supp}\widehat{\theta} \subset (-\pi, \pi)$ *and* $\widehat{\theta}(x) \equiv 1$ *on an open set containing* $\operatorname{supp} T$. *Then,*

$$T = \sum_{n=-\infty}^{\infty} f(n)\,\mathrm{e}^{-in\omega}\widehat{\theta} \quad \text{in } \mathcal{E}'(\mathbb{R})$$

*As usual,* $\mathcal{D}(\mathbb{R})$ *denotes the space of compactly supported infinitely differentiable functions on* $\mathbb{R}$.

***Proof*** First, since $\widehat{\theta}(x) \equiv 1$ on an open set containing $\operatorname{supp} T$, one has $\widehat{\theta}\, T_{per} = T$. Multiplication by $\widehat{\theta}$ is a continuous operation in $\mathcal{S}'(\mathbb{R})$, so

$$T = \sum_{n=-\infty}^{\infty} c_n \mathrm{e}^{-in\omega}\widehat{\theta}$$

in $\mathcal{S}'(\mathbb{R})$. Furthermore, the convergence is also in $\mathcal{E}'(\mathbb{R})$, since for any $\varphi \in \mathcal{E}(\mathbb{R})$ we have $\langle T, \varphi\rangle = \langle \widehat{\theta}\, T_{per}, \varphi\rangle = \langle T, \widehat{\theta}\,\varphi\rangle$. This implies

$$\langle T, \varphi\rangle = \lim_{N\to\infty} \Big\langle \sum_{n=-N}^{N} c_n\, \mathrm{e}^{-in\omega}, \varphi\widehat{\theta}\Big\rangle = \lim_{N\to\infty} \Big\langle \sum_{n=-N}^{N} c_n\, \mathrm{e}^{-in\omega}\widehat{\theta}, \varphi\Big\rangle,$$

which concludes the proof. □

We are now in the position of proving the *Campbell sampling theorem* (see Ref. [7]):

**Theorem 2.6** *Let* $f$ *be a function in* $PW$ *with* $T = \mathcal{F}(f) \in \mathcal{E}'(\mathbb{R})$ *supported in* $(-\pi, \pi)$, *and* $\widehat{\theta} \in \mathcal{D}(\mathbb{R})$ *as in the statement of Lemma 2.2. Then,*

$$f(z) = \sum_{n=-\infty}^{\infty} f(n)\,\theta(z-n)\,, \quad z \in \mathbb{C}\,, \tag{2.24}$$

*with uniform convergence on compact sets of* $\mathbb{C}$.

***Proof*** Applying the inverse Fourier transform to the equality in Lemma 2.2 we obtain

$$f(z) = \sum_{n=-\infty}^{\infty} f(n)\,\mathcal{F}^{-1}(\mathrm{e}^{-in\omega}\widehat{\theta}(\omega))(z) = \sum_{n=-\infty}^{\infty} f(n)\,\theta(z-n) \quad \text{in } PW\,,$$

in particular, with uniform convergence on compact subsets of $\mathbb{C}$. □

We end with two comments related to the expansion (2.24). The first one concerns the convergence factor $\widehat{\theta}$ introduced in Lemma 2.2 to obtain convergence in $\mathcal{E}'(\mathbb{R})$. The price we have to pay for this convergence is that it forces us to oversample, i.e. to sample $f(z)$ at frequency 1, which is larger than the Nyquist frequency $m(\operatorname{supp} T)/2\pi$, where $m(\cdot)$ denotes the Lebesgue measure in $\mathbb{R}$. Note, finally, that (2.24) lacks the appearance of a cardinal series. However, we may introduce the characteristic function $\chi_{[-\pi,\pi]}$ in the above expansion, obtaining

$$\begin{aligned} f(z) &= \sum_{n=-\infty}^{\infty} f(n)\mathcal{F}^{-1}(e^{-in\omega}\chi_{[-\pi,\pi]}\widehat{\theta}(\omega))(z) \\ &= \sum_{n=-\infty}^{\infty} f(n)\left[\mathcal{F}^{-1}(e^{-in\omega}\chi_{[-\pi,\pi]}) * \theta\right](z) \\ &= \sum_{n=-\infty}^{\infty} f(n)\left(\tau_n \operatorname{sinc} * \theta\right)(z), \qquad z \in \mathbb{C}, \end{aligned}$$

where $\tau_n \operatorname{sinc}(t) = \operatorname{sinc}(t-n)$. Thus, the sampling functions of (2.24) are given through a convolution involving the inverse Fourier transform of the convergence factor.

**Example 2.5 (Bessel-Neumann Expansions in $PW_1$)** It is well-known that $\{\sqrt{n+\frac{1}{2}}\,P_n(x)\}_{n=0}^{\infty}$ is an orthonormal basis for $L^2[-1,1]$, where $\{P_n(x)\}_{n=0}^{\infty}$ denotes the sequence of Legendre orthogonal polynomials. Moreover, for any $n \in \mathbb{N}_0$, we have the Fourier dual (see, for instance, Ref. [1, 10.1.14])

$$i^n t^{-1/2} J_{n+\frac{1}{2}}(t) = \frac{1}{\sqrt{2\pi}} \int_{-1}^{1} P_n(w)\, e^{itw} dw, \quad t \in \mathbb{R},$$

where $J_{n+\frac{1}{2}}(t)$ is the Bessel function of half odd integer order given by the formula

$$J_{n+\frac{1}{2}}(t) = (-1)^n \sqrt{\frac{2}{\pi}}\, t^{n+\frac{1}{2}} \left(\frac{1}{t}\frac{d}{dt}\right)^n \frac{\sin t}{t}.$$

Since the Fourier transform is a unitary operator and $|i^n| = 1$, we obtain that the sequence

$$\left\{\sqrt{n+\frac{1}{2}}\, t^{-1/2} J_{n+\frac{1}{2}}(t)\right\}_{n=0}^{\infty}$$

forms an orthonormal basis for $PW_1 = \{f \in L^2(\mathbb{R}) \,:\, \operatorname{supp} \widehat{f} \subseteq [-1,1]\}$ giving the Bessel-Neumann expansion:

**Proposition 2.11** *Any function $f$ in $PW_1$ can be expanded as*

$$f(z) = \sum_{n=0}^{\infty} a_n \, i^n \, z^{-1/2} J_{n+\frac{1}{2}}(z) \,, \quad z \in \mathbb{C} \,, \tag{2.25}$$

*with uniform convergence on compact subsets of $\mathbb{C}$, where*

$$a_n = (n + \frac{1}{2}) \, \langle \widehat{f}, P_n \rangle_{L^2[-1,1]} \,, \quad n \in \mathbb{N}_0 \,.$$

***Proof*** We expand the Fourier transform $\widehat{f}$ of $f$ in the orthonormalized Legendre basis of $L^2[-1, 1]$ as

$$\widehat{f}(x) = \sum_{n=0}^{\infty} a_n P_n(x) \quad \text{in } L^2[-1, 1] \,.$$

The result is obtained applying the inverse Fourier transform. □

**Example 2.6 (Bessel-Neumann Expansions in PW)** By using the same technique one can prove the Bessel-Neumann expansion of a function $f$ whose Fourier transform is a distribution $T \in \mathcal{E}'(\mathbb{R})$. To obtain an analogue of the Bessel-Neumann expansion (2.25) for a function $f$ whose Fourier transform is a distribution $T$ of $\mathcal{E}'(\mathbb{R})$ with $\operatorname{supp} T \subset (-1, 1)$ we need to expand $T$ with respect to the Legendre polynomials with convergence in $\mathcal{E}'(\mathbb{R})$. Again, this forces us to introduce a convergence factor $\widehat{\theta}$ as before.

**Proposition 2.12** *Let $T \in \mathcal{E}'(\mathbb{R})$ with $\operatorname{supp} T \subset (-1, 1)$ and $\widehat{\theta} \in \mathcal{D}(\mathbb{R})$ with $\operatorname{supp} \widehat{\theta} \subset (-1, 1)$ and $\widehat{\theta}(x) \equiv 1$ on an open set containing $\operatorname{supp} T$. Then, the function $f = \mathcal{F}^{-1}(T) \in PW$ can be expanded as*

$$f(z) = \sum_{n=0}^{\infty} b_n (\Gamma_n * \theta)(z) \,, \quad z \in \mathbb{C} \,,$$

*with uniform convergence on compact sets of $\mathbb{C}$, where $b_n = \langle T, P_n \rangle$, and the function $\Gamma_n$ is given by $\Gamma_n(z) = i^n z^{-1/2} J_{n+\frac{1}{2}}(z)$ for $n \in \mathbb{N}_0$.*

***Proof*** We may mirror the proof in Ref. [34, Proposition 6.3] to ensure that

$$T = \sum_{n=0}^{\infty} b_n P_n \chi_{[-1,1]} \widehat{\theta} \quad \text{in } \mathcal{E}'(\mathbb{R}) \,.$$

Using again the inverse Fourier transform we obtain the desired result. □

Similar results can be obtained by using the Hermite functions; see, for instance, Refs. [16, 34].

## References

1. Abramowitz, G., Stegun, I.: Handbook of Mathematical Functions. Dover, New York (1972)
2. Al-Hammali, H., Faridani, A.: Uniform and non-uniform sampling of bandlimited functions at minimal density with a few additional samples. Sampl. Theory Signal Process. Data Anal. **21**, 2 (2023)
3. Bachman, G., Narici, L., Beckenstein, E.: Fourier and Wavelet Analysis. Springer, New York (2000)
4. Benedetto, J.J.: Irregular frames and sampling. In: Chui, C.K. (ed.) Wavelets-A Tutorial in Theory and Applications, pp. 445–507. Academic Press, San Diego, CA (1992)
5. Berenstein, C.A., Gay, R.: Complex Analysis and Special Topics in Harmonic Analysis. Springer, New York (1995)
6. Bond, F.E., Chan, C.R.: On sampling the zeros of bandwidth limited signals. IRE Trans. Inform. Theory **IT-4**, 110–113 (1958)
7. Campbell, L.: Sampling theorems for the Fourier transform of a distribution with bounded support. SIAM J. Appl. Math. **16**(3), 626–636 (1968)
8. Cauchy, A.L.: Mémoire sur diverses formules d'analyse. C. R. Acad. Sci. Paris **12**, 283–298 (1841)
9. Duffin, R., Schaeffer, A.: Some properties of functions of exponential type. Bull. Amer. Math. Soc. **44**, 236–240 (1938)
10. Duffin, R.J., Shaeffer, A.C.: A class of nonharmonic Fourier series. Trans. Amer. Math. Soc. **72**, 341–366 (1952)
11. Duren, P.L.: Theory of the $H^p$ Spaces. Dover Publications, New York (2000)
12. Feichtinger, H.G., Gröchenig, K.: Theory and practice of irregular sampling. In: Benedetto, J., Frazier, M. (eds.) Wavelets: Mathematics and Applications, pp. 305–363. CRC Press, Boca Raton, FL (1994)
13. Führ, H.: Abstract Harmonic Analysis of Continuous Wavelet Transform. Springer, Berlin-Heidelberg (2005)
14. García, A.G.: Sampling theory and reproducing kernel Hilbert spaces. In: Alpay, D. (ed.) Handbook of Operator Theory. Springer, Basel (2015)
15. García, A.G., Pérez-Villalón, G.: Multivariate generalized sampling in shift-invariant spaces and its approximation properties. J. Math. Anal. Appl. **355**, 397–413 (2009)
16. García, A.G., Moro, J., Hernández-Medina, M.A.: On the distributional Fourier duality and its applications. J. Math. Anal. Appl. **227**, 43–54 (1998)
17. García, A.G., Kim, J.M., Kwon, K.H., Yoon, G.J.: Multi-channel sampling on shift-invariant spaces with frame generators. Int. J. Wavelets Multiresolut. Inf. Process. **10**(1), 41–60 (2012)
18. Gasquet, C., Witomski, P.: Fourier Analysis and Applications. Springer, New York (1998)
19. Hamming, R.W.: Numerical Methods for Scientists and Engineers. Dover Publications, New York (1973)
20. Hardy, G.H.: Notes on special systems of orthogonal functions, IV: The Whittaker's cardinal series. Proc. Camb. Philos. Soc. **37**, 331–348 (1941)
21. Higgins, J.R.: Sampling Theory in Fourier and Signal Analysis: Foundations. Oxford University Press, Oxford (1996)
22. Higgins, J.R., Stens, R.L. (eds.): Sampling Theory in Fourier and Signal Analysis: Advanced Topics. Oxford University Press, Oxford (1999)
23. Levinson, N.: Gap and Density Theorems. AMS Colloq. Publ., vol. 26. AMS, New York (1940)
24. Mallat, S.: A Wavelet Tour of Signal Processing. Academic Press, San Diego, CA (1999)
25. Marvasti, F.: A Unified Approach to Zero Crossings and Nonuniform Sampling. Department of Electrical Engineering, Institute of Technology, Illinois (1987)
26. Paley, R.E.A.C., Wiener, N.: Fourier Transforms in the Complex Domain. AMS Colloq. Publ., vol. 19. AMS, New York (1934)
27. Partington, J.R.: Interpolation, Identification and Sampling. Clarendon Press, Oxford (1997)

28. Rudin, W.: Functional Analysis, 2nd edn. McGraw-Hill, New York (1991)
29. Sansone, G.: Orthogonal Functions. Dover Publications, New York (1991)
30. Szegö, G.: Orthogonal Polynomials. AMS Colloq. Publ., vol. 23. AMS, Providence RI (1991)
31. Titchmarsh, E.C.: The zeros of certain integral functions. Proc. London Math. Soc. **25**, 283–302 (1926)
32. Treves, F.: Topological Vector Spaces, Distributions and Kernels. Academic Press, San Diego (1967)
33. Walker, W.J.: Oscillatory properties of Paley-Wiener functions. Ind. J. Pure Appl. Math. **25**, 1253–1258 (1994)
34. Walter, G.G.: Wavelets and Other Orthogonal Systems with Applications. CRC Press, Boca Raton FL (1994)
35. Young, R.: An Introduction to Nonharmonic Fourier Series. Academic Press, San Diego (2001)

# Chapter 3
# Sampling in Shift-Invariant Subspaces

Although Shannon's sampling theory has had an enormous impact, it has a number of problems, as pointed out by Unser in Ref. [16]:

1. It relies on the use of *ideal filters*. In other words, in Fig. 2.1, $\widehat{f}$ can be obtained from $\widehat{f_p}$ multiplying by the characteristic function $\chi_{[-\pi,\pi]}$;
2. The bandlimited hypothesis is in contradiction with the idea of a finite duration signal since $f$ is an entire function (it has isolated zeros);
3. The band-limiting operation generates *Gibbs oscillations* and, finally,
4. The sinc function has a very slow decay at infinity which makes computation in the time domain very inefficient.

Moreover, many applied problems impose different a priori constraints on the type of signals. For this reason, sampling and reconstruction problems have been investigated in spline spaces, wavelet spaces, and general shift-invariant spaces. Signals are assumed to belong to some shift-invariant space of the form:

$$V_\varphi^2 := \overline{\operatorname{span}}_{L^2}\{\varphi(t-n)\}_{n\in\mathbb{Z}}\,,$$

where the function $\varphi$ in $L^2(\mathbb{R})$ is called the generator of $V_\varphi^2$.

## 3.1 The Basic Theory: Shannon-Type Sampling Formulas

Let $V_\varphi^2 := \overline{\operatorname{span}}\big\{\varphi(\cdot-n)\big\}_{n\in\mathbb{Z}}$ be the *shift-invariant space* generated by the function $\varphi \in L^2(\mathbb{R})$. If the sequence $\{\varphi(\cdot-n)\}_{n\in\mathbb{Z}}$ is a Riesz basis for $V_\varphi^2$, we say that $\varphi$ is a *stable generator* for $V_\varphi^2$. A characterization of stable generators is given in the following result:

 
A. García García, *The Use of Frames in Sampling Theory*,
RSME Springer Series 14, https://doi.org/10.1007/978-3-031-63242-6_3

**Proposition 3.1** *The sequence $\{\varphi(\cdot - n)\}_{n\in\mathbb{Z}}$ is a Riesz sequence in $L^2(\mathbb{R})$, i.e., a Riesz basis for $V_\varphi^2$ if and only if there exist two positive constants $0 < A \le B$ such that*

$$A \le \sum_{k\in\mathbb{Z}} |\widehat{\varphi}(w + 2\pi k)|^2 \le B\,, \quad a.e.\ w \in (-\pi, \pi)\,, \tag{3.1}$$

*where $\widehat{\varphi}$ stands for the Fourier transform of $\varphi$.*

Later on, in Sect. 3.3, we will give the proof of Theorem 3.5, a more general result.

Observe that, for any function $\varphi \in L^2(\mathbb{R})$, the sequence $\{\varphi(t-n)\}_{n\in\mathbb{Z}}$ is not a complete system in $L^2(\mathbb{R})$.

Indeed, consider the sets $A := \{x \in \mathbb{R} \mid x - [x] < 1/2\}$, where $[x]$ denotes the integer part of $x$, its complement $B = \mathbb{R} \setminus A$, and the function $\psi$ in $L^2(\mathbb{R})$ defined by

$$\psi(t) := \big(\chi_A(t) - \chi_B(t)\big)\overline{\varphi(-t)} \quad \text{a.e. in } \mathbb{R}\,.$$

Since

$$\begin{aligned}\int_{\mathbb{R}} \chi_A(t)\,\overline{\varphi(-t)}\,\overline{\varphi(t-n)}\,dt &= \int_{\mathbb{R}} \chi_A(-u+n)\,\overline{\varphi(u-n)}\,\overline{\varphi(-u)}\,du \\ &= \int_{\mathbb{R}} \chi_B(u)\,\overline{\varphi(u-n)}\,\overline{\varphi(-u)}\,du\,,\end{aligned}$$

we obtain $\langle \psi, \varphi(\cdot - n)\rangle_{L^2(\mathbb{R})} = 0$ for all $n \in \mathbb{Z}$. However, the function $\psi$ is not the zero function in $L^2(\mathbb{R})$.

As a consequence, there is not a Riesz (orthonormal) basis (neither a frame) for $L^2(\mathbb{R})$ having the form $\{\varphi(t-n)\}_{n\in\mathbb{Z}}$.

In case $\varphi$ is a stable generator, the shift-invariant subspace $V_\varphi^2$ can be expressed as

$$V_\varphi^2 = \Big\{\sum_{n\in\mathbb{Z}} a_n\,\varphi(\cdot - n)\ :\ \{a_n\} \in \ell^2(\mathbb{Z})\Big\} \subset L^2(\mathbb{R})\,.$$

For sampling purposes we require that the functions in $V_\varphi^2$ are continuous on $\mathbb{R}$. As proved by Zhou and Sun in Ref. [19], this is equivalent to saying that the generator $\varphi$ is continuous on $\mathbb{R}$ and the function $t \longmapsto \sum_{n\in\mathbb{Z}} |\varphi(t-n)|^2$ is bounded on $\mathbb{R}$:

**Theorem 3.1** *The shift-invariant subspace $V_\varphi^2$ with stable generator consists of continuous functions on $\mathbb{R}$ if and only if the generator $\varphi$ is continuous on $\mathbb{R}$ and the series $\sum_{n\in\mathbb{Z}} |\varphi(t-n)|^2$ is uniformly bounded on $\mathbb{R}$.*

***Proof*** The Cauchy-Schwarz inequality for $f = \sum_{n\in\mathbb{Z}} a_n\, \varphi(\cdot - n)$ and the Weierstrass $M$-test give that the series $\sum_{n\in\mathbb{Z}} a_n\, \varphi(t-n)$ converges, uniformly on $\mathbb{R}$, to a continuous function that coincides with $f$. This proves the sufficient condition.

For the necessary condition, fix $t \in [0, 1)$ and define the linear functional

$$\begin{aligned} \Lambda_{t,N} :\quad & \ell^2(\mathbb{Z}) \longrightarrow \mathbb{C} \\ & c := \{c_n\}_{n\in\mathbb{Z}} \longmapsto \sum\nolimits_{|n|<N} c_n \varphi(t-n)\,. \end{aligned}$$

For each $c \in \ell^2(\mathbb{Z})$ we have $\sup_{N\in\mathbb{N}} |\Lambda_{t,N}(c)| \le \sum_{n\in\mathbb{Z}} |c_n \varphi(t-n)| < \infty$. Thus, the *Banach-Steinhaus theorem* gives

$$\sup_{N\in\mathbb{N}} \|\Lambda_{t,N}\| = \Big(\sum_{n\in\mathbb{Z}} |\varphi(t-n)|^2\Big)^{1/2} < \infty\,.$$

Next we define the linear functional

$$\begin{aligned} \Lambda_t :\quad & \ell^2(\mathbb{Z}) \longrightarrow \mathbb{C} \\ & c := \{c_n\}_{n\in\mathbb{Z}} \longmapsto \sum\nolimits_{n\in\mathbb{Z}} c_n \varphi(t-n)\,. \end{aligned}$$

For each $c \in \ell^2(\mathbb{Z})$ we have that $\sup_{[0,1)} \big|\sum_{n\in\mathbb{Z}} c_n \varphi(t-n)\big| < \infty$. Applying again the Banach-Steinhaus theorem we finally get

$$\sup_{t\in[0,1)} \|\Lambda_t\| = \sup_{t\in[0,1)} \big\|\{\varphi(t-n)\}_{n\in\mathbb{Z}}\big\|_2 = \sup_{t\in\mathbb{R}} \Big(\sum_{n\in\mathbb{Z}} |\varphi(t-n)|^2\Big)^{1/2} < \infty\,.$$

□

Thus, any $f \in V_\varphi^2$ is defined on $\mathbb{R}$ as the pointwise sum

$$f(t) = \sum_{n\in\mathbb{Z}} a_n \varphi(t-n) \quad \text{for each } t \in \mathbb{R}\,.$$

On the other hand, the space $V_\varphi^2$ is the image of the Hilbert space $L^2[0, 1]$ by means of the isomorphism

$$\begin{aligned} \mathcal{T}_\varphi :\quad & L^2[0,1] \longrightarrow V_\varphi^2 \\ & \{e^{-2\pi i n x}\}_{n\in\mathbb{Z}} \longmapsto \{\varphi(t-n)\}_{n\in\mathbb{Z}}\,, \end{aligned}$$

which maps the orthonormal basis $\{e^{-2\pi inx}\}_{n\in\mathbb{Z}}$ for $L^2[0,1]$ onto the Riesz basis $\{\varphi(t-n)\}_{n\in\mathbb{Z}}$ for $V_\varphi^2$. For any $f \in V_\varphi^2$ there exists $F \in L^2[0,1]$ such that

$$\begin{aligned} f(t) = \mathcal{T}_\varphi F(t) &= \sum_{n\in\mathbb{Z}} \langle F, e^{-2\pi inx}\rangle\, \varphi(t-n) \\ &= \Big\langle F, \sum_{n\in\mathbb{Z}} \overline{\varphi(t-n)} e^{-2\pi inx} \Big\rangle = \big\langle F, K_t \big\rangle, \qquad t \in \mathbb{R}, \end{aligned} \tag{3.2}$$

where, for each $t \in \mathbb{R}$, the function $K_t \in L^2[0,1]$ is given by

$$K_t(x) := \sum_{n\in\mathbb{Z}} \overline{\varphi(t-n)} e^{-2\pi inx} = \overline{\sum_{n\in\mathbb{Z}} \varphi(t+n) e^{-2\pi inx}} = \overline{Z\varphi(t,x)}\,. \tag{3.3}$$

Here,

$$Z\varphi(t,x) := \sum_{n\in\mathbb{Z}} \varphi(t+n) e^{-2\pi inx} \quad \text{a.e. in } [0,1) \times [0,1)\,,$$

is just the *Zak transform* of the function $\varphi \in L^2(\mathbb{R})$. See Ref. [6] for properties and uses of the Zak transform.

As a consequence of Eq. (3.2), the shift-invariant space $V_\varphi^2$ is a RKHS in $L^2(\mathbb{R})$ and the mapping $\mathcal{T}_\varphi$ satisfies the *shifting property*

$$\mathcal{T}_\varphi(e^{-2\pi imx} F)(t) = (\mathcal{T}_\varphi F)(t-m)\,, \quad t \in \mathbb{R} \text{ and } m \in \mathbb{Z} \tag{3.4}$$

From Eq. (3.2), for $a \in [0,1)$ fixed and $m \in \mathbb{Z}$ we have

$$f(a+m) = \langle F, K_{a+m}\rangle_{L^2[0,1]} = \langle F, e^{-2\pi imx} K_a\rangle_{L^2[0,1]} \quad \text{with} \quad F = \mathcal{T}_\varphi^{-1} f\,.$$

In order to obtain a sampling formula in $V_\varphi^2$, we look for sampling points of the form $t_m := a + m$, $m \in \mathbb{Z}$, such that the sequence $\big\{e^{-2\pi imx} K_a(x)\big\}_{m\in\mathbb{Z}}$ is a Riesz basis for $L^2[0,1]$. We will need the following result about multiplication operators in $L^2[0,1]$:

**Lemma 3.1** *The multiplication operator $m_g : L^2[0,1] \to L^2[0,1]$ defined as the product $m_g(f) = gf$ is well-defined if and only if $g \in L^\infty[0,1]$, and then, it is bounded with operator norm $\|m_g\| = \|g\|_\infty$.*

***Proof*** Whenever $g \in L^\infty[0,1]$, the function $gf \in L^2[0,1]$ for every $f \in L^2[0,1]$. Assume that $gf \in L^2[0,1]$ for all $f \in L^2[0,1]$ with $g \notin L^\infty[0,1]$. In this case, for each $k > 0$, the set

$$A_k := \big\{t \in [0,1] : |g(t)| \geq k\big\}$$

has positive Lebesgue measure $m(A_k)$. Since $g \notin L^\infty[0,1]$, there exists a sequence $\{t_n\}_{n\in\mathbb{N}}$ in $\mathbb{R}$ with $t_n \geq n$ such that the subsets defined by $E_n := A_{t_n} \setminus A_{t_{n+1}}$ have positive Lebesgue measure $m(E_n) > 0$ and they are mutually disjoint; besides we have $|g(t)| \geq n$ in $E_n$.

Consider the sequence of functions $\{f_n\}_{n\in\mathbb{N}}$ in $L^2[0,1]$ where $f_n$ is defined as

$$f_n(t) := \begin{cases} \frac{1}{\sqrt{m(E_n)}} & \text{if } t \in E_n\,, \\ \\ 0 & \text{if } t \notin E_n\,. \end{cases}$$

This sequence forms an orthonormal set in $L^2[0,1]$ and as a consequence, the function $h := \sum_{n=1}^\infty \frac{1}{n} f_n$ belongs to $L^2[0,1]$.

Having in mind that $f_n \overline{f}_k = 0$ if $n \neq k$, for each $N \in \mathbb{N}$ we have

$$\begin{aligned}\int_0^1 |g(t)h(t)|^2 dt &\geq \int_0^1 |g(t)|^2 \Big|\sum_{n=1}^N \frac{f_n(t)}{n}\Big|^2 dt \\ &= \sum_{n=1}^N \int_{E_n} \frac{|g(t)|^2}{n^2\, m(E_n)} dt \geq \sum_{n=1}^N 1 = N\,,\end{aligned}$$

from which $gh \notin L^2[0,1]$, and a contradiction follows. Therefore, the function $g$ belongs to $L^\infty[0,1]$.

Obviously $\|m_g\| \leq \|g\|_\infty$. For the equality, suppose that $\|g\|_\infty > \|m_g\|$. Then the set

$$\big\{t \in [0,1] : |g(t)| > \|m_g\|\big\} = \bigcup_{n=1}^\infty \Big\{t \in [0,1] : |g(t)| \geq \|m_g\| + \frac{1}{n}\Big\}$$

has positive measure. In particular, there exist $\epsilon > 0$ and a set $E \subset [0,1]$ with positive measure such that $|g(t)| \geq \|m_g\| + \epsilon$ in $E$. Define the function

$$k(t) := \begin{cases} \frac{1}{\sqrt{m(E)}} & \text{if } t \in E\,, \\ \\ 0 & \text{if } t \notin E\,, \end{cases}$$

having unit norm. We get

$$\|m_g\|^2 \geq \|m_g k\|^2 = \int_E \frac{|g(t)|^2}{m(E)} dt \geq (\|m_g\| + \epsilon)^2\,,$$

a contradiction. Therefore, $\|m_g\| = \|g\|_\infty$. □

Essentially, the above lemma characterizes all bounded convolution operators

$$\ell^2(\mathbb{Z}) \ni x \longmapsto h * x \in \ell^2(\mathbb{Z})$$

as those whose *transfer function* $\widehat{h}(\omega) = \sum_{n\in\mathbb{Z}} h(n)\, e^{-2\pi i n\omega}$ belongs to $L^\infty[0, 1]$.

Bounded convolution systems will be treated in a more general setting in Sect. 5.2 of Chap. 5.

As a consequence of the above lemma, we obtain the following result:

**Proposition 3.2** *The sequence of functions* $\{e^{-2\pi i m x} K_a(x)\}_{m\in\mathbb{Z}}$ *is a Riesz basis for* $L^2[0, 1]$ *if and only if the inequalities* $0 < \|K_a\|_0 \le \|K_a\|_\infty < \infty$ *hold, where*

$$\|K_a\|_0 := \operatorname*{ess\,inf}_{x\in[0,1]} |K_a(x)| \quad \textit{and} \quad \|K_a\|_\infty := \operatorname*{ess\,sup}_{x\in[0,1]} |K_a(x)|\,.$$

*Moreover, its dual Riesz basis is* $\{e^{-2\pi i m x}/\overline{K_a(x)}\}_{m\in\mathbb{Z}}$.

Recall that the *essential supremum* $\|f\|_\infty$ of a function $f$ on an interval $I$ is defined by

$$\|f\|_\infty = \operatorname*{ess\,sup}_{t\in I} |f(t)| := \inf\big\{C \,:\, m(\{t \in I \,:\, |f(t)| > C\}) = 0\big\}\,.$$

Similarly, the *essential infimum* $\|f\|_0$ of a function $f$ on an interval $I$ is defined by

$$\|f\|_0 = \operatorname*{ess\,inf}_{t\in I} |f(t)| := \sup\big\{c \,:\, m(\{t \in I \,:\, |f(t)| < c\}) = 0\big\}\,.$$

Notice that the Riesz basis $\{e^{-2\pi i m x} K_a(x)\}_{m\in\mathbb{Z}}$ for $L^2[0, 1]$ has optimal Riesz bounds $A_{opt} = \|K_a\|_0^{-1}$ and $B_{opt} = \|K_a\|_\infty$.

In particular, the sequence $\{e^{-2\pi i m x} K_a(x)\}_{m\in\mathbb{Z}}$ is an orthonormal basis in $L^2[0, 1]$ if and only if $|K_a(x)| = 1$ a.e. in $[0, 1]$.

Let $a$ be a real number in $[0, 1)$ such that $0 < \|K_a\|_0 \le \|K_a\|_\infty < \infty$. Any $F \in L^2[0, 1]$ can be expanded as

$$\begin{aligned} F &= \sum_{m\in\mathbb{Z}} \langle F, e^{-2\pi i m x} K_a\rangle \frac{e^{-2\pi i m x}}{\overline{K_a(x)}} \\ &= \sum_{m\in\mathbb{Z}} f(a+m) \frac{e^{-2\pi i m x}}{\overline{K_a(x)}} \qquad \text{in } L^2[0, 1]\,. \end{aligned} \tag{3.5}$$

Having in mind the shifting property of $\mathcal{T}_\varphi$,

$$\Big\langle \frac{e^{-2\pi i m x}}{\overline{K}_a(x)}, K_t(x) \Big\rangle_{L^2[0,1]} = \mathcal{T}_\varphi\Big(\frac{e^{-2\pi i m x}}{\overline{K}_a(x)}\Big)(t) = S_a(t-m)\,, \quad t \in \mathbb{R}\,,$$

where $S_a := \mathcal{T}_\varphi\big(1/\overline{K}_a\big) \in V_\varphi^2$. Thus, the isomorphism $\mathcal{T}_\varphi$ acting in formula (3.5) gives the sampling result in $V_\varphi^2$:

**Theorem 3.2 (Shannon-Type Sampling Formula)** *Let $a \in [0,1)$ be such that $0 < \|K_a\|_0 \le \|K_a\|_\infty < \infty$. Then, any function $f \in V_\varphi^2$ can be expanded as the Shannon-type sampling series*

$$f(t) = \sum_{n=-\infty}^{\infty} f(a+n) S_a(t-n)\,, \quad t \in \mathbb{R}\,. \tag{3.6}$$

*The convergence of the series in (3.6) is absolute and uniform on $\mathbb{R}$ since the function $t \longmapsto \|K_t\|^2 = \sum_{n\in\mathbb{Z}} |\varphi(t-n)|^2$ is bounded on $\mathbb{R}$.*

The above Shannon-type sampling formula for the shift-invariant subspace $V_\varphi^2$ was proved for the first time by G. Walter in Ref. [18]. The $d$-dimensional case goes in the same manner by using the $d$-dimensional version of the Zak transform.

### 3.1.1 Some Sampling Examples Involving B-Splines

Consider the space $V_\varphi^2$ for the generator $\varphi := N_m$ where $N_m$ is the $B$-spline of order $m-1$. This $B$-spline is defined as $N_m := N_1 * N_1 * \cdots * N_1$ ($m$ times) and $N_1 := \chi_{[0,1]}$, i.e., the characteristic function of the interval $[0,1]$. In Fig. 3.1 they are depicted the first $B$-splines.

**Proposition 3.3** *For each fixed $m \ge 2$, the sequence $\big\{N_m(t-n)\big\}_{n\in\mathbb{Z}}$ is a Riesz sequence for $L^2(\mathbb{R})$, i.e., a stable generator for $V_{N_m}^2$.*

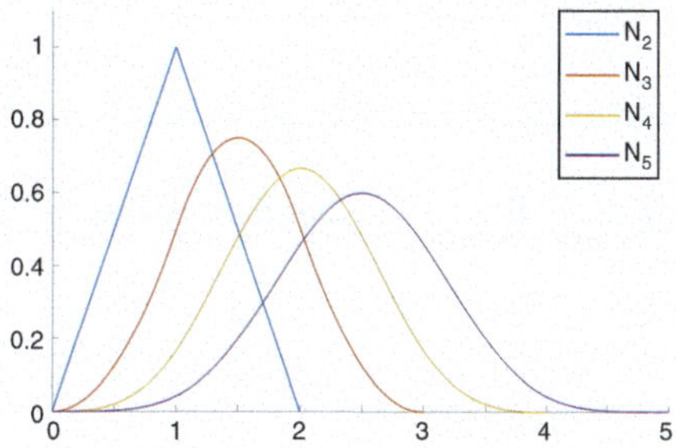


**Fig. 3.1** Linear $N_2$, quadratic $N_3$, cubic $N_4$ and quartic $N_5$ $B$-splines

***Proof*** For the proof we use Proposition 3.1. Firstly, from $\widehat{N}_1(w) = \frac{e^{-iw/2}}{\sqrt{2\pi}}\frac{\sin(w/2)}{w/2}$, we obtain that

$$\widehat{N}_m(w) = (2\pi)^{(m-1)/2}\big[\widehat{N}_1(w)\big]^m = \frac{e^{-imw/2}}{\sqrt{2\pi}}\Big(\frac{\sin(w/2)}{w/2}\Big)^m .$$

Since $|\widehat{N}_1(w)| \le 1$ and $\{N_1(t-n)\}_{n\in\mathbb{Z}}$ is an orthonormal system in $L^2(\mathbb{R})$, the equality

$$\sum_{p\in\mathbb{Z}} |\widehat{N}_1(w+2\pi p)|^2 = \frac{1}{2\pi} \quad \text{a.e. in } [0, 2\pi]$$

holds, and we get

$$\sum_{p\in\mathbb{Z}} |\widehat{N}_m(w+2\pi p)|^2 \le (2\pi)^{m-1} \sum_{p\in\mathbb{Z}} |\widehat{N}_1(w+2\pi p)|^2 = (2\pi)^{m-2} \quad \text{a.e. in } [0, 2\pi]\,.$$

Secondly,

$$\sum_{p\in\mathbb{Z}} |\widehat{N}_m(w+2\pi p)|^2 \ge \inf_{w\in[-\pi,\pi]} |\widehat{N}_m(w)|^2 = \frac{1}{2\pi}\Big(\frac{\sin\pi/2}{\pi/2}\Big)^{2m} = \frac{1}{2\pi}\Big(\frac{2}{\pi}\Big)^{2m} .$$

□

Some sampling examples in spline shift-invariant subspaces:

**Example 3.1 (Sampling Formula Associated with the Quadratic Spline $N_3$)** For the quadratic spline $N_3$ we have $ZN_3(t,x) = \frac{t^2}{2} + \big[\frac{3}{4} - (t-\frac{1}{2})^2\big]z + \frac{(1-t)^2}{2}z^2$ where $z = e^{-2\pi i x}$. Thus, for $t=0$ we have $ZN_3(0,x) = \frac{z}{2}(1+z)$ which vanishes at $x = 1/2$. However, for $t = 1/2$ we have $ZN_3(1/2,x) = \frac{1}{8}(1+6z+z^2)$.

According to (3.3) we deduce $0 < \|K_{1/2}\|_0 \le \|K_{1/2}\|_\infty < \infty$. Hence, for any $f \in V^2_{N_3}$, we have

$$f(t) = \sum_{n=-\infty}^{\infty} f(n+\frac{1}{2})\, S^q_{1/2}(t-n)\,, \quad t\in\mathbb{R}\,,$$

where $S^q_{1/2}(t) = \sqrt{2}\sum_{n=-\infty}^{\infty}(2\sqrt{2}-3)^{|n+1|}\, N_3(t-n)$. This function has been obtained from the Laurent expansion of the function $8(1+6z+z^2)^{-1}$ in the annulus $3-2\sqrt{2} < |z| < 3+2\sqrt{2}$.

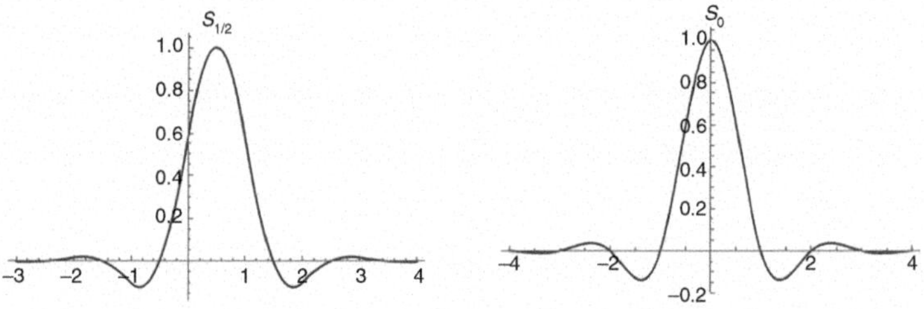


**Fig. 3.2** Quadratic $S^q_{1/2}$ and cubic $S^c_0$ sampling functions

**Example 3.2 (Sampling Formula Associated with the Cubic Spline $N_4$)** We have that $ZN_4(0, x) = \frac{z}{6}\big(1+4z+z^2\big) = \frac{z}{6}(z-\lambda)(z-1/\lambda)$, where $z = \mathrm{e}^{-2\pi i x}$ and $\lambda = \sqrt{3}-2$.

According to (3.3) we deduce that $0 < \|K_0\|_0 \le \|K_0\|_\infty < \infty$. Thus, for any $f \in V^2_{N_4}$ we have

$$f(t) = \sum_{n=-\infty}^{\infty} f(n)\, S^c_0(t-n)\,, \quad t \in \mathbb{R}\,,$$

where $S^c_0(t) = \sqrt{3} \sum_{n=-\infty}^{\infty} (-1)^n (2-\sqrt{3})^{|n|}\, N_4(t-n+2)$. We have used the Laurent expansion of the function $6(z+4z^2+z^3)^{-1}$ in the annulus $2-\sqrt{3} < |z| < 2+\sqrt{3}$ for obtaining the function $S^c_0$.

In Fig. 3.2 they are depicted the quadratic sampling function $S^q_{1/2}$ and the cubic sampling function $S^c_0$.

### 3.1.2 *Sampling Formulas for Two-Dimensional Shift-Invariant Subspaces*

Here we deal with the two-dimensional case which will be used in Sect. 3.2.3. We will distinguish two cases: the first one where the stable generator $\Phi$ is a function in $L^2(\mathbb{R}^2)$, and the second one where the stable generator $\Phi$ has the particular form $\Phi(t, s) = \varphi(t)\, \psi(s)$ with $\varphi, \psi \in L^2(\mathbb{R})$. Just to say that the case where $V^2_\Phi$ is a shift-invariant subspace of $L^2(\mathbb{R}^d)$ goes in the same manner.

**The General Two-Dimensional Case**

Consider a generator $\Phi \in L^2(\mathbb{R}^2)$ such that the sequence $\{\Phi(t-n, s-m)\}_{n,m\in\mathbb{Z}}$ forms a Riesz sequence for $L^2(\mathbb{R}^2)$. Thus,

$$\begin{aligned} V_\Phi^2 :=& \overline{\text{span}}\{\Phi(t-n, s-m)\}_{n,m\in\mathbb{Z}} \\ =& \Big\{ \sum_{n,m\in\mathbb{Z}} a_{nm}\, \Phi(t-n, s-m) \ :\ \{a_{nm}\} \in \ell^2(\mathbb{Z}^2)\Big\}. \end{aligned}$$

Assuming that $\Phi$ is continuous in $\mathbb{R}^2$ and that $\sum_{n,m\in\mathbb{Z}} |\Phi(t-n, s-m)|^2$ is bounded in $[0,1]^2$ one can easily deduce the same results that in Sect. 3.1 for the one dimensional case. Firstly, the isomorphism

$$\begin{aligned} \mathcal{T}_\Phi : \qquad & L^2(0,1)^2 && \longrightarrow V_\Phi^2 \\ & \{e^{-2\pi inx}\, e^{-2\pi imy}\}_{n,m\in\mathbb{Z}} && \longmapsto \{\Phi(t-n, s-m)\}_{n,m\in\mathbb{Z}}\,, \end{aligned}$$

which maps the orthonormal basis $\{e^{-2\pi inx}\, e^{-2\pi imy}\}_{n,m\in\mathbb{Z}}$ of $L^2(0,1)^2$ onto the Riesz basis $\{\Phi(t-n, s-m)\}_{n,m\in\mathbb{Z}}$ for $V_\Phi^2$, satisfies, for any $F \in L^2(0,1)^2$, the *shifting property*:

$$\mathcal{T}_\Phi\big[e^{-2\pi inx} e^{-2\pi imy} F(x,y)\big](t,s) = \mathcal{T}_\Phi[F](t-n, s-m)\,, \quad t,s \in \mathbb{R}\,,\ n,m \in \mathbb{Z}\,.$$

Moreover, any $f = \mathcal{T}_\Phi F \in V_\varphi^2$ can be expressed as

$$f(t,s) = \big\langle F(x,y), K_{t,s}(x,y)\big\rangle_{L^2(0,1)^2}\,, \quad t,s \in \mathbb{R}\,, \tag{3.7}$$

where

$$K_{t,s}(x,y) = \sum_{n,m\in\mathbb{Z}} \overline{\Phi(t-n, s-m)}\, e^{-2\pi inx}\, e^{-2\pi imy} \in L^2(0,1)^2\,.$$

Thus, for fixed $a, b \in [0,1)$ and $n, m \in \mathbb{Z}$ we get the expression for the samples

$$\begin{aligned} f(a+n, b+m) &= \langle F, K_{a+n,b+m}\rangle_{L^2(0,1)^2} \\ &= \langle F, e^{-2\pi inx} e^{-2\pi imy} K_{a,b}(x,y)\rangle_{L^2(0,1)^2}\,. \end{aligned}$$

The sequence $\{e^{-2\pi inx}\, e^{-2\pi imy} K_{a,b}(x,y)\}_{n,m\in\mathbb{Z}}$ is a Riesz basis for $L^2(0,1)^2$ if and only if the inequalities $0 < \|K_{a,b}\|_0 \le \|K_{a,b}\|_\infty < \infty$ hold.

Besides, its dual Riesz basis is the sequence $\{e^{-2\pi inx}\, e^{-2\pi imy}/\overline{K}_{a,b}(x,y)\}_{n,m\in\mathbb{Z}}$. Proceeding as in the former section we obtain:

**Proposition 3.4** *For any $f \in V^2_\Phi$ the sampling formula*

$$f(t,s) = \sum_{n,m\in\mathbb{Z}} f(a+n, b+m)\, S_{a,b}(t-n, s-m)\,, \quad t,s \in \mathbb{R}\,, \tag{3.8}$$

*holds, where $S_{a,b} := \mathcal{T}_\Phi\big(1/\overline{K}_{a,b}\big) \in V^2_\Phi$. The convergence of the series is absolute due to the unconditional convergence of a Riesz basis expansion and it is also uniform on $\mathbb{R}^2$ since the shift-invariant space $V^2_\Phi$ is a RKHS contained in $L^2(\mathbb{R}^2)$ and the function*

$$(t,s) \longmapsto \sum_{n,m\in\mathbb{Z}} \big|\Phi(t-n, s-m)\big|^2$$

*is bounded on $\mathbb{R}^2$.*

### The Case of Two Separable Variables

Consider two generators $\varphi(x)$ and $\psi(y)$ in $L^2(\mathbb{R})$ satisfying the same hypotheses as in Sect. 3.1, and define $\Phi(x,y) := \varphi(x)\psi(y)$ in $L^2(\mathbb{R}^2)$ and its associated shift-invariant subspace $V^2_{\varphi\psi} := V^2_\Phi$. It can be seen as the tensor product $V^2_\varphi \otimes V^2_\psi$. For an overview on tensor products see, for instance, Ref. [9] and references therein.

It is straightforward to show that $\mathcal{T}_{\varphi\psi} := \mathcal{T}_\Phi = \mathcal{T}_\varphi \otimes \mathcal{T}_\psi$, and

$$\mathcal{T}_\Phi\big(F(x)G(y)\big) = \mathcal{T}_\varphi\big(F(x)\big)\,\mathcal{T}_\psi\big(G(y)\big)\,, \quad F, G \in L^2(0,1)$$

holds.

Consequently, for $H, G \in L^2(0,1)$ and $n, m \in \mathbb{Z}$, the *shifting property* reads:

$$\begin{aligned}
&\mathcal{T}_{\varphi\psi}\big[e^{-2\pi inx} e^{-2\pi imy} H(x)G(y)\big](t,s) \\
&\quad = \mathcal{T}_\varphi[e^{-2\pi inx} H(x)](t)\mathcal{T}_\psi[e^{-2\pi imy} G(y)](s) \\
&\quad = \mathcal{T}_\varphi[H](t-n)\mathcal{T}_\psi[G](s-m)\,, \qquad t,s \in \mathbb{R}\,.
\end{aligned}$$

Besides, any $f = \mathcal{T}_{\varphi\psi} F \in V^2_{\varphi\psi}$, where $F \in L^2(0,1)^2$, can be expressed as

$$f(t,s) = \Big\langle F(x,y), K_t(x)\widetilde{K}_s(y)\Big\rangle_{L^2(0,1)^2}\,, \quad t,s \in \mathbb{R}\,, \tag{3.9}$$

where

$$K_t(x) = \sum_{n\in\mathbb{Z}} \overline{\varphi(t-n)}\, \mathrm{e}^{-2\pi i n x} \quad \text{and} \quad \widetilde{K}_s(y) = \sum_{m\in\mathbb{Z}} \overline{\psi(s-m)}\, \mathrm{e}^{-2\pi i m y}$$

belong to $L^2(0,1)$. Thus, for fixed $a, b \in [0,1)$ and $n, m \in \mathbb{Z}$ we get the following expression for the samples

$$\begin{aligned} f(a+n, b+m) &= \langle F, K_{a+n}\, \widetilde{K}_{b+m}\rangle_{L^2(0,1)^2} \\ &= \langle F, \mathrm{e}^{-2\pi i n x} K_a(x)\, \mathrm{e}^{-2\pi i m y} \widetilde{K}_b(y)\rangle_{L^2(0,1)^2} \,. \end{aligned}$$

Assuming that $0 < \|K_a\|_0 \le \|K_a\|_\infty < \infty$ and $0 < \|\widetilde{K}_b\|_0 \le \|\widetilde{K}_b\|_\infty < \infty$, the sequences

$$\big\{\mathrm{e}^{-2\pi i n x} \mathrm{e}^{-2\pi i m y} K_a(x) \widetilde{K}_b(y)\big\}_{n,m\in\mathbb{Z}} \quad \text{and} \quad \left\{\frac{\mathrm{e}^{-2\pi i n x}\mathrm{e}^{-2\pi i m}}{(\overline{K_a(x)}\, \overline{\widetilde{K}_b(y)})}\right\}_{n,m\in\mathbb{Z}}$$

form a pair of dual Riesz bases $L^2(0,1)^2$.

In tensor notation, the sequences $\{x_n \otimes \widetilde{x}_m\}_{n,m\in\mathbb{Z}}$ and $\{y_n \otimes \widetilde{y}_m\}_{n,m\in\mathbb{Z}}$ form a pair of dual Riesz bases in $L^2(0,1)^2$, where

$$x_n = \mathrm{e}^{-2\pi i n x} K_a(x)\,; \quad \widetilde{x}_m = \mathrm{e}^{-2\pi i m y} \widetilde{K}_b(y)$$

$$y_n = \frac{\mathrm{e}^{-2\pi i n x}}{\overline{K_a(x)}}\,; \quad \widetilde{y}_m = \frac{\mathrm{e}^{-2\pi i m y}}{\overline{\widetilde{K}_b(y)}}\,.$$

Proceeding as in the former section we obtain:

**Proposition 3.5** *For any $f \in V^2_{\varphi\psi}$ the sampling formula*

$$f(t,s) = \sum_{n,m\in\mathbb{Z}} f(a+n, b+m)\, S_a(t-n)\, \widetilde{S}_b(s-m)\,, \quad t, s \in \mathbb{R}\,, \tag{3.10}$$

*holds, where $S_a := \mathcal{T}_\varphi\big(1/\overline{K_a}\big) \in V^2_\varphi$ and $\widetilde{S}_b := \mathcal{T}_\psi\big(1/\overline{\widetilde{K}_b}\big) \in V^2_\psi$. The convergence of the series is absolute due to the unconditional convergence of a Riesz basis expansion and it is also uniform on $\mathbb{R}^2$ since the shift-invariant space $V^2_{\varphi\psi}$ is a RKHS contained in $L^2(\mathbb{R}^2)$ and the function*

$$(t,s) \longmapsto \sum_{n\in\mathbb{Z}} |\varphi(t-n)|^2 \sum_{m\in\mathbb{Z}} |\psi(s-m)|^2$$

*is bounded on $\mathbb{R}^2$.*

## 3.2 Sampling Formulas Involving Samples Differences

We start this section with an easy motivating example. Consider the shift-invariant space $V_\varphi^2 = \big\{ \sum_{n\in\mathbb{Z}} a_n\, \varphi(t-n) \ :\ \{a_n\} \in \ell^2(\mathbb{Z}) \big\}$ of $L^2(\mathbb{R})$ with stable generator $\varphi \in L^2(\mathbb{R})$. The goal in this section is to investigate the existence of stable sampling formulas involving differences (or averages) obtained from the regular sequence of the samples $\{f(a+n)\}_{n\in\mathbb{Z}}$ of any $f \in V_\varphi^2$.

In general, the involved samples are $p$ subsequences taken, for instance, from

$$\big\{f(a+pn)\big\}_{n\in\mathbb{Z}}; \quad \big\{\Delta_+^k f(a+pn)\big\}_{n\in\mathbb{Z}}; \quad \text{and/or} \quad \big\{\Delta_-^k f(a+pn)\big\}_{n\in\mathbb{Z}}. \tag{3.11}$$

In the second sequence in (3.11),

$$\Delta_+^k f(a+pn) := \Delta_+^{k-1} f(a+pn+1) - \Delta_+^{k-1} f(a+pn)$$

denotes the $k$-th forward difference operator; here,

$$\Delta_+ f(a+pn) = f(a+pn+1) - f(a+pn) \quad \text{and} \quad \Delta_+^0 = I\,.$$

Analogously,

$$\Delta_-^k f(a+pn) := \Delta_-^{k-1} f(a+pn) - \Delta_-^{k-1} f(a+pn-1)$$

denotes the $k$-th backward difference operator, $1 \le k \le p-1$, and $a$ is a fixed number in $[0, 1)$; here,

$$\Delta_- f(a+pn) = f(a+pn) - f(a+pn-1) \quad \text{and} \quad \Delta_-^0 = I\,.$$

Successive differences on samples $\big\{f(a+n)\big\}_{n\in\mathbb{Z}}$ can help us to discover some smoothness features on the original function $f$.

In the above section we have proved that, under some appropriate hypotheses, a Shannon-type sampling formula as

$$f(t) = \sum_{n\in\mathbb{Z}} f(a+n)\, S_a(t-n)\,, \quad t \in \mathbb{R}\,, \tag{3.12}$$

holds in $V_\varphi^2$ for some sampling function $S_a \in V_\varphi^2$ satisfying that the shifted sequence $\big\{S_a(t-n)\big\}_{n\in\mathbb{Z}}$ is a Riesz basis for $V_\varphi^2$.

As a consequence, the above sampling formula is an interpolation formula. Indeed, the function $S_a$ satisfies the interpolation property

$$S_a(a+n) = \delta_n \quad \text{for each} \quad n \in \mathbb{Z}\,.$$

Roughly speaking, the aim is to rewrite formula (3.12) by replacing some samples by any other information given by differences, averages, etc.

Let us consider an easy example. Initially, we write formula (3.12) as

$$f(t)=\sum_{n\in\mathbb{Z}}\big\{f(a+2n)\,S_a(t-2n)+f(a+2n+1)\,S_a(t-2n-1)\big\}\,,\quad t\in\mathbb{R}\,.$$

Then, adding and subtracting $f(a+2n)\,S_a(t-2n-1)$ in each summand, for each $t\in\mathbb{R}$, yields

$$f(t)=\sum_{n\in\mathbb{Z}}\big\{f(a+2n)\big[S_a(t-2n)+S_a(t-2n-1)\big]+\Delta_+f(a+2n)\,S_a(t-2n-1)\big\}.$$

The stable recovery of any $f\in V_\varphi^2$ from the samples

$$\big\{f(a+2n)\big\}_{n\in\mathbb{Z}}\cup\big\{\Delta_+f(a+2n)\big\}_{n\in\mathbb{Z}}$$

follows from the fact that the sequence

$$\big\{S_a(t-2n)+S_a(t-2n-1)\big\}_{n\in\mathbb{Z}}\cup\big\{S_a(t-2n-1)\big\}_{n\in\mathbb{Z}}$$

turns out to be a Riesz basis for $V_\varphi^2$, that is, there exist constants $0<A\le B$ such that

$$A\|f\|^2\le\sum_{n\in\mathbb{Z}}\big\{|f(a+2n)|^2+|\Delta_+f(a+2n)|^2\big\}\le B\|f\|^2\quad\text{for each } f\in V_\varphi^2\,.$$

In this section, stable sampling formulas involving forward and/or backward differences of arbitrary order of the sequence of samples $\{f(a+n)\}_{n\in\mathbb{Z}}$ are derived in a systematic way. The used technique will allow to deal also with central differences and/or averages.

Thus we obtain a sort of counterpart to the family of finite interpolation formulas in the infinite setting of sampling in shift-invariant subspaces of $L^2(\mathbb{R})$.

The existence of an isomorphism $\mathcal{T}_\varphi$ between $L^2(0,1)$ and $V_\varphi^2$ and an associated representation of the functions in $V_\varphi^2$ make the mathematical technique very simple: we identify the available samples as inner products with respect to a suitable Riesz basis in $L^2(0,1)$ for which we obtain its corresponding dual Riesz basis. Finally, we derive the corresponding sampling formula in $V_\varphi^2$ by using the isomorphism $\mathcal{T}_\varphi$.

In Ref. [17], sampling using, for instance, the samples $\{x(2n)\}_{n\in\mathbb{Z}}\cup\{\Delta x(2n)\}_{n\in\mathbb{Z}}$ goes under the name of *unconventional sampling*.

### 3.2.1 The Involved Riesz Bases in $L^2(0,1)$ and Their Dual Bases

Having in mind the expression $f(t) = \langle F, K_t\rangle$, $t \in \mathbb{R}$, for the *forward difference operator* $\Delta_+$ of $f \in V_\varphi^2$, i.e.,

$$\Delta_+ f(t) := f(t+1) - f(t)\,, \quad t \in \mathbb{R}\,,$$

we have

$$\Delta_+ f(t) = \big\langle F, K_{t+1} - K_t\big\rangle_{L^2(0,1)} = \big\langle F, (\mathrm{e}^{-2\pi i x} - 1)\, K_t\big\rangle_{L^2(0,1)}\,, \quad t \in \mathbb{R}\,.$$

In general, for the $k$-th forward difference operator we have the expression

$$\Delta_+^k f(t) = \big\langle F, (\mathrm{e}^{-2\pi i x} - 1)^k\, K_t\big\rangle_{L^2(0,1)}\,, \quad t \in \mathbb{R}\,. \tag{3.13}$$

For the *backward difference operator* $\Delta_-$ of $f \in V_\varphi^2$, i.e.,

$$\Delta_- f(t) := f(t) - f(t-1)\,, \quad t \in \mathbb{R}\,,$$

we have

$$\Delta_- f(t) = \big\langle F, K_t - K_{t-1}\big\rangle_{L^2(0,1)} = \big\langle F, (1 - \mathrm{e}^{2\pi i x})\, K_t\big\rangle_{L^2(0,1)}\,, \quad t \in \mathbb{R}\,.$$

In general, for the $k$-th backward difference operator we have the expression

$$\Delta_-^k f(t) = \big\langle F, (1 - \mathrm{e}^{2\pi i x})^k\, K_t\big\rangle_{L^2(0,1)}\,, \quad t \in \mathbb{R}\,. \tag{3.14}$$

Similar expressions can be also obtained from iterates of the *central difference operator* $\Delta_0$ , i.e.,

$$\Delta_0 f(t) := f(t+1) - f(t-1)\,, \qquad t \in \mathbb{R}\,,$$

or from the *average operators* $\mu_+$, $\mu_-$ or $\mu_0$ defined as:

$$\mu_+ f(t) := \frac{1}{2}[f(t+1) + f(t)]\,,$$

$$\mu_- f(t) := \frac{1}{2}[f(t) + f(t-1)]\,,$$

$$\mu_0 f(t) := \frac{1}{2}[f(t+1) + f(t-1)]\,, \qquad t \in \mathbb{R}\,.$$

All these cases can be put under a common umbrella by considering *generalized finite differences* defined as

$$\Delta^{\mathbf{a}}_{M,N} f(t) := \sum_{k=M}^{N} a_k f(t+k)\,, \quad \text{where } \mathbf{a} = (a_k)_{M\le k\le N} \subset \mathbb{C} \text{ and } M, N \in \mathbb{Z}\,.$$

Going back to our previous example we have for the samples $f(a+2n)$ and $\Delta_+ f(a+2n)$, $n \in \mathbb{Z}$:

$$f(a+2n) = \big\langle F, \mathrm{e}^{-2\pi i(2n)x} K_a \big\rangle_{L^2(0,1)}\,,$$
$$\Delta_+ f(a+2n) = \big\langle F, \mathrm{e}^{-2\pi i(2n+1)x} K_a - \mathrm{e}^{-2\pi i(2n)x} K_a \big\rangle_{L^2(0,1)}\,.$$

In other words, denoting $x_n := \mathrm{e}^{-2\pi i n x} K_a(x)$, $n \in \mathbb{Z}$, the arising questions are:

- Is the sequence $\{x_{2n}\}_{n\in\mathbb{Z}} \cup \{x_{2n+1} - x_{2n}\}_{n\in\mathbb{Z}}$ a Riesz basis for $L^2(0,1)$?
- How to obtain, in the affirmative case, its dual basis?

As was mentioned in the introduction section, both are true. Indeed, denoting $y_n := \mathrm{e}^{-2\pi i n x}/\overline{K_a(x)}$, $n \in \mathbb{Z}$, the dual Riesz basis of $\{x_n\}_{n\in\mathbb{Z}}$, the dual Riesz basis of $\{x_{2n}\}_{n\in\mathbb{Z}} \cup \{x_{2n+1} - x_{2n}\}_{n\in\mathbb{Z}}$ is $\{y_{2n} + y_{2n+1}\}_{n\in\mathbb{Z}} \cup \{y_{2n+1}\}_{n\in\mathbb{Z}}$.

A general answer which involves the aforementioned generalized finite differences is given in the following simple result borrowed from Ref. [8]:

**Theorem 3.3** *Let $\{x_n\}_{n\in\mathbb{Z}}$ be a Riesz basis for a separable Hilbert space $\mathcal{H}$ with dual Riesz basis $\{y_n\}_{n\in\mathbb{Z}}$. For a fixed $p \in \mathbb{N}$, consider a partition*

$$\{x_{1n}\}_{n\in\mathbb{Z}} \cup \{x_{2n}\}_{n\in\mathbb{Z}} \cup \cdots \cup \{x_{pn}\}_{n\in\mathbb{Z}}$$

*of the basis $\{x_n\}_{n\in\mathbb{Z}}$, the same partition $\{y_{1n}\}_{n\in\mathbb{Z}} \cup \{y_{2n}\}_{n\in\mathbb{Z}} \cup \cdots \cup \{y_{pn}\}_{n\in\mathbb{Z}}$ for its dual basis $\{y_n\}_{n\in\mathbb{Z}}$, and let $M = [a_{jk}] \in \mathbb{C}^{p\times p}$ be a matrix of scalars.*

*Then, the new sequence defined by using the rows of $M$ as*

$$\Big\{\sum_{k=1}^{p} a_{1k} x_{kn}\Big\}_{n\in\mathbb{Z}} \cup \Big\{\sum_{k=1}^{p} a_{2k} x_{kn}\Big\}_{n\in\mathbb{Z}} \cup \cdots \cup \Big\{\sum_{k=1}^{p} a_{pk} x_{kn}\Big\}_{n\in\mathbb{Z}} \tag{3.15}$$

*is a Riesz basis for $\mathcal{H}$ if and only if* $\det M \neq 0$.

*In this case, its dual Riesz basis is given by*

$$\Big\{\sum_{k=1}^{p} \overline{b}_{k1} y_{kn}\Big\}_{n\in\mathbb{Z}} \cup \Big\{\sum_{k=1}^{p} \overline{b}_{k2} y_{kn}\Big\}_{n\in\mathbb{Z}} \cup \cdots \cup \Big\{\sum_{k=1}^{p} \overline{b}_{kp} y_{kn}\Big\}_{n\in\mathbb{Z}}\,,$$

*where $\big(b_{1k}\ b_{2k}\ \ldots\ b_{pk}\big)^\top$ denotes the $k$-th column, $k = 1, 2, \ldots, p$, of the inverse matrix $M^{-1}$.*

*Moreover, if a sequence as* (3.15) *with* $\det M \neq 0$ *is a Riesz basis for* $\mathcal{H}$*, then the sequence* $\{x_n\}_{n\in\mathbb{Z}}$ *is also a Riesz basis for* $\mathcal{H}$*.*

***Proof*** If $\det M = 0$, then $\left\{\sum_{k=1}^{p} a_{1k}x_{km}, \sum_{k=1}^{p} a_{2k}x_{km}, \dots, \sum_{k=1}^{p} a_{pk}x_{km}\right\}$ is a linearly dependent system for each $m \in \mathbb{Z}$, so that the sequence defined in (3.15) is not a Riesz basis.

Conversely, assume that $\det M \neq 0$. Denoting the new sequence defined in (3.15) as $\{z_n\}_{n\in\mathbb{Z}}$ and its partition as $\{z_{1n}\}_{n\in\mathbb{Z}} \cup \{z_{2n}\}_{n\in\mathbb{Z}} \cup \cdots \cup \{z_{pn}\}_{n\in\mathbb{Z}}$, we rewrite (3.15) using the symbolic notation

$$M\big(x_{1n}, x_{2n}, \dots, x_{pn}\big)^{\top} = \big(z_{1n}, z_{2n}, \dots, z_{pn}\big)^{\top}, \quad n \in \mathbb{Z}.$$

It is easily deduced what we also symbolically write as

$$\big(x_{1n}, x_{2n}, \dots, x_{pn}\big)^{\top} = M^{-1}\big(z_{1n}, z_{2n}, \dots, z_{pn}\big)^{\top}, \quad n \in \mathbb{Z}.$$

Thus $\operatorname{span}\{z_n\}_{n\in\mathbb{Z}} = \operatorname{span}\{x_n\}_{n\in\mathbb{Z}}$ and consequently the sequence $\{z_n\}_{n\in\mathbb{Z}}$ is complete in $\mathcal{H}$.

In addition, for every finite scalar sequence $\{c_m\}$ and for every finite sequence $\{z_m\} = \left\{\sum_{k=1}^{p} a_{j_m k}x_{kn_m}\right\}$ with $n_m \in \mathbb{Z}$ and $j_m \in \{1, 2, \dots, p\}$, one has

$$\begin{aligned}\Big\|\sum_m c_m\, z_m\Big\|^2 &= \Big\|\sum_m c_m\Big(\sum_{k=1}^{p} a_{j_m k}x_{kn_m}\Big)\Big\|^2 \le B\sum_m\sum_{k=1}^{p} |c_m a_{j_m k}|^2 \\ &= B\sum_m |c_m|^2 \sum_{k=1}^{p} |a_{j_m k}|^2 \le BD\sum_m |c_m|^2,\end{aligned}$$

where $B$ is an upper Riesz bound of $\{x_n\}_{n\in\mathbb{Z}}$ and $D = \max\limits_{1\le j\le p} \sum_{k=1}^{p} |a_{jk}|^2$.

Similarly,

$$\begin{aligned}\Big\|\sum_m c_m\, z_m\Big\|^2 &= \Big\|\sum_m c_m\Big(\sum_{k=1}^{p} a_{j_m k}x_{kn_m}\Big)\Big\|^2 \ge A\sum_m\sum_{k=1}^{p} |c_m a_{j_m k}|^2 \\ &= A\sum_m |c_m|^2 \sum_{k=1}^{p} |a_{j_m k}|^2 \ge AC\sum_m |c_m|^2,\end{aligned}$$

where $A$ is a lower Riesz bound of the Riesz basis $\{x_n\}_{n\in\mathbb{Z}}$ and $C = \min\limits_{1\le j\le p} \sum_{k=1}^{p} |a_{jk}|^2$.

Observe that $C > 0$ since $\det M \neq 0$ and, consequently, the sequence $\{z_n\}_{n\in\mathbb{Z}}$, in addition to $\overline{\operatorname{span}}\{z_n\}_{n\in\mathbb{Z}} = \mathcal{H}$, satisfies the Riesz sequence condition (1.6), i.e.,

$$A' \sum_m |c_m|^2 \le \Big\| \sum_m c_m\, z_m \Big\|^2 \le B' \sum_m |c_m|^2 ,$$

with constants $A' = AC > 0$ and $B' = BD > 0$.

Finally, since the sequences $\{x_n\}_{n\in\mathbb{Z}}$ and $\{y_n\}_{n\in\mathbb{Z}}$ are biorthogonal and $M\,M^{-1} = I_p$, we have

$$\langle x_q, y_{q'} \rangle = \delta_{q,q'} \quad \text{and} \quad \sum_{k=1}^{p} a_{jk} b_{kj'} = \delta_{j,j'}$$

from which we obtain

$$\begin{aligned} \Big\langle \sum_{k=1}^{p} a_{jk} x_{kn}, \sum_{k'=1}^{p} \overline{b}_{k'j'} y_{k'm} \Big\rangle &= \sum_{k,k'=1}^{p} a_{jk} b_{k'j'} \langle x_{kn}, y_{k'm} \rangle \\ &= \begin{cases} 1 & \text{if } n \neq m \text{ and } j = j' \\ 0 & \text{otherwise} \end{cases}, \end{aligned}$$

proving that the sequences

$$\Big\{ \sum_{k=1}^{p} a_{jk} x_{kn} \Big\}_{j=1,2,\ldots,p;\, n\in\mathbb{Z}} \quad \text{and} \quad \Big\{ \sum_{k=1}^{p} \overline{b}_{kj} y_{kn} \Big\}_{j=1,2,\ldots p;\, n\in\mathbb{Z}}$$

are a dual pair of Riesz bases for $\mathcal{H}$. □

Observe that in the case of one forward difference, the partition of the Riesz basis $\{x_n\}_{n\in\mathbb{Z}}$ is $\{x_{2n}\}_{n\in\mathbb{Z}} \cup \{x_{2n+1}\}_{n\in\mathbb{Z}}$. Then, the associated matrix to the Riesz basis $\{x_{2n}\}_{n\in\mathbb{Z}} \cup \{x_{2n+1} - x_{2n}\}_{n\in\mathbb{Z}}$ is $M = \begin{pmatrix} 1 & 0 \\ -1 & 1 \end{pmatrix}$. Meanwhile, the columns of its inverse $M^{-1} = \begin{pmatrix} 1 & 0 \\ 1 & 1 \end{pmatrix}$ give the dual Riesz basis

$$\big\{ y_{2n} + y_{2n+1} \big\}_{n\in\mathbb{Z}} \cup \big\{ y_{2n+1} \big\}_{n\in\mathbb{Z}} .$$

Although we will not derive overcomplete expansions in the sequel, it is worth mentioning the frame counterpart to the above theorem. Under the circumstances of Theorem 3.3 we have:

**Theorem 3.4** *Under the same assumptions of Theorem 3.3, let $M = [a_{jk}] \in \mathbb{C}^{q\times p}$ be a matrix of scalars such that $q > p$ and rank $M = p$. By using the rows of $M$ we define*

$$z_{jn} = \sum_{k=1}^{p} a_{jk}\, x_{kn}\,, \quad \text{where } 1 \le j \le q \text{ and } n \in \mathbb{Z}\,.$$

*Then the sequence*

$$\{z_{1n}\}_{n\in\mathbb{Z}} \cup \{z_{2n}\}_{n\in\mathbb{Z}} \cup \cdots \cup \{z_{qn}\}_{n\in\mathbb{Z}}$$

*is a frame for $\mathcal{H}$. Moreover, a dual frame $\{\tilde{z}_{1n}\}_{n\in\mathbb{Z}} \cup \{\tilde{z}_{2n}\}_{n\in\mathbb{Z}} \cup \cdots \cup \{\tilde{z}_{qn}\}_{n\in\mathbb{Z}}$ is obtained from the columns of a left-inverse matrix $N = [b_{kj}] \in \mathbb{C}^{p\times q}$ of $M$ as*

$$\tilde{z}_{jn} = \sum_{k=1}^{p} \overline{b}_{kj}\, y_{kn}\,, \quad \text{where } 1 \le j \le q \text{ and } n \in \mathbb{Z}\,.$$

***Proof*** Clearly, both sequences are Bessel sequences. Following Proposition 1.11, it suffices to prove that

$$x = \sum_{n\in\mathbb{Z}} \sum_{j=1}^{q} \langle x, \tilde{z}_{jn}\rangle\, z_{jn} \quad \text{for each } x \in \mathcal{H}\,.$$

Indeed,

$$\sum_{n\in\mathbb{Z}} \sum_{j=1}^{q} \Big\langle x, \sum_{k=1}^{p} \overline{b}_{kj}\, y_{kn}\Big\rangle \sum_{k=1}^{p} a_{jk}\, x_{kn} = \sum_{n\in\mathbb{Z}} \sum_{k,k'=1}^{p} \Big(\sum_{j=1}^{q} b_{kj} a_{jk'}\Big) \langle x, y_{kn}\rangle x_{k'n}$$

$$= \sum_{n\in\mathbb{Z}} \sum_{k,k'=1}^{p} \delta_{k,k'} \langle x, y_{kn}\rangle x_{k'n} = \sum_{n\in\mathbb{Z}} \sum_{k=1}^{p} \langle x, y_{kn}\rangle x_{kn} = \sum_{n\in\mathbb{Z}} \langle x, y_n\rangle x_n = x\,,$$

since $\sum_{j=1}^{q} b_{kj} a_{jk'} = \delta_{k,k'}$, and the sequences $\{x_n\}_{n\in\mathbb{Z}}$ and $\{y_n\}_{n\in\mathbb{Z}}$ form a pair of dual Riesz bases. □

Observe that the proof of the above lemma is also valid in case the sequences $\{x_n\}_{n\in\mathbb{Z}}$ and $\{y_n\}_{n\in\mathbb{Z}}$ form a pair of dual frames for $\mathcal{H}$.

Next we include a necessary reminder on matrix analysis needed in the sequel:

(*a*) Let $M$ be a $q \times p$ matrix with $q \geq p$ and $\operatorname{rank} M = p$. Then, the $p \times p$ matrix $M^*M$ is invertible, and its Moore-Penrose pseudo-inverse $M^\dagger$ given by the $p \times q$ matrix

$$M^\dagger = \left(M^*M\right)^{-1}M^*$$

is a left-inverse of $M$, i.e., $M^\dagger M = I_p$.

(*b*) In this case, the set of all left-inverse matrices $N$ of the matrix $M$ can be expressed as

$$N = M^\dagger + U\big[I_q - M\,M^\dagger\big], \tag{3.16}$$

where $U$ denotes an arbitrary $p \times q$ matrix.

Indeed, any matrix defined as (3.16) is a left-inverse of $M$

$$\big(M^\dagger + U\big[I_q - M\,M^\dagger\big]\big)M = M^\dagger M + U\big[M - M\big] = I_q\,.$$

Conversely, given a left-inverse $P$ of $M$, i.e., $P\,M = I_p$, it belongs to the set defined in (3.16) by taking $U = P$,

$$M^\dagger + P\big[I_q - M\,M^\dagger\big] = M^\dagger + P - M^\dagger = P\,.$$

(*c*) Note that, conversely, for a matrix $M$ of order $q \times p$ with $q \geq p$, the existence of a matrix $N$ of order $p \times q$ such that $N\,M = I_p$ implies that $\operatorname{rank} M = p$.

### 3.2.2 Sampling Results in the One-Dimensional Case

Throughout this section we fix $a \in [0, 1)$ such that $0 < \|K_a\|_0 \leq \|K_a\|_\infty < \infty$, and we will denote by

$$x_n = \mathrm{e}^{-2\pi i n x} K_a(x) \quad \text{and} \quad y_n = \frac{\mathrm{e}^{-2\pi i n x}}{\overline{K_a(x)}}\,, \qquad n \in \mathbb{Z}\,,$$

the associated pair of dual Riesz bases for $L^2(0, 1)$ which appeared in Sect. 3.1. The followed general procedure is easy:

(*i*) For any $f \in V_\varphi^2$, let $F \in L^2(0, 1)$ be such that $\mathcal{T}_\varphi F = f$; once we identify the Riesz basis for $L^2(0, 1)$ associated with the successive differences considered on the data samples, we compute its dual Riesz basis by using Theorem 3.3. Then we expand $F$ with respect to this dual basis in $L^2(0, 1)$. Finally, by applying the isomorphism $\mathcal{T}_\varphi$ we obtain a sampling formula in $V_\varphi^2$ with $L^2(\mathbb{R})$-convergence.

(*ii*) Since $V_\varphi^2$ is a RKHS we obtain the pointwise convergence properties of the sampling expansion: the convergence will be uniform on $\mathbb{R}$. It will be also absolute due the unconditional convergence of an expansion with respect to a Riesz basis.

Next we will exhibit a variety of examples involving different types of samples obtained from *forward, backward or central difference operators*, or from *average operators*.

**Example 3.3 (Using First and Second Forward Differences)** For the samples $f(a+3n)$, $\Delta_+ f(a+3n)$ and $\Delta_+^2 f(a+3n)$, $n \in \mathbb{Z}$, we have (see Eq. (3.13)):

$$\begin{aligned} f(a+3n) &= \big\langle F, \mathrm{e}^{-2\pi i(3n)x} K_a \big\rangle_{L^2(0,1)}, \\ \Delta_+ f(a+3n) &= \big\langle F, \mathrm{e}^{-2\pi i(3n+1)x} K_a - \mathrm{e}^{-2\pi i(3n)x} K_a \big\rangle_{L^2(0,1)}, \\ \Delta_+^2 f(a+3n) &= \big\langle F, \mathrm{e}^{-2\pi i(3n+2)x} K_a - 2\mathrm{e}^{-2\pi i(3n+1)x} K_a + \mathrm{e}^{-2\pi i(3n)x} K_a \big\rangle_{L^2(0,1)}. \end{aligned}$$

If we consider the partition $\{x_{3n}\}_{n\in\mathbb{Z}} \cup \{x_{3n+1}\}_{n\in\mathbb{Z}} \cup \{x_{3n+2}\}_{n\in\mathbb{Z}}$ of $\{x_n\}_{n\in\mathbb{Z}}$, according to Theorem 3.3 the sequence

$$\{x_{3n}\}_{n\in\mathbb{Z}} \cup \{x_{3n+1} - x_{3n}\}_{n\in\mathbb{Z}} \cup \{x_{3n+2} - 2x_{3n+1} + x_{3n}\}_{n\in\mathbb{Z}}$$

is a Riesz basis for $L^2(0,1)$ and its dual Riesz basis is

$$\{y_{3n} + y_{3n+1} + y_{3n+2}\}_{n\in\mathbb{Z}} \cup \{y_{3n+1} + 2y_{3n+2}\}_{n\in\mathbb{Z}} \cup \{y_{3n+2}\}_{n\in\mathbb{Z}},$$

since

$$M = \begin{pmatrix} 1 & 0 & 0 \\ -1 & 1 & 0 \\ 1 & -2 & 1 \end{pmatrix} \quad \text{and} \quad M^{-1} = \begin{pmatrix} 1 & 0 & 0 \\ 1 & 1 & 0 \\ 1 & 2 & 1 \end{pmatrix}.$$

Expanding $F$ with respect to the above dual basis we obtain

$$\begin{aligned} F = \sum_{n\in\mathbb{Z}} \Big\{ & f(a+3n)\big[y_{3n} + y_{3n+1} + y_{3n+2}\big] + \Delta_+ f(a+3n)\big[y_{3n+1} + 2y_{3n+2}\big] + \\ & + \Delta_+^2 f(a+3n)\, y_{3n+2} \Big\} \qquad \text{in } L^2(0,1). \end{aligned}$$

Applying the isomorphism $\mathcal{T}_\varphi$ and its shifting property (3.4), we obtain in $V_\varphi^2$ the sampling formula:

$$f(t)=\sum_{n\in\mathbb{Z}}\Big\{f(a+3n)\big[S_a(t-3n)+S_a(t-3n-1)+S_a(t-3n-2)\big]$$
$$+\Delta_+ f(a+3n)\big[S_a(t-3n-1)+2S_a(t-3n-2)\big]$$
$$+\Delta_+^2 f(a+3n)\,S_a(t-3n-2)\Big\}\,,\qquad t\in\mathbb{R}\,.$$

Observe that, due to the shifting property (3.4), we have

$$\mathcal{T}_\varphi\big(y_{pn+k}\big)(t)=S_a\big(t-(pn+k)\big)\,.$$

**Example 3.4 (Using $p-1$ Forward Differences)** Proceeding as in the above example, for the set of samples

$$f(a+pn),\ \Delta_+ f(a+pn),\ldots,\ \Delta_+^{p-1} f(a+pn),\quad n\in\mathbb{Z}\,,$$

we have (see Eq. (3.13)):

$$f(a+pn)=\big\langle F,\mathrm{e}^{-2\pi i(pn)x}K_a\big\rangle_{L^2(0,1)}\,,$$
$$\Delta_+ f(a+pn)=\big\langle F,\mathrm{e}^{-2\pi i(pn+1)x}K_a-\mathrm{e}^{-2\pi i(pn)x}K_a\big\rangle_{L^2(0,1)}\,,$$
$$\Delta_+^2 f(a+pn)=\big\langle F,\mathrm{e}^{-2\pi i(pn+2)x}K_a-2\mathrm{e}^{-2\pi i(pn+1)x}K_a+\mathrm{e}^{-2\pi i(pn)x}K_a\big\rangle_{L^2(0,1)}\,,$$
$$\cdots\cdots$$
$$\Delta_+^{p-1} f(a+pn)=\Big\langle F,\mathrm{e}^{-2\pi i(pn+p-1)x}K_a-(p-1)\mathrm{e}^{-2\pi i(pn+p-2)x}K_a$$
$$+\binom{p-1}{2}\mathrm{e}^{-2\pi i(pn+p-3)x}K_a+\ldots$$
$$+(-1)^{p-2}\binom{p-1}{p-2}\mathrm{e}^{-2\pi i(pn+1)x}K_a$$
$$+(-1)^{p-1}\mathrm{e}^{-2\pi i(pn)x}K_a\Big\rangle_{L^2(0,1)}\,.$$

If we consider the partition $\{x_{pn}\}_{n\in\mathbb{Z}}\cup\{x_{pn+1}\}_{n\in\mathbb{Z}}\cup\cdots\cup\{x_{pn+p-1}\}_{n\in\mathbb{Z}}$ of $\{x_n\}_{n\in\mathbb{Z}}$, according to Theorem 3.3 the sequence defined by using the rows of the lower triangular matrix $M$ given by

$$\begin{pmatrix} 1 & & & & & & \\ -1 & 1 & & & & & \\ 1 & -\binom{2}{1} & 1 & & & & \\ -1 & \binom{3}{1} & -\binom{3}{2} & 1 & & & \\ \vdots & \vdots & \vdots & \vdots & \ddots & & \\ (-1)^{p-2} & (-1)^{p-3}\binom{p-2}{1} & (-1)^{p-4}\binom{p-2}{2} & (-1)^{p-5}\binom{p-2}{3} & \dots & 1 & \\ (-1)^{p-1} & (-1)^{p-2}\binom{p-1}{1} & (-1)^{p-3}\binom{p-1}{2} & (-1)^{p-4}\binom{p-1}{3} & \dots & -\binom{p-1}{p-2} & 1 \end{pmatrix}$$

is a Riesz basis for $L^2(0, 1)$ and its dual Riesz basis and, consequently, the sampling functions are obtained from the columns of its inverse $M^{-1}$ which is:

$$M^{-1} = \begin{pmatrix} 1 & & & & & & \\ 1 & 1 & & & & & \\ 1 & \binom{2}{1} & 1 & & & & \\ 1 & \binom{3}{1} & \binom{3}{2} & 1 & & & \\ \vdots & \vdots & \vdots & \vdots & \ddots & & \\ 1 & \binom{p-2}{1} & \binom{p-2}{2} & \binom{p-2}{3} & \dots & 1 & \\ 1 & \binom{p-1}{1} & \binom{p-1}{2} & \binom{p-1}{3} & \dots & \binom{p-1}{p-2} & 1 \end{pmatrix}.$$

The resulting sampling formula in $V_\varphi^2$ reads:

$$\begin{aligned} f(t) = \sum_{n\in\mathbb{Z}} \Big\{ & f(a+pn)\, T_{a,1}(t-pn) + \Delta_+ f(a+pn)\, T_{a,2}(t-pn) \\ & + \Delta_+^2 f(a+pn)\, T_{a,3}(t-pn) + \cdots \\ & + \Delta_+^{p-1} f(a+pn)\, T_{a,p}(t-pn) \Big\}, \quad t\in\mathbb{R}, \end{aligned}$$

where, for $1 \le k \le p$, the sampling function $T_{a,k}$ is given by

$$\begin{aligned} T_{a,k}(t) = & \binom{k-1}{k-1} S_a(t-(k-1)) + \binom{k}{k-1} S_a(t-k) + \\ & + \binom{k+1}{k-1} S_a(t-(k+1)) + \cdots + \binom{p-1}{k-1} S_a(t-(p-1)), \quad t\in\mathbb{R}. \end{aligned}$$

Observe that the sampling function $T_{a,k}$, where $1 \le k \le p$, has $p-k+1$ summands.

**Example 3.5 (The General Case Involving $p-1$ Backward Differences)** Here, for the set of samples

$$f(a+pn),\ \Delta_- f(a+pn), \dots,\ \Delta_-^{p-1} f(a+pn), \quad n \in \mathbb{Z},$$

we have (see Eq. (3.14)):

$$\begin{aligned}
f(a+pn) &= \big\langle F, \mathrm{e}^{-2\pi i (pn)x} K_a \big\rangle_{L^2(0,1)}, \\
\Delta_- f(a+pn) &= \big\langle F, \mathrm{e}^{-2\pi i (pn)x} K_a - \mathrm{e}^{-2\pi i (pn-1)x} K_a \big\rangle_{L^2(0,1)}, \\
\Delta_-^2 f(a+pn) &= \big\langle F, \mathrm{e}^{-2\pi i (pn)x} K_a - 2\mathrm{e}^{-2\pi i (pn-1)x} K_a + \mathrm{e}^{-2\pi i (pn-2)x} K_a \big\rangle_{L^2(0,1)}, \\
&\cdots\cdots \\
\Delta_-^{p-1} f(a+pn) &= \Big\langle F, \mathrm{e}^{-2\pi i (pn)x} K_a - (p-1)\mathrm{e}^{-2\pi i (pn-1)x} K_a + \\
&\quad + \binom{p-1}{2} \mathrm{e}^{-2\pi i (pn-2)x} K_a + \cdots\cdots + \\
&\quad + (-1)^{p-2}\binom{p-1}{p-2} \mathrm{e}^{-2\pi i (pn-p+2)x} K_a + (-1)^{p-1} \mathrm{e}^{-2\pi i (pn-p+1)x} K_a \Big\rangle_{L^2(0,1)}.
\end{aligned}$$

If we consider the partition

$$\{x_{pn-p+1}\}_{n\in\mathbb{Z}} \cup \{x_{pn-p+2}\}_{n\in\mathbb{Z}} \cup \cdots \cup \{x_{pn-1}\}_{n\in\mathbb{Z}} \cup \{x_{pn}\}_{n\in\mathbb{Z}}$$

of $\{x_n\}_{n\in\mathbb{Z}}$, according to Theorem 3.3 the sequence defined by using the rows of the upper triangular matrix

$$M = \begin{pmatrix}
(-1)^{p-1} & (-1)^{p-2}\binom{p-1}{1} & (-1)^{p-3}\binom{p-1}{2} & \dots & -\binom{p-1}{3} & \binom{p-1}{2} & -\binom{p-1}{1} & 1 \\
 & (-1)^{p-2} & (-1)^{p-3}\binom{p-2}{1} & \dots & -\binom{p-2}{3} & \binom{p-2}{2} & -\binom{p-2}{1} & 1 \\
 & & (-1)^{p-3} & \dots & -\binom{p-3}{3} & \binom{p-3}{2} & -\binom{p-3}{1} & 1 \\
 & & & \ddots & \vdots & \vdots & \vdots & \vdots \\
 & & & & -1 & \binom{3}{2} & -\binom{3}{1} & 1 \\
 & & & & & 1 & -\binom{2}{1} & 1 \\
 & & & & & & -1 & 1 \\
 & & & & & & & 1
\end{pmatrix}$$

is a Riesz basis for $L^2(0,1)$ and its dual Riesz basis and, consequently, the sampling functions are obtained from the columns of its inverse which here coincides with itself.

Therefore, writing from the last column to the first one, the resulting sampling formula in $V_\varphi^2$ reads:

$$f(t) = \sum_{n\in\mathbb{Z}} \Big\{ f(a+pn)\,\widetilde{T}_{a,1}(t-pn) + \Delta_- f(a+pn)\,\widetilde{T}_{a,2}(t-pn) + \Delta_-^2 f(a+pn)\,\widetilde{T}_{a,3}(t-pn) + \cdots + \Delta_-^{p-1} f(a+pn)\,\widetilde{T}_{a,p}(t-pn)\Big\}, \quad t\in\mathbb{R},$$

where, for $1\le k\le p$, the sampling function $\widetilde{T}_{a,k}$ is given by

$$\widetilde{T}_{a,k}(t) = (-1)^{k-1}\Big[\binom{p-1}{k-1} S_a(t+p-1) + \binom{p-2}{k-1} S_a(t+p-2) + \binom{p-3}{k-1} S_a(t+p-3) + \cdots + \binom{k-1}{k-1} S_a(t+k-1)\Big], \quad t\in\mathbb{R}.$$

Observe that the sampling function $\widetilde{T}_{a,k}$, where $1\le k\le p$, has $p-k+1$ summands.

**Example 3.6 (Using First Forward and Backward Differences)** For the set of samples $f(a+3n)$, $\Delta_- f(a+3n)$ and $\Delta_+ f(a+3n)$, $n\in\mathbb{Z}$, we have:

$$\begin{aligned}
\Delta_- f(a+3n) &= \big\langle F, \mathrm{e}^{-2\pi i(3n)x} K_a - \mathrm{e}^{-2\pi i(3n-1)x} K_a\big\rangle_{L^2(0,1)},\\
f(a+3n) &= \big\langle F, \mathrm{e}^{-2\pi i(3n)x} K_a\big\rangle_{L^2(0,1)},\\
\Delta_+ f(a+3n) &= \big\langle F, \mathrm{e}^{-2\pi i(3n+1)x} K_a - \mathrm{e}^{-2\pi i(3n)x} K_a\big\rangle_{L^2(0,1)}.
\end{aligned}$$

If we consider the partition $\{x_{3n-1}\}_{n\in\mathbb{Z}} \cup \{x_{3n}\}_{n\in\mathbb{Z}} \cup \{x_{3n+1}\}_{n\in\mathbb{Z}}$ of $\{x_n\}_{n\in\mathbb{Z}}$, according to Theorem 3.3 the sequence

$$\{-x_{3n-1}+x_{3n}\}_{n\in\mathbb{Z}} \cup \{x_{3n}\}_{n\in\mathbb{Z}} \cup \{-x_{3n}+x_{3n+1}\}_{n\in\mathbb{Z}}$$

is a Riesz basis for $L^2(0,1)$ and its dual Riesz basis is

$$\{-y_{3n-1}\}_{n\in\mathbb{Z}} \cup \{y_{3n-1}+y_{3n}+y_{3n+1}\}_{n\in\mathbb{Z}} \cup \{y_{3n+1}\}_{n\in\mathbb{Z}}$$

since $\det M \neq 0$, where

$$M = \begin{pmatrix} -1 & 1 & 0\\ 0 & 1 & 0\\ 0 & -1 & 1\end{pmatrix} \quad\text{and}\quad M^{-1} = \begin{pmatrix} -1 & 1 & 0\\ 0 & 1 & 0\\ 0 & 1 & 1\end{pmatrix}.$$

The resulting sampling formula in $V_\varphi^2$ reads:

$$f(t)=\sum_{n\in\mathbb{Z}}\Big\{-\Delta_- f(a+3n)\,S_a(t-3n+1)$$
$$+f(a+3n)\big[S_a(t-3n+1)+S_a(t-3n)+S_a(t-3n-1)\big]$$
$$+\Delta_+ f(a+3n)\,S_a(t-3n-1)\Big\},\qquad t\in\mathbb{R}.$$

**Example 3.7 (Using First Backward and First and Second Forward Differences)** For the set of samples $f(a+4n)$, $\Delta_+ f(a+4n)$, $\Delta_+^2 f(a+4n)$ and $\Delta_- f(a+4n)$, $n\in\mathbb{Z}$, we have (see Eqs. (3.13) and (3.14)):

$$\Delta_- f(a+4n)=\big\langle F, e^{-2\pi i(4n)x}K_a-e^{-2\pi i(4n-1)x}K_a\big\rangle_{L^2(0,1)},$$
$$f(a+4n)=\big\langle F, e^{-2\pi i(4n)x}K_a\big\rangle_{L^2(0,1)},$$
$$\Delta_+ f(a+4n)=\big\langle F, e^{-2\pi i(4n+1)x}K_a-e^{-2\pi i(4n)x}K_a\big\rangle_{L^2(0,1)},$$
$$\Delta_+^2 f(a+4n)=\big\langle F, e^{-2\pi i(4n+2)x}K_a-2e^{-2\pi i(4n+1)x}K_a+e^{-2\pi i(4n)x}K_a\big\rangle_{L^2(0,1)}.$$

If we consider the partition $\{x_{4n-1}\}_{n\in\mathbb{Z}}\cup\{x_{4n}\}_{n\in\mathbb{Z}}\cup\{x_{4n+1}\}_{n\in\mathbb{Z}}\cup\{x_{4n+2}\}_{n\in\mathbb{Z}}$ of $\{x_n\}_{n\in\mathbb{Z}}$, according to Theorem 3.3 the sequence

$$\{-x_{4n-1}+x_{4n}\}_{n\in\mathbb{Z}}\cup\{x_{4n}\}_{n\in\mathbb{Z}}\cup\{-x_{4n}+x_{4n+1}\}_{n\in\mathbb{Z}}\cup\{x_{4n}-2x_{4n+1}+x_{4n+2}\}_{n\in\mathbb{Z}}$$

is a Riesz basis for $L^2(0,1)$ and its dual Riesz basis is

$$\{-y_{4n-1}\}_{n\in\mathbb{Z}}\cup\{y_{4n-1}+y_{4n}+y_{4n+1}+y_{4n+2}\}_{n\in\mathbb{Z}}\cup\{y_{4n+1}+2y_{4n+2}\}_{n\in\mathbb{Z}}\cup\{y_{4n+2}\}_{n\in\mathbb{Z}}$$

since $\det M\neq 0$, where

$$M=\begin{pmatrix}-1&1&0&0\\0&1&0&0\\0&-1&1&0\\0&1&-2&1\end{pmatrix}\quad\text{and}\quad M^{-1}=\begin{pmatrix}-1&1&0&0\\0&1&0&0\\0&1&1&0\\0&1&2&1\end{pmatrix}.$$

The resulting sampling formula in $V_\varphi^2$ reads:

$$f(t)=\sum_{n\in\mathbb{Z}}\Big\{-\Delta_- f(a+4n)S_a(t-4n+1)$$
$$+f(a+4n)\big[S_a(t-4n+1)+S_a(t-4n)+S_a(t-4n-1)+S_a(t-4n-2)\big]$$
$$+\Delta_+ f(a+4n)\big[S_a(t-4n-1)+2S_a(t-4n-2)\big]$$
$$+\Delta_+^2 f(a+4n)\,S_a(t-4n-2)\Big\},\qquad t\in\mathbb{R}.$$

From the above example, the general case involving $p$ forward and $q-1$ backward differences with sampling period $p+q$ easily comes out.

**Example 3.8 (Using Forward Averages and Differences)** For the set of samples $\mu_+ f(a+2n)$ and $\Delta_+ f(a+2n)$, $n \in \mathbb{Z}$, we have:

$$\mu_+ f(a+2n) = \big\langle F, \frac{1}{2}(\mathrm{e}^{-2\pi i(2n+1)x} K_a + \mathrm{e}^{-2\pi i(2n)x} K_a\big\rangle_{L^2(0,1)},$$

$$\Delta_+ f(a+2n) = \big\langle F, \mathrm{e}^{-2\pi i(2n+1)x} K_a - \mathrm{e}^{-2\pi i(2n)x} K_a\big\rangle_{L^2(0,1)}.$$

If we consider the partition $\{x_{2n}\}_{n\in\mathbb{Z}} \cup \{x_{2n+1}\}_{n\in\mathbb{Z}}$ of $\{x_n\}_{n\in\mathbb{Z}}$, according to Theorem 3.3 the sequence

$$\big\{\frac{1}{2}x_{2n} + \frac{1}{2}x_{2n+1}\big\}_{n\in\mathbb{Z}} \cup \big\{-x_{2n} + x_{2n+1}\big\}_{n\in\mathbb{Z}}$$

is a Riesz basis for $L^2(0,1)$ and its dual Riesz basis is

$$\big\{y_{2n} + y_{2n+1}\big\}_{n\in\mathbb{Z}} \cup \big\{-\frac{1}{2}y_{2n} + \frac{1}{2}y_{2n+1}\big\}_{n\in\mathbb{Z}}$$

since $\det M \neq 0$, where

$$M = \begin{pmatrix} \frac{1}{2} & \frac{1}{2} \\ -1 & 1 \end{pmatrix} \quad \text{and} \quad M^{-1} = \begin{pmatrix} 1 & \frac{-1}{2} \\ 1 & \frac{1}{2} \end{pmatrix}.$$

The resulting sampling formula in $V_\varphi^2$ reads:

$$\begin{aligned} f(t) = \sum_{n\in\mathbb{Z}} \Big\{ & \mu_+ f(a+2n)\big[S_a(t-2n) + S_a(t-2n-1)\big] \\ & + \Delta_+ f(a+2n)\frac{1}{2}\big[-S_a(t-2n) + S_a(t-2n-1)\big]\Big\}, \qquad t \in \mathbb{R}. \end{aligned}$$

**Example 3.9 (Using Central Averages and Differences)** For the set of samples $f(a+3n)$, $\mu_0 f(a+3n)$ and $\Delta_0 f(a+3n)$, $n \in \mathbb{Z}$, we have:

$$\mu_0 f(a+3n) = \big\langle F, \frac{1}{2}(\mathrm{e}^{-2\pi i(3n+1)x} K_a + \mathrm{e}^{-2\pi i(3n-1)x} K_a\big\rangle_{L^2(0,1)},$$

$$f(a+3n) = \big\langle F, \mathrm{e}^{-2\pi i(3n)x} K_a\big\rangle_{L^2(0,1)},$$

$$\Delta_0 f(a+3n) = \big\langle F, \mathrm{e}^{-2\pi i(3n+1)x} K_a - \mathrm{e}^{-2\pi i(3n-1)x} K_a\big\rangle_{L^2(0,1)}.$$

If we consider the partition $\{x_{3n-1}\}_{n\in\mathbb{Z}} \cup \{x_{3n}\}_{n\in\mathbb{Z}} \cup \{x_{3n+1}\}_{n\in\mathbb{Z}}$ of $\{x_n\}_{n\in\mathbb{Z}}$, according to Theorem 3.3 the sequence

$$\{x_{3n}\}_{n\in\mathbb{Z}} \cup \{\frac{1}{2}x_{3n-1} + \frac{1}{2}x_{3n+1}\}_{n\in\mathbb{Z}} \cup \{-x_{3n-1} + x_{3n+1}\}_{n\in\mathbb{Z}}$$

is a Riesz basis for $L^2(0,1)$ and its dual Riesz basis is

$$\{y_{3n}\}_{n\in\mathbb{Z}} \cup \{y_{3n-1} + y_{3n+1}\}_{n\in\mathbb{Z}} \cup \{-\frac{1}{2}y_{3n-1} + \frac{1}{2}y_{3n+1}\}_{n\in\mathbb{Z}}$$

since $\det M \neq 0$, where

$$M = \begin{pmatrix} 0 & 1 & 0 \\ \frac{1}{2} & 0 & \frac{1}{2} \\ -1 & 0 & 1 \end{pmatrix} \quad \text{and} \quad M^{-1} = \begin{pmatrix} 0 & 1 & \frac{-1}{2} \\ 1 & 0 & 0 \\ 0 & 1 & \frac{1}{2} \end{pmatrix}.$$

The resulting sampling formula in $V_\varphi^2$ reads:

$$\begin{aligned} f(t) = \sum_{n\in\mathbb{Z}} \Big[ & f(a+3n)\, S_a(t-3n) \\ & + \mu_0 f(a+3n)\big[S_a(t-3n+1) + S_a(t-3n-1)\big] \\ & + \Delta_0 f(a+3n)\frac{1}{2}\big[-S_a(t-3n+1) + S_a(t-3n-1)\big]\Big], \qquad t\in\mathbb{R}. \end{aligned}$$

### 3.2.3 *Sampling Results in the Two-Dimensional Case*

Throughout this section we will use the *forward differences* defined as:

$$\Delta_+^{k,k'} f = \Delta_+^{1,0}(\Delta_+^{k-1,k'} f) = \Delta_+^{0,1}(\Delta_+^{k,k'-1} f),$$

where

$$\Delta_+^{1,0} f(t,s) = f(t+1,s) - f(t,s) \quad \text{and} \quad \Delta_+^{0,1} f(t,s) = f(t,s+1) - f(t,s);$$

we adopt the convention $\Delta_+^{0,0} = I$.

For any $f \in V^2_{\varphi\psi}$ (such that $\mathcal{T}_{\varphi\psi} F = f$) or $f \in V^2_{\Phi}$ (such that $\mathcal{T}_{\Phi} F = f$), we respectively get

$$\Delta_+^{k,k'} f(t,s) = \big\langle F, (\mathrm{e}^{-2\pi i x} - 1)^k\, (\mathrm{e}^{-2\pi i y} - 1)^{k'}\, K_t(x)\, \tilde{K}_s(y)\big\rangle_{L^2(0,1)^2}, \tag{3.17}$$

or

$$\Delta_+^{k,k'} f(t,s) = \big\langle F, (\mathrm{e}^{-2\pi i x} - 1)^k\, (\mathrm{e}^{-2\pi i y} - 1)^{k'}\, K_{t,s}(x,y)\big\rangle_{L^2(0,1)^2}. \tag{3.18}$$

Next we exhibit a couple of examples involving forward differences up to second order. The obtained results could be generalized to high order forward differences and even to higher dimensions. Besides, backward differences or averages could be used as in the previous section.

**Example 3.10 (Using the Samples $\Delta_+^{k,k'} f(a+2n, b+3m)$ $k = 0, 1$ and $k' = 0, 1, 2$ in $V^2_{\varphi\psi}$)** For any $f = \mathcal{T}_{\varphi\psi} F$ in $V^2_{\varphi\psi}$ we have the expression for the samples (see Eq. (3.17)):

$$\begin{aligned}
&\Delta_+^{k,k'} f(a+2n, b+3m)\\
&\quad = \Big\langle F, (\mathrm{e}^{-2\pi i x} - 1)^k (\mathrm{e}^{-2\pi i y} - 1)^{k'} \mathrm{e}^{-2\pi i (2n) x} K_a(x) \mathrm{e}^{-2\pi i (3m) y} \widetilde{K}_b(y)\Big\rangle,
\end{aligned}$$

where $n, m \in \mathbb{Z}$ and $k = 0, 1$ ; $k' = 0, 1, 2$.

Consider the Riesz basis $\{x_n \otimes \widetilde{x}_m\}_{n,m\in\mathbb{Z}}$ for $L^2(0,1)^2$, where

$$x_n := \mathrm{e}^{-2\pi i n x} K_a(x) \quad \text{and} \quad \widetilde{x}_m := \mathrm{e}^{-2\pi i m y} \widetilde{K}_b(y)\,, \qquad n, m \in \mathbb{Z}\,.$$

According to Theorem 3.3, for the partition

$$\begin{aligned}
&\{x_{2n} \otimes \widetilde{x}_{3m}\}_{n,m\in\mathbb{Z}} \cup \{x_{2n} \otimes \widetilde{x}_{3m+1}\}_{n,m\in\mathbb{Z}} \cup \{x_{2n} \otimes \widetilde{x}_{3m+2}\}_{n,m\in\mathbb{Z}}\\
&\cup \{x_{2n+1} \otimes \widetilde{x}_{3m}\}_{n,m\in\mathbb{Z}} \cup \{x_{2n+1} \otimes \widetilde{x}_{3m+1}\}_{n,m\in\mathbb{Z}} \cup \{x_{2n+1} \otimes \widetilde{x}_{3m+2}\}_{n,m\in\mathbb{Z}}\,,
\end{aligned}$$

the matrix $M$ of the Riesz basis associated with the above samples will be the *Kronecker product* (tensor product) of the corresponding matrices derived in Sect. 3.1 for the one-dimensional case.

Recall that the Kronecker product $A \otimes B$ of the matrices $A \in \mathbb{R}^{m\times n}$ and $B \in \mathbb{R}^{p\times q}$ is defined as the matrix:

$$A \otimes B = \begin{pmatrix} a_{11}B & \dots & a_{1n}B \\ \vdots & \ddots & \vdots \\ a_{m1}B & \dots & a_{mn}B \end{pmatrix} \in \mathbb{R}^{mp\times nq}\,.$$

In our case we obtain:

$$M=\begin{pmatrix}1&0\\-1&1\end{pmatrix}\otimes\begin{pmatrix}1&0&0\\-1&1&0\\1&-2&1\end{pmatrix}=\begin{pmatrix}1&0&0&0&0&0\\-1&1&0&0&0&0\\1&-2&1&0&0&0\\-1&0&0&1&0&0\\1&-1&0&-1&1&0\\-1&2&-1&1&-2&1\end{pmatrix}.$$

In case $A$ and $B$ are nonsingular matrices we have $(A\otimes B)^{-1}=A^{-1}\otimes B^{-1}$. See, for instance, Ref. [12] for the Kronecker product of matrices and its properties.

As a consequence, its inverse is the Kronecker product of their respective inverses; that is:

$$M^{-1}=\begin{pmatrix}1&0\\1&1\end{pmatrix}\otimes\begin{pmatrix}1&0&0\\1&1&0\\1&2&1\end{pmatrix}=\begin{pmatrix}1&0&0&0&0&0\\1&1&0&0&0&0\\1&2&1&0&0&0\\1&0&0&1&0&0\\1&1&0&1&1&0\\1&2&1&1&2&1\end{pmatrix}.$$

Thus, the corresponding sampling formula for each $f\in V^2_{\varphi\psi}$ reads:

$$f(t,s)=\sum_{n,m\in\mathbb{Z}}\Big\{\sum_{k=0}^{1}\sum_{k'=0}^{2}\Delta_+^{k,k'}f(a+2n,b+3m)\,\widetilde{T}^{k,k'}_{a,b}(t-2n,s-3m)\Big\},\quad t,s\in\mathbb{R},$$

where the sampling functions $\widetilde{T}^{k,k'}_{a,b}(t,s)$ are obtained from the columns of the inverse matrix $M^{-1}$ as

$$\begin{aligned}
\widetilde{T}^{0,0}_{a,b}(t,s)&=S_a(t)\,\widetilde{S}_b(s)+S_a(t)\,\widetilde{S}_b(s-1)+S_a(t)\,\widetilde{S}_b(s-2)+S_a(t-1)\,\widetilde{S}_b(s)\\
&\quad+S_a(t-1)\,\widetilde{S}_b(s-1)+S_a(t-1)\,\widetilde{S}_b(s-2)\,;\\
\widetilde{T}^{0,1}_{a,b}(t,s)&=S_a(t)\,\widetilde{S}_b(s-1)+2S_a(t)\,\widetilde{S}_b(s-2)+S_a(t-1)\,\widetilde{S}_b(s-1)\\
&\quad+2S_a(t-1)\,\widetilde{S}_b(s-2)\,;\\
\widetilde{T}^{0,2}_{a,b}(t,s)&=S_a(t)\,\widetilde{S}_b(s-2)+S_a(t-1)\,\widetilde{S}_b(s-2)\,;\\
\widetilde{T}^{1,0}_{a,b}(t,s)&=S_a(t-1)\,\widetilde{S}_b(s)+S_a(t-1)\,\widetilde{S}_b(s-1)+S_a(t-1)\,\widetilde{S}_b(s-2)\,;\\
\widetilde{T}^{1,1}_{a,b}(t,s)&=S_a(t-1)\,\widetilde{S}_b(s-1)+2S_a(t-1)\,\widetilde{S}_b(s-2)\,;\\
\widetilde{T}^{1,2}_{a,b}(t,s)&=S_a(t-1)\,\widetilde{S}_b(s-2)\,,\quad t,s\in\mathbb{R}\,.
\end{aligned}$$

Observe that

$$\mathcal{T}_{\varphi\psi}\big(y_{2n+k}\otimes\widetilde{y}_{3m+k'}\big)(t,s)=S_a(t-(2n+k))\widetilde{S}_b(s-(3m+k')),$$

where

$$y_n=\frac{\mathrm{e}^{-2\pi inx}}{\overline{K_a(x)}}\quad\text{and}\quad\widetilde{y}_m=\frac{\mathrm{e}^{-2\pi imy}}{\overline{\widetilde{K}_b(y)}},\qquad n,m\in\mathbb{Z}.$$

**Example 3.11 (Using the Samples $\Delta_+^{k,k'}f(a+3n,b+3m)$ with $k,k'=0,1,2$ in $V_\Phi^2$)** For any $f=\mathcal{T}_\Phi F$ in $V_\Phi^2$ we have the expression for the samples (see Eq. (3.18)):

$$\begin{aligned}\Delta_+^{k,k'}&f(a+3n,b+3m)\\&=\Big\langle F,(\mathrm{e}^{-2\pi ix}-1)^k(\mathrm{e}^{-2\pi iy}-1)^{k'}\mathrm{e}^{-2\pi i(3n)x}\,\mathrm{e}^{-2\pi i(3m)y}K_{a,b}(x,y)\Big\rangle,\end{aligned}$$

where $n,m\in\mathbb{Z}$ and $k,k'=0,1,2$. Consider the Riesz basis $\big\{x_{n,m}\big\}_{n,m\in\mathbb{Z}}$ for $L^2(0,1)^2$, where

$$x_{n,m}:=\mathrm{e}^{-2\pi inx}\,\mathrm{e}^{-2\pi imy}K_{a,b}(x,y),\quad n,m\in\mathbb{Z}.$$

According to Theorem 3.3, for the partition

$$\begin{aligned}&\{x_{3n,3m}\}_{n,m\in\mathbb{Z}}\cup\{x_{3n,3m+1}\}_{n,m\in\mathbb{Z}}\cup\{x_{3n,3m+2}\}_{n,m\in\mathbb{Z}}\\&\cup\{x_{3n+1,3m}\}_{n,m\in\mathbb{Z}}\cup\{x_{3n+1,3m+1}\}_{n,m\in\mathbb{Z}}\cup\{x_{3n+1,3m+2}\}_{n,m\in\mathbb{Z}}\\&\cup\{x_{3n+2,3m}\}_{n,m\in\mathbb{Z}}\cup\{x_{3n+2,3m+1}\}_{n,m\in\mathbb{Z}}\cup\{x_{3n+2,3m+2}\}_{n,m\in\mathbb{Z}},\end{aligned}$$

the Riesz basis associated to the above samples has as matrix $M$ the Kronecker product:

$$M=\begin{pmatrix}1&0&0\\-1&1&0\\1&-2&1\end{pmatrix}\otimes\begin{pmatrix}1&0&0\\-1&1&0\\1&-2&1\end{pmatrix}=\begin{pmatrix}1&0&0&0&0&0&0&0&0\\-1&1&0&0&0&0&0&0&0\\1&-2&1&0&0&0&0&0&0\\-1&0&0&1&0&0&0&0&0\\1&-1&0&-1&1&0&0&0&0\\-1&2&-1&1&-2&1&0&0&0\\1&0&0&-2&0&0&1&0&0\\-1&1&0&2&-2&0&-1&1&0\\1&-2&1&-2&4&-2&1&-2&1\end{pmatrix},$$

whose inverse is:

$$M^{-1} = \begin{pmatrix} 1 & 0 & 0 \\ 1 & 1 & 0 \\ 1 & 2 & 1 \end{pmatrix} \otimes \begin{pmatrix} 1 & 0 & 0 \\ 1 & 1 & 0 \\ 1 & 2 & 1 \end{pmatrix} = \begin{pmatrix} 1&0&0&0&0&0&0&0&0 \\ 1&1&0&0&0&0&0&0&0 \\ 1&2&1&0&0&0&0&0&0 \\ 1&0&0&1&0&0&0&0&0 \\ 1&1&0&1&1&0&0&0&0 \\ 1&2&1&1&2&1&0&0&0 \\ 1&0&0&2&0&0&1&0&0 \\ 1&1&0&2&2&0&1&1&0 \\ 1&2&1&2&4&2&1&2&1 \end{pmatrix}.$$

Thus, the corresponding sampling formula for each $f \in V_\Phi^2$ reads:

$$f(t,s) = \sum_{n,m\in\mathbb{Z}} \Big\{ \sum_{k,k'=0}^{2} \Delta_+^{k,k'} f(a+3n, b+3m)\, T_{a,b}^{k,k'}(t-3n, s-3m) \Big\}, \quad t,s \in \mathbb{R},$$

where the sampling functions $T_{a,b}^{k,k'}(t,s)$ are obtained from the columns of the inverse matrix (4.2) as:

$$\begin{aligned}
T_{a,b}^{0,0}(t,s) &= S_{a,b}(t,s) + S_{a,b}(t,s-1) + S_{a,b}(t,s-2) + S_{a,b}(t-1,s) \\
&\quad + S_{a,b}(t-1,s-1) + S_{a,b}(t-1,s-2) + S_{a,b}(t-2,s) \\
&\quad + S_{a,b}(t-2,s-1) + S_{a,b}(t-2,s-2);\\
T_{a,b}^{0,1}(t,s) &= S_{a,b}(t,s-1) + 2S_{a,b}(t,s-2) + S_{a,b}(t-1,s-1) \\
&\quad + 2S_{a,b}(t-1,s-2) + S_{a,b}(t-2,s-1) + 2S_{a,b}(t-2,s-2);\\
T_{a,b}^{0,2}(t,s) &= S_{a,b}(t,s-2) + S_{a,b}(t-1,s-2) + S_{a,b}(t-2,s-2);\\
T_{a,b}^{1,0}(t,s) &= S_{a,b}(t-1,s) + S_{a,b}(t-1,s-1) + S_{a,b}(t-1,s-2) \\
&\quad + 2S_{a,b}(t-2,s-2) + 2S_{a,b}(t-2,s-1) + 2S_{a,b}(t-2,s-2);
\end{aligned}$$

$$\begin{aligned}
T_{a,b}^{1,1}(t,s) &= S_{a,b}(t-1,s-1) + 2S_{a,b}(t-1,s-2) + 2S_{a,b}(t-2,s-1) \\
&\quad + 4S_{a,b}(t-2,s-2);\\
T_{a,b}^{1,2}(t,s) &= S_{a,b}(t-1,s-2) + 2S_{a,b}(t-2,s-2);\\
T_{a,b}^{2,0}(t,s) &= S_{a,b}(t-2,s) + S_{a,b}(t-2,s-1) + S_{a,b}(t-2,s-2);\\
T_{a,b}^{2,1}(t,s) &= S_{a,b}(t-2,s-1) + 2S_{a,b}(t-2,s-2);\\
T_{a,b}^{2,2}(t,s) &= S_{a,b}(t-2,s-2), \quad t,s \in \mathbb{R}.
\end{aligned}$$

Observe that

$$\mathcal{T}_{\Phi}\big(y_{3n+k,3m+k'}\big)(t,s) = S_{a,b}(t-(3n+k), s-(3m+k'))\,,$$

where

$$y_{n,m} := \frac{\mathrm{e}^{-2\pi i n x}\,\mathrm{e}^{-2\pi i m y}}{\overline{K_{a,b}(x,y)}}\,, \qquad n, m \in \mathbb{Z}\,.$$

## 3.3 From Classical Shift-Invariant to $U$-Invariant Subspaces

Let $U$ be a unitary operator in a separable Hilbert space $\mathcal{H}$. We can consider the group $\{U^n\}_{n\in\mathbb{Z}}$ generated by $U$. It satisfies

- $U^n U^m = U^{n+m}$ for $n, m \in \mathbb{Z}$,
- $U^0 = I_{\mathcal{H}}, \quad U^{-1} = U^*$

As we will see at Chap. 5, defining $n \longmapsto U^n$ we obtain a unitary representation of the group $(\mathbb{Z}, +)$ on $\mathcal{H}$.

For a fixed $a \in \mathcal{H}$, consider the closed subspace given by $\mathcal{V}_a := \overline{\text{span}}\big\{U^n a\big\}_{n\in\mathbb{Z}}$. In case that the sequence $\{U^n a\}_{n\in\mathbb{Z}}$ is a Riesz sequence in $\mathcal{H}$ we have

$$\mathcal{V}_a = \Big\{\sum_{n\in\mathbb{Z}} \alpha_n\, U^n a \,:\, \{\alpha_n\}_{n\in\mathbb{Z}} \in \ell^2(\mathbb{Z})\Big\}\,,$$

that is, $\mathcal{V}_a$ is the $U$-invariant subspace in $\mathcal{H}$ with stable generator $a \in \mathcal{H}$.

Thus we obtain a generalization of shift-invariant subspaces in $L^2(\mathbb{R})$ where $U$ is nothing but the *shift operator* $T : f(u) \mapsto f(u-1)$ in $L^2(\mathbb{R})$.

### 3.3.1 Stationary Sequences in a Hilbert Space

For simplicity and ease of notation we are considering the one-generator setting. The sequence $\{U^n a\}_{n\in\mathbb{Z}}$ is a *stationary sequence* in $\mathcal{H}$ since the inner product

$$\langle U^n a, U^m a\rangle_{\mathcal{H}} \quad \text{depends only on the difference } n-m \in \mathbb{Z}\,.$$

The concept of stationary sequence was introduced by Kolmogorov (see, Ref. [11]). Moreover, the *auto-covariance* $R_a$ of the sequence $\{U^n a\}_{n\in\mathbb{Z}}$ admits the integral representation

$$R_a(k) := \langle U^k a, a\rangle_{\mathcal{H}} = \int_{-\pi}^{\pi} \mathrm{e}^{ik\theta}\, d\mu_a(\theta)\,, \qquad k \in \mathbb{Z}\,,$$

in terms of a *positive Borel measure* $\mu_a$ on $(-\pi, \pi)$ called the *spectral measure* of the sequence (see Ref. [11, Theorem 2, p. 236]). This result is obtained from the integral representation of the unitary operator $U$ on $\mathcal{H}$ as

$$U = \int_{-\pi}^{\pi} e^{i\theta} dE_\theta \,,$$

where $\{E_\theta \ : \ -\pi \le \theta \le \pi\}$ denotes a *resolution of the identity* for $U$. Indeed, in short,

$$R_a(k) = \langle U^k a, a\rangle = \Big\langle \int_{-\pi}^{\pi} e^{ik\theta} dE_\theta \, a, a\Big\rangle = \int_{-\pi}^{\pi} e^{ik\theta} d\langle E_\theta \, a, a\rangle$$
$$= \int_{-\pi}^{\pi} e^{ik\theta} d\mu_a(\theta) \,, \qquad k \in \mathbb{Z} \,,$$

where the function $\mu_a(\theta) = \langle E_\theta \, a, a\rangle$ is nondecreasing, has bounded variation, is continuous from the left and, $\mu_a(-\pi) = 0$ and $\mu_a(\pi) = \|a\|^2$. The interested reader can find all the details, for instance, in Ref. [1, Vol. 2, pp. 16–21].

The spectral measure $\mu_a$ can be decomposed into an absolutely continuous and a singular part as

$$d\mu_a(\theta) = \phi_a(\theta) d\theta + d\mu_a^s(\theta) \,.$$

See, for instance, Ref. [15]. A necessary and sufficient condition in order for the sequence $\{U^n a\}_{n\in\mathbb{Z}}$ to be a Riesz sequence for $\mathcal{H}$ is given in the next theorem in terms of the decomposition of the spectral measure $\mu_a$:

**Theorem 3.5** *Let $\{U^n a\}_{n\in\mathbb{Z}}$ be a sequence obtained from a unitary operator in a separable Hilbert space $\mathcal{H}$ with spectral measure $d\mu_a(\theta) = \phi_a(\theta)d\theta + d\mu_a^s(\theta)$, and let $\mathcal{V}_a$ be the closed subspace spanned by $\{U^n a\}_{n\in\mathbb{Z}}$. Then the sequence $\{U^n a\}_{n\in\mathbb{Z}}$ is a Riesz basis for $\mathcal{V}_a$ if and only if the singular part $\mu_a^s \equiv 0$ and*

$$0 < \operatorname*{ess\,inf}_{\theta\in(-\pi,\pi)} \phi_a(\theta) \le \operatorname*{ess\,sup}_{\theta\in(-\pi,\pi)} \phi_a(\theta) < \infty \,.$$

Theorem 3.5 is just the one-generator case ($L = 1$) of Theorem 3.6 proved in next section.

An straightforward computation gives the dual Riesz basis of $\{U^n a\}_{n\in\mathbb{Z}}$ in $\mathcal{V}_a$. Namely:

**Proposition 3.6** *The dual Riesz basis of $\{U^n a\}_{n\in\mathbb{Z}}$ is given by $\{U^n b\}_{n\in\mathbb{Z}}$ with $b = \sum_{k\in\mathbb{Z}} b_k \, U^k a \in \mathcal{V}_a$, where the terms of the sequence $\{b_k\}_{k\in\mathbb{Z}} \in \ell^2(\mathbb{Z})$ are the Fourier coefficients of the function $\frac{1}{2\pi\phi_a(\theta)} \in L^2(-\pi, \pi)$. Moreover,*

$$\phi_b(\theta) = \frac{1}{4\pi^2 \phi_a(\theta)} \quad \text{a.e. in } (-\pi, \pi) \,.$$

***Proof*** Indeed, for $b = \sum_{k\in\mathbb{Z}} b_k\, U^k a$ in $\mathcal{V}_a$, the biorthogonality between the sequences $\{U^n a\}_{n\in\mathbb{Z}}$ and $\{U^n b\}_{n\in\mathbb{Z}}$ means

$$\begin{aligned}\delta_{m,0} = \langle U^m a, b\rangle_{\mathcal{H}} &= \langle U^m a, \sum_{k\in\mathbb{Z}} b_k\, U^k a\rangle_{\mathcal{H}} = \sum_{k\in\mathbb{Z}} \overline{b}_k \int_{-\pi}^{\pi} \mathrm{e}^{i(m-k)\theta}\phi_a(\theta)d\theta \\ &= \int_{-\pi}^{\pi} \big(\sum_{k\in\mathbb{Z}} \overline{b}_k \mathrm{e}^{-ik\theta}\big)\phi_a(\theta)\mathrm{e}^{im\theta}\, d\theta = \int_{-\pi}^{\pi} B(\theta)\phi_a(\theta)\mathrm{e}^{-im\theta}\, d\theta\,,\end{aligned}$$

where $B(\theta) := \sum_{k\in\mathbb{Z}} b_k\, \mathrm{e}^{ik\theta}$; in other words, we have $B(\theta)\phi_a(\theta) = \frac{1}{2\pi}$ in $L^2(-\pi,\pi)$.

To deduce $\phi_b$ we have,

$$\begin{aligned}\langle U^k b, b\rangle &= \big\langle \sum_m b_m U^{m+k} a, \sum_n b_n U^n a\big\rangle = \sum_{m,n} b_m \overline{b}_n\, \langle U^{m+k} a, U^n a\rangle \\ &= \int_{-\pi}^{\pi} \big(\sum_{m,n} b_m \overline{b}_n\, \mathrm{e}^{i(m-n)\theta}\big)\, \mathrm{e}^{ik\theta}\phi_a(\theta)d\theta = \int_{-\pi}^{\pi} |B(\theta)|^2\, \mathrm{e}^{ik\theta}\phi_a(\theta)d\theta \\ &= \int_{-\pi}^{\pi} \mathrm{e}^{ik\theta}\phi_b(\theta)d\theta\,,\end{aligned}$$

where $\phi_b(\theta) = |B(\theta)|^2\, \phi_a(\theta) = \frac{1}{4\pi^2\phi_a(\theta)}$ a.e. in $(-\pi,\pi)$. □

For the shift operator $T : f(u) \longmapsto f(u-1)$ in $L^2(\mathbb{R})$, Theorem 3.5 allows to recover the classical necessary and sufficient condition on the function $\varphi \in L^2(\mathbb{R})$ for the sequence $\{\varphi(t-n)\}_{n\in\mathbb{Z}}$ to be a Riesz basis for the corresponding shift-invariant subspace $V_\varphi^2$ in $L^2(\mathbb{R})$.

Indeed, using the Parseval equality (1.7) one easily gets

$$\begin{aligned}\langle T^k\varphi, \varphi\rangle_{L^2(\mathbb{R})} &= \int_{-\infty}^{\infty} \varphi(u-k)\,\overline{\varphi(u)}\, du = \int_{-\infty}^{\infty} \widehat{\varphi(u-k)}(\theta)\,\overline{\widehat{\varphi}(\theta)}\, d\theta \\ &= \int_{-\infty}^{\infty} |\widehat{\varphi}(\theta)|^2\, \mathrm{e}^{-ik\theta}\, d\theta = \int_{-\pi}^{\pi} \sum_{n\in\mathbb{Z}} |\widehat{\varphi}(\theta+2\pi n)|^2\, \mathrm{e}^{-ik\theta}\, d\theta \\ &= \int_{-\pi}^{\pi} \mathrm{e}^{ik\theta} \sum_{n\in\mathbb{Z}} |\widehat{\varphi}(-\theta+2\pi n)|^2\, d\theta\,,\end{aligned}$$

that is, $\phi_\varphi(\theta) = \sum_{n\in\mathbb{Z}} |\widehat{\varphi}(-\theta+2\pi n)|^2$, a.e. $\theta \in (-\pi,\pi)$.

Thus, Theorem 3.5 yields the classical condition in Proposition 3.1:

$$0 < \operatorname*{ess\,inf}_{\theta\in(-\pi,\pi)} \sum_{n\in\mathbb{Z}} |\widehat{\varphi}(\theta+2\pi n)|^2 \le \operatorname*{ess\,sup}_{\theta\in(-\pi,\pi)} \sum_{n\in\mathbb{Z}} |\widehat{\varphi}(\theta+2\pi n)|^2 < \infty\,.$$

### 3.3.2 The Case of Multiple Generators

The case of $L$ generators can be analogously derived. Indeed, consider the $U$-invariant subspace generated by the set $\mathbf{a} = \{a_1, a_2, \dots, a_L\} \subset \mathcal{H}$, i.e.,

$$\mathcal{V}_{\mathbf{a}} := \overline{\text{span}}\big\{U^n a_l \ : \ n \in \mathbb{Z};\, l = 1, 2\dots, L\big\}.$$

Assuming that the sequence $\{U^n a_l\}_{n\in\mathbb{Z};\, l=1,2,\dots,L}$ is a Riesz sequence in $\mathcal{H}$, the $U$-invariant subspace $\mathcal{V}_{\mathbf{a}}$ can be expressed as

$$\mathcal{V}_{\mathbf{a}} = \Big\{\sum_{l=1}^{L}\sum_{n\in\mathbb{Z}} \alpha_n^l\, U^n a : \{\alpha_n^l\}_{n\in\mathbb{Z}} \in \ell^2(\mathbb{Z});\, l = 1, 2\dots, L\Big\}.$$

The sequence $\{U^n a_l\}_{n\in\mathbb{Z};\, l=1,2,\dots,L}$ can be thought as an $L$-dimensional stationary sequence. Its *covariance matrix* $\mathbf{R}_{\mathbf{a}}(k)$ is the $L \times L$ matrix

$$\mathbf{R}_{\mathbf{a}}(k) := \Big[\langle U^k a_m, a_n\rangle_{\mathcal{H}}\Big]_{m,n=1,2,\dots,L}, \qquad k \in \mathbb{Z}.$$

Its admits the spectral representation Ref. [11]:

$$\mathbf{R}_{\mathbf{a}}(k) = \int_{-\pi}^{\pi} e^{ik\theta} d\boldsymbol{\mu}_{\mathbf{a}}(\theta), \quad k \in \mathbb{Z}.$$

The *spectral measure* $\boldsymbol{\mu}_{\mathbf{a}}$ is an $L \times L$ matrix whose entries are the spectral measures associated with the cross-correlation functions

$$R_{m,n}(k) := \langle U^k a_m, a_n\rangle_{\mathcal{H}}. \quad k \in \mathbb{Z}.$$

It can be decomposed into an absolutely continuous part and its singular part. Thus we can write

$$d\boldsymbol{\mu}_{\mathbf{a}}(\theta) = \boldsymbol{\Phi}_{\mathbf{a}}(\theta)d\theta + d\boldsymbol{\mu}_{\mathbf{a}}^s(\theta).$$

In case that the singular part $\boldsymbol{\mu}_{\mathbf{a}}^s \equiv 0$, the hermitian $L \times L$ matrix $\boldsymbol{\Phi}_{\mathbf{a}}(\theta)$ is called the *spectral density* of the sequence $\{U^n a_l\}_{n\in\mathbb{Z};\, l=1,2,\dots,L}$. The following theorem holds:

**Theorem 3.6** *Let $\{U^n a_l\}_{n\in\mathbb{Z};\, l=1,2,\dots,L}$ be a stationary sequence obtained from a unitary operator $U$ in a separable Hilbert space $\mathcal{H}$ with spectral measure*

$$d\boldsymbol{\mu}_{\mathbf{a}}(\theta) = \boldsymbol{\Phi}_{\mathbf{a}}(\theta)d\theta + d\boldsymbol{\mu}_{\mathbf{a}}^s(\theta),$$

*and let $\mathcal{V}_{\mathbf{a}}$ be the closed subspace spanned by $\{U^n a_l\}_{n\in\mathbb{Z};\, l=1,2,\dots,L}$.*

*Then the sequence* $\{U^n a_l\}_{n\in\mathbb{Z};\, l=1,2,\dots,L}$ *is a Riesz basis for* $\mathcal{V}_\mathbf{a}$ *if and only if the singular part* $\boldsymbol{\mu}_\mathbf{a}^s \equiv 0$ *and*

$$0 < \operatorname*{ess\,inf}_{\theta\in(-\pi,\pi)} \lambda_{\min}\big[\boldsymbol{\Phi}_\mathbf{a}(\theta)\big] \le \operatorname*{ess\,sup}_{\theta\in(-\pi,\pi)} \lambda_{\max}\big[\boldsymbol{\Phi}_\mathbf{a}(\theta)\big] < \infty\,, \tag{3.19}$$

*where* $\lambda_{\min}$ *(respectively* $\lambda_{\max}$*) denotes the smallest (respectively the largest) eigenvalue of the hermitian matrix* $\boldsymbol{\Phi}_\mathbf{a}(\theta)$.

***Proof*** For a fixed $\ell^2_L$-sequence $c := \{c_n^l\}_{n\in\mathbb{Z};\, l=1,2,\dots,L}$ we have

$$\begin{aligned}
\Big\|\sum_{l=1}^{L}\sum_{k\in\mathbb{Z}} c_k^l U^k a_l\Big\|^2 &= \sum_{i,j=1}^{L}\sum_{m\in\mathbb{Z}}\sum_{n\in\mathbb{Z}} c_m^i \overline{c}_m^j \langle U^m a_i, U^n a_j\rangle \\
&= \sum_{i,j=1}^{L}\sum_{m\in\mathbb{Z}}\sum_{n\in\mathbb{Z}} c_m^i \overline{c}_n^j \int_{-\pi}^{\pi} e^{im\theta} e^{-in\theta}\, d\mu_{a_i,a_j}(\theta) \qquad (3.20)\\
&= \int_{-\pi}^{\pi} \sum_{m\in\mathbb{Z}}\sum_{n\in\mathbb{Z}} (\mathbf{c}_m\, e^{im\theta})^\top d\boldsymbol{\mu}_\mathbf{a}(\theta)\overline{\mathbf{c}}_n\, e^{-in\theta}\,,
\end{aligned}$$

where $\mathbf{c}_k = (c_k^1, c_k^2, \dots, c_k^L)^\top$ for every $k \in \mathbb{Z}$.

First we show that if the measure $\boldsymbol{\mu}_\mathbf{a}$ is not absolutely continuous with respect to Lebesgue measure $m$, then $\{U^n a_l\}_{n\in\mathbb{Z};\, l=1,2,\dots,L}$ is not a Riesz basis for $\mathcal{V}_\mathbf{a}$.

Indeed, if the spectral measure $\boldsymbol{\mu}_\mathbf{a}$ is not absolutely continuous with respect to Lebesgue measure $m$, then there exists $i \in \{1, 2, \dots, L\}$ such that the positive spectral measure $\mu_{a_i,a_i}$ is not absolutely continuous with respect to Lebesgue measure.

This comes from the fact that, if any spectral measure in the diagonal $\mu_{a_j,a_j}$ is absolutely continuous with respect to Lebesgue measure, the same occurs for each measure $\mu_{a_j,a_k}$ with $k \neq j$ (see Ref. [5, p. 137]).

Then, $\mu_{a_i,a_i}(B) > 0$ for a (Lebesgue) measurable set $B \subset (-\pi, \pi)$ of Lebesgue measure zero. Bearing in mind that every measurable set is included in a Borel set, actually an intersection of a countable collection of open sets, having the same Lebesgue measure (see Ref. [15, p. 63]), we take $B$ to be a Borel set.

Moreover, since every finite Borel measure on $(-\pi, \pi)$ is inner regular (see [15, p. 340]) we may also assume that $B$ is a compact set. For any $\varepsilon > 0$ there exists a sequence of disjoint open intervals $I_j \subset (-\pi, \pi)$ such that

$$B \subset \bigcup_{j=1}^{\infty} I_j \ \text{ and } \ \sum_{j=1}^{\infty} m(I_j) \le m(B) + \varepsilon = \varepsilon\,,$$

(see [15, pp. 58 and 42]). Since $B$ is compact we may take the sequence to be finite. Hence, for every $N \in \mathbb{N}$ there exist open disjoint intervals $I_1^N, I_2^N, \dots, I_{j_N}^N$

in $(-\pi,\pi)$ such that

$$B \subset \bigcup_{j=1}^{j_N} I_j^N \quad\text{and}\quad \sum_{j=1}^{j_N} m(I_j^N) \le \frac{1}{3^N}\,.$$

Besides, $\sum_{j=1}^{j_N} \mu_{a_i,a_i}(I_j^N) \ge \mu_{a_i,a_i}(B)$. Consider the function $g_N\colon (-\pi,\pi) \to \mathbb{R}$, where $g_N = 2^{N/2}\chi_{\bigcup_{j=1}^{j_N} I_j^N}$, that satisfies

$$\|g_N\|_2^2 = 2^N \sum_{j=1}^{j_N} m(I_j^N) \le \frac{2^N}{3^N} < 1\,.$$

We modify and extend each $g_N$ to obtain a $2\pi$-periodic function $f_N\colon \mathbb{R} \longrightarrow \mathbb{R}$ such that $f_N$ and its derivative are continuous on $\mathbb{R}$, $\|f_N\|_2^2 \le 1$ and $f_N(\theta) = g_N(\theta)$ for every $\theta \in \bigcup_{j=1}^{j_N} I_j^N$.

Let $\sum_k c_k^N \mathrm{e}^{ik\theta}$ be the Fourier series of $f_N$. First, by using Parseval's identity we have

$$\|c_k^N\|_2^2 = \frac{1}{2\pi}\|f_N\|_2^2 \le \frac{1}{2\pi} \quad \text{for every } N \in \mathbb{N}\,,$$

so that $\{c^N\}_{N=1}^{\infty}$ is a bounded sequence in $\ell^2(\mathbb{Z})$. Besides, the regularity of each $f_N$ ensures that each Fourier series converges uniformly to $f_N$. Therefore each series $\sum_k c_k^N \mathrm{e}^{ik\theta}$ converges to $f_N$ in $L^2_{\mu_{a_i,a_i}(-\pi,\pi)}$ and, consequently,

$$\begin{aligned}\Big\|\sum_k c_k^N \mathrm{e}^{ik\theta}\Big\|^2_{L^2_{\mu_{a_i,a_i}}(-\pi,\pi)} &= \int_{-\pi}^{\pi} |f_N|^2 d\mu_{a_i,a_i} \ge \int_{-\pi}^{\pi} |g_N|^2 d\mu_{a_i,a_i} \\ &= 2^N \sum_{j=1}^{j_N} \mu_{a_i,a_i}(I_j^N) \ge 2^N \mu_{a_i,a_i}(B)\,.\end{aligned}$$

For every $c^N \in \ell^2(\mathbb{Z})$ we consider the $\ell^2_L$-sequence $\{c_n^{Nl}\}_{n\in\mathbb{Z};\, l=1,2,\dots,L}$ given by $c_n^{Ni} = c_n^N$ and $c_n^{Nl} = 0$ if $l \ne i$. Substituting each $\{c_n^{Nl}\}_{n\in\mathbb{Z};\, l=1,2,\dots,L}$ in (3.20) we have that

$$\Big\|\sum_{l=1}^{L}\sum_{k\in\mathbb{Z}} c_k^{Nl} U^k a_l\Big\|^2 = \int_{-\pi}^{\pi} \Big|\sum_{k\in\mathbb{Z}} c_k^N \mathrm{e}^{ik\theta}\Big|^2 d\mu_{a_i,a_i}(\theta)$$

tends to infinity with $N$, so $\{U^n a_l\}_{n\in\mathbb{Z};\, l=1,2,\dots,L}$ cannot be a Bessel sequence, therefore, not a Riesz basis.

For the remainder of the proof we assume that the singular part $\mu_{\mathbf{a}}^s \equiv 0$ and that $d\mu_{\mathbf{a}}(\theta) = \mathbf{\Phi}_{\mathbf{a}}(\theta)d\theta$. Then (3.20) yields

$$\Big\|\sum_{l=1}^{L}\sum_{k\in\mathbb{Z}} c_k^l U^k a_l\Big\|^2 = \int_{-\pi}^{\pi} \Big(\sum_{m\in\mathbb{Z}} \mathbf{c}_m\, \mathrm{e}^{im\theta}\Big)^\top \mathbf{\Phi}_{\mathbf{a}}(\theta)\overline{\sum_{n\in\mathbb{Z}} \mathbf{c}_n\, \mathrm{e}^{in\theta}}\, d\theta\,. \tag{3.21}$$

We have to show that $\{U^n a_l\}_{n\in\mathbb{Z};\, l=1,2,\dots,L}$ is a Riesz basis for $\mathcal{V}_{\mathbf{a}}$ if and only if (3.19) holds. The Rayleigh-Ritz theorem (see [10, p. 176]) provides the inequalities

$$\begin{aligned}\lambda_{\min}\big[\mathbf{\Phi}_{\mathbf{a}}(\theta)\big]\Big|\sum_{k\in\mathbb{Z}} \mathbf{c}_k\, \mathrm{e}^{ik\theta}\Big|^2 &\le \Big(\sum_{m\in\mathbb{Z}} \mathbf{c}_m\, \mathrm{e}^{im\theta}\Big)^\top \mathbf{\Phi}_{\mathbf{a}}(\theta)\overline{\sum_{n\in\mathbb{Z}} \mathbf{c}_n\, \mathrm{e}^{in\theta}}\\ &\le \lambda_{\max}\big[\mathbf{\Phi}_{\mathbf{a}}(\theta)\big]\Big|\sum_{k\in\mathbb{Z}} \mathbf{c}_k\, \mathrm{e}^{ik\theta}\Big|^2\end{aligned}$$

and taking into account (3.21) we have

$$\begin{aligned}\int_{-\pi}^{\pi} \lambda_{\min}\big[\mathbf{\Phi}_{\mathbf{a}}(\theta)\big]\Big|\sum_{k\in\mathbb{Z}} \mathbf{c}_k\, \mathrm{e}^{ik\theta}\Big|^2 d\theta &\le \Big\|\sum_{l=1}^{L}\sum_{k\in\mathbb{Z}} c_k^l U^k a_l\Big\|^2\\ &\le \int_{-\pi}^{\pi} \lambda_{\max}\big[\mathbf{\Phi}_{\mathbf{a}}(\theta)\big]\Big|\sum_{k\in\mathbb{Z}} \mathbf{c}_k\, \mathrm{e}^{ik\theta}\Big|^2 d\theta\,,\end{aligned}$$

so that

$$\begin{aligned}\operatorname*{ess\,inf}_{\theta\in(-\pi,\pi)} \lambda_{\min}\big[\mathbf{\Phi}_{\mathbf{a}}(\theta)\big]\, 2\pi \sum_{l=1}^{L}\sum_{k\in\mathbb{Z}} |c_k^l|^2 &\le \Big\|\sum_{l=1}^{L}\sum_{k\in\mathbb{Z}} c_k^l U^k a_l\Big\|^2\\ &\le \operatorname*{ess\,sup}_{\theta\in(-\pi,\pi)} \lambda_{\max}\big[\mathbf{\Phi}_{\mathbf{a}}(\theta)\big]\, 2\pi \sum_{l=1}^{L}\sum_{k\in\mathbb{Z}} |c_k^l|^2\,.\end{aligned}$$

Therefore (3.19) implies that $\{U^n a_l\}_{n\in\mathbb{Z};\, l=1,2,\dots,L}$ is a Riesz basis for $\mathcal{V}_{\mathbf{a}}$.

Conversely, if $\{U^n a_l\}_{n\in\mathbb{Z};\, l=1,2,\dots,L}$ is a Riesz basis for $\mathcal{V}_{\mathbf{a}}$, then there exist constants $0 < A \le B < \infty$ such that

$$A\sum_{l=1}^{L}\sum_{k\in\mathbb{Z}} |c_k^l|^2 \le \Big\|\sum_{l=1}^{L}\sum_{k\in\mathbb{Z}} c_k^l U^k a_l\Big\|^2 \le B\sum_{l=1}^{L}\sum_{k\in\mathbb{Z}} |c_k^l|^2 \tag{3.22}$$

for every $\ell^2_L$-sequence $\{c^l_n\}_{n\in\mathbb{Z};\, l=1,2,\dots,L}$. Let us prove that

$$A \le \operatorname*{ess\,inf}_{\theta\in(-\pi,\pi)} \lambda_{\min}\big[\boldsymbol{\Phi}_{\mathbf{a}}(\theta)\big] \le \operatorname*{ess\,sup}_{\theta\in(-\pi,\pi)} \lambda_{\max}\big[\boldsymbol{\Phi}_{\mathbf{a}}(\theta)\big] \le B\,. \tag{3.23}$$

Proceeding by contradiction, if (3.23) would not hold, then

$$A \le \lambda_{\min}\big[\boldsymbol{\Phi}_{\mathbf{a}}(\theta)\big] \le \lambda_{\max}\big[\boldsymbol{\Phi}_{\mathbf{a}}(\theta)\big] \le B$$

does not hold on a subset of $(-\pi,\pi)$ with positive Lebesgue measure.

In case the set $\Gamma_B := \big\{\theta \in (-\pi,\pi) \colon \lambda_{\max}\big[\boldsymbol{\Phi}_{\mathbf{a}}(\theta)\big] > B\big\}$ has positive Lebesgue measure we introduce the Fourier expansion of the function $F \in L^2_L(-\pi,\pi)$ ($L^2_L(-\pi,\pi)$ denotes the usual product Hilbert space $L^2(-\pi,\pi)\times \cdots \times L^2(-\pi,\pi)$ ($L$ times)) in (3.21), where $F(\theta) = \mathbf{X}(\theta)\,\chi_{\Gamma_B}(\theta)$ and $\mathbf{X}(\theta)$ is an eigenvector of norm 1 associated with the biggest eigenvalue of $\boldsymbol{\Phi}_{\mathbf{a}}(\theta)$. We get

$$\Big\|\sum_{l=1}^{L}\sum_{k\in\mathbb{Z}} c^l_k U^k a_l\Big\|^2 = \frac{1}{2\pi}\int_{\Gamma_B} \lambda_{\max}\big[\boldsymbol{\Phi}_{\mathbf{a}}(\theta)\big]d\theta > \frac{1}{2\pi}\int_{\Gamma_B} B d\theta$$

which contradicts the right inequality in (3.22) for such a Fourier expansion.

Whenever Lebesgue measure of the set $\Gamma_B$ is zero, then we proceed in a similar way with the set of positive Lebesgue measure

$$\Gamma_A := \{\theta \in (-\pi,\pi) \colon \lambda_{\min}\big[\boldsymbol{\Phi}_{\mathbf{a}}(\theta)\big] < A\}\,.$$

□

In particular, for $L=1$ we deduce Theorem 3.5. The above proof is similar to that of Lemma 2 in Ref. [14], except we do not exclude the case in which the singular measure is atomless. Another characterization for being $\{U^n a_l\}_{n\in\mathbb{Z};\, l=1,2,\dots,L}$ a Riesz basis for $\mathcal{V}_{\mathbf{a}}$ can be found in Ref. [2].

Whenever $U$ is the shift operator, the spectral density matrix $\boldsymbol{\Phi}_{\mathbf{a}}(\theta)$ is nothing but than the so-called *Gramian matrix-function* $G_{\boldsymbol{\Phi}}(w)$

$$G_{\boldsymbol{\Phi}}(w) = \sum_{n\in\mathbb{Z}} \widehat{\Phi}(w+2\pi n)\,\overline{\widehat{\Phi}(w+2\pi n)}^{\top} \quad \text{a.e. in } (-\pi,\pi)\,,$$

where $\widehat{\Phi} := (\widehat{\varphi}_1, \widehat{\varphi}_2, \dots, \widehat{\varphi}_L)^{\top}$.

We deduce the well-known necessary and sufficient condition for the sequence $\{\varphi_l(\cdot - n)\}_{n\in\mathbb{Z};\, l=1,2\dots,L}$ to be a Riesz basis for the shift-invariant subspace $V^2_{\Phi}$:

Let $\Phi := (\varphi_1, \varphi_2, \dots, \varphi_L)^{\top}$, where $\varphi_l \in L^2(\mathbb{R})$, $l = 1, 2, \dots, L$, such that the sequence $\big\{\varphi_l(t-n)\big\}_{n\in\mathbb{Z};\, l=1,2\dots,L}$ is a Riesz basis ($\Phi$ is a stable set of generators) for the shift-invariant space

$$V^2_{\Phi} := \Big\{\sum_{l=1}^{L}\sum_{n\in\mathbb{Z}} d_l(n)\,\varphi_l(t-n) \,:\, d_l \in \ell^2(\mathbb{Z});\ l = 1, 2\dots, L\Big\} \subset L^2(\mathbb{R})\,.$$

There exists a necessary and sufficient condition, involving the *Gramian matrix-function* $G_{\mathbf{\Phi}}(w)$ which assures to be the sequence $\{\varphi_l(t-n)\}_{n\in\mathbb{Z};\, l=1,2...,L}$ a Riesz basis for $V_\Phi$. Namely:

**Corollary 3.1** *The sequence* $\{\varphi_l(t-n)\}_{n\in\mathbb{Z};\, l=1,2...,L}$ *is a Riesz basis for* $V_\Phi$ *if and only if there exist two positive constants* $m$ *and* $M$ *such that*

$$m\,\mathbb{I}_L \le G_{\mathbf{\Phi}}(w) \le M\,\mathbb{I}_L \quad a.e.\ in\ (-\pi,\pi)\,. \tag{3.24}$$

Recall that the above inequalities mean that for a.e. $w \in (0,1)$ we have

$$m\,\|\mathbf{x}\|^2 \le \big\langle G_{\mathbf{\Phi}}(w)\mathbf{x},\mathbf{x}\big\rangle_{\mathbb{C}^L} \le M\,\|\mathbf{x}\|^2 \quad \text{for all } \mathbf{x}\in\mathbb{C}^L\,.$$

The above corollary is a consequence of (3.19) and the *Rayleigh-Ritz theorem* applied to the hermitian matrix $G_{\mathbf{\Phi}}(w)$. See also, for instance, Refs. [2–4].

Whenever $V_\Phi^2$ is a shift-invariant subspace of $L^2(\mathbb{R}^d)$, that is

$$V_\Phi^2 := \Big\{\sum_{l=1}^{L}\sum_{\alpha\in\mathbb{Z}^d} d_l(\alpha)\,\varphi_l(t-\alpha)\ :\ d_l\in\ell^2(\mathbb{Z}^d);\ l=1,2\ldots,L\Big\}\subset L^2(\mathbb{R}^d)\,,$$

condition (3.24) on the Gramian matrix-function

$$G_{\mathbf{\Phi}}(w) := \sum_{\alpha\in\mathbb{Z}^d}\widehat{\Phi}(w+2\pi\alpha)\,\overline{\widehat{\Phi}(w+2\pi\alpha)}^{\top} \quad \text{a.e. in } (-\pi,\pi)^d\,,$$

reads:

$$m\,\mathbb{I}_L \le G_{\mathbf{\Phi}}(w) \le M\,\mathbb{I}_L \quad \text{a.e. in } (-\pi,\pi)^d\,.$$

### 3.3.3 Generalized $U$-Average Sampling

Suppose we have a generator set $\mathbf{a} := \{a_1, a_2, \ldots, a_L\} \subset \mathcal{H}$ with $L$ elements, and assume that the sequence $\{U^n a_l\}_{n\in\mathbb{Z};\, l=1,2,...,L}$ is a Riesz sequence for $\mathcal{H}$. Then, the $U$-invariant subspace $\mathcal{V}_{\mathbf{a}}$ generated by $\mathbf{a}$ can be expressed as

$$\mathcal{V}_{\mathbf{a}} = \Big\{\sum_{l=1}^{L}\sum_{n\in\mathbb{Z}}\alpha_l(n)\,U^n a_l\ :\ \{\alpha_l(n)\}_{n\in\mathbb{Z}}\in\ell^2(\mathbb{Z});\ l=1,2\ldots,L\Big\}\,.$$

For $s$ vectors $h_j \in \mathcal{H}$, $j = 1, 2, \ldots, s$, not necessarily in $\mathcal{V}_{\mathbf{a}}$, we define the generalized samples of $x = \sum_{l=1}^{L} \sum_{n\in\mathbb{Z}} \alpha_l(n)\, U^n a_l$ in $\mathcal{V}_{\mathbf{a}}$ as

$$\mathcal{L}_j x(k) := \langle x, U^k h_j \rangle_{\mathcal{H}}, \quad \text{for } k \in \mathbb{Z} \text{ and } j = 1, 2, \ldots, s\,.$$

These samples can be expressed as a discrete convolution. Indeed:

$$\begin{aligned}\mathcal{L}_j x(k) &= \langle x, U^k h_j \rangle_{\mathcal{H}} = \Big\langle \sum_{l=1}^{L} \sum_{n\in\mathbb{Z}} \alpha_l(n)\, U^n a_l, U^k h_j \Big\rangle_{\mathcal{H}} \\ &= \sum_{l=1}^{L} \sum_{n\in\mathbb{Z}} \alpha_l(n) \langle U^n a_l, U^k h_j \rangle_{\mathcal{H}} = \sum_{l=1}^{L} \sum_{n\in\mathbb{Z}} \alpha_l(n) \langle a_l, U^{k-n} h_j \rangle_{\mathcal{H}} \\ &= \sum_{l=1}^{L} \sum_{n\in\mathbb{Z}} \alpha_l(n)\, c_{j,l}(k-n) = \sum_{l=1}^{L} (\boldsymbol{\alpha}_l * \mathbf{c}_{j,l})(k)\,, \quad k \in \mathbb{Z}\,,\end{aligned}$$

where $\mathbf{c}_{j,l} = \{c_{j,l}(n)\}_{n\in\mathbb{Z}}$ with $c_{j,l}(n) = \langle a_l, U^n h_j \rangle$ and $\boldsymbol{\alpha}_l = \{\alpha_l(n)\}_{n\in\mathbb{Z}}$, $n \in \mathbb{Z}$, $l = 1, 2, \ldots, L$ and $j = 1, 2, \ldots, s$.

As we will prove in Chap. 5 in a more general setting, under appropriate hypotheses on the Fourier transform in $\ell^2(\mathbb{Z})$ of the entries of the $s \times L$ matrix $A = [\mathbf{c}_{j,l}]$, there exist $s$ elements $b_j$ in $\mathcal{V}_{\mathbf{a}}$, $j = 1, 2, \ldots, s$, where, necessarily $s \geq L$, such that the sequence $\big\{U^k b_j\big\}_{k\in\mathbb{Z};\, j=1,2,\ldots,s}$ is a frame for $\mathcal{V}_{\mathbf{a}}$ and, for any $x \in \mathcal{V}_{\mathbf{a}}$, the sampling formula

$$x = \sum_{j=1}^{s} \sum_{k\in\mathbb{Z}} \mathcal{L}_j x(k)\, U^k b_j \quad \text{in } \mathcal{H}\,,$$

holds.

We can also take a sampling period $r \in \mathbb{N}$ and consider the sequence of samples $\{\mathcal{L}_j x(rk)\}_{k\in\mathbb{Z};\, j=1,2,\ldots,s}$. By using the theory in Sect. 5.3.3 of Chap. 5, under appropriate hypotheses on the Fourier transform in $\ell^2(r\mathbb{Z})$ of the entries of the $s \times Lr$ matrix $A = [\mathbf{c}_{j,li}]$, with entries

$$c_{j,li}(rk) = \langle a_l, U^{rk-i} h_j \rangle\,, \quad k \in \mathbb{Z}\,,$$

where $j = 1, 2, \ldots, s$, and the new index $li$ runs as the lexicographic order

$$10, 11, \ldots, 1(r-1), 20, 21, \ldots, 2(r-1), \ldots, L0, L1, \ldots, L(r-1)\,,$$

there exist $s$ elements $b_j$ in $\mathcal{V}_\mathbf{a}$, $j = 1, 2, \ldots, s$, where necessarily $s \geq Lr$, such that the sequence $\{U^{rk} b_j\}_{k\in\mathbb{Z};\, j=1,2,\ldots,s}$ is a frame for $\mathcal{V}_\mathbf{a}$ and, for any $x \in \mathcal{V}_\mathbf{a}$, the sampling formula

$$x = \sum_{j=1}^{s} \sum_{k\in\mathbb{Z}} \mathcal{L}_j x(rk)\, U^{rk} b_j \quad \text{in } \mathcal{H},$$

holds.

The $U$-sampling problem was considered, for the first time, in Refs. [7, 13, 14].

In the particular case where $U$ is the shift operator $T : f(t) \mapsto f(t-1)$ in $L^2(\mathbb{R})$ and $V_\Phi^2$ the shift-invariant subspace in $L^2(\mathbb{R})$ generated by $\Phi = \{\varphi_1, \varphi_2, \cdots, \varphi_L\}$

$$V_\Phi^2 := \Big\{ \sum_{l=1}^{L} \sum_{n\in\mathbb{Z}} c_l(n)\, \varphi_l(t-n) \ : \ c_l \in \ell^2(\mathbb{Z});\ l = 1, 2 \ldots, L \Big\}$$

we are dealing with the so-called *average sampling* in $L^2(\mathbb{R})$ where the corresponding samples for $f \in V_\Phi^2$ are given by

$$\mathcal{L}_j f(n) = \langle f, \psi_j(\cdot - n)\rangle_{L^2(\mathbb{R})} = (f * \widetilde{\psi}_j)(n)\,, \quad n \in \mathbb{Z}\,,$$

and $\widetilde{\psi}_j(t) = \overline{\psi_j(-t)}$, $j = 1, 2, \ldots, s$, are the average functions.

The above sampling formulas in this case read:

$$f(t) = \sum_{j=1}^{s} \sum_{n\in\mathbb{Z}} \mathcal{L}_j f(n)\, S_j(t-n) \quad \text{in } L^2(\mathbb{R})\,,$$

or

$$f(t) = \sum_{j=1}^{s} \sum_{n\in\mathbb{Z}} \mathcal{L}_j f(rn)\, S_j(t-rn) \quad \text{in } L^2(\mathbb{R})\,,$$

for some sampling functions $S_j$, $j = 1, 2, \ldots, s$, in $V_\Phi^2$.

In the first case, $s \geq L$ and the sequence $\{S_j(t-n)\}_{n\in\mathbb{Z};\, j=1,2,\ldots,s}$ is a frame for $V_\Phi^2$; in the second case, $s \geq Lr$ and the sequence $\{S_j(t-rn)\}_{n\in\mathbb{Z};\, j=1,2,\ldots,s}$ is a frame for $V_\Phi^2$.

Moreover, the convergence of both sampling series is also absolute since it comes out from a frame expansion, and uniform on $\mathbb{R}$ as the shift-invariant subspace $V_\Phi^2$ is a RKHS.

As we will see, all these examples are particular examples of the sampling theory developed in Chap. 5.

## References

1. Akhiezer, N.I., Glazman, I.M.: Theory of Linear Operators in Hilbert Space. Dover Publications, New York (1993)
2. Aldroubi, A.: Oblique projections in atomic spaces. Proc. Amer. Math. Soc. **124**, 2051–2060 (1996)
3. Aldroubi, A.: Oblique and hierarchical multiwavelet bases. Appl. Comput. Harmon. Anal. **4**, 231–263 (1997)
4. Aldroubi, A., Sun, Q., Tang, W.S.: Convolution, average sampling, and a Calderon resolution of the identity for shift-invariant spaces. J. Fourier Anal. Appl. **11**(2), 215–244 (2005)
5. Birman, M.S., Solomjak, M.Z.: Spectral Theory of Self-adjoint Operators in Hilbert Space. Springer, New York (1987)
6. Christensen, O.: An Introduction to Frames and Riesz Bases, 2nd edn. Birkhäuser, Boston (2016)
7. Fernández-Morales, H.R., García, A.G., Hernández-Medina, M.A., Muñoz-Bouzo, M.J.: Generalized sampling: from shift-invariant to $U$-invariant spaces. Anal. Appl. **13**(3), 303–329 (2015)
8. García, A.G., Muñoz-Bouzo, M.J.: Sampling formulas involving differences in shift-invariant subspaces: a unified approach. Numer. Funct. Anal. Optim. **39**(6), 677–688 (2018)
9. García, A.G., Ibort, A., Muñoz-Bouzo, M.J.: Modeling sampling in tensor products of unitary invariant subspaces. J. Funct. Spaces **2016**, Article ID 4573940, 14 pp. (2016)
10. Horn, R.A., Johnson, C.R.: Matrix Analysis. Cambridge University Press, Cambridge UK (1999)
11. Kolmogorov, A.N.: Stationary sequences in Hilbert space. In: Shiryayev, A.N. (ed.) Selected Works of A.N. Kolmogorov, Vol. II, pp. 228–271. Kluwer Academic Publ., Dordrecht (1992)
12. Laub, A.J.: Matrix Analysis. SIAM, Philadelphia (2005)
13. Michaeli, T., Pohl, V., Eldar, Y.C.: $U$-invariant sampling: extrapolation and causal interpolation from generalized samples. IEEE Trans. Signal Process. **59**(5), 2085–2100 (2011)
14. Pohl, V., Boche, H.: $U$-invariant sampling and reconstruction in atomic spaces with multiple generators. IEEE Trans. Signal Process. **60**(7), 3506–3519 (2012)
15. Royden, H.L.: Real Analysis, 3rd edn. Macmillan Publishing Company, New York (1988)
16. Unser, M.: Sampling 50 Years After Shannon. Proc. IEEE **88**(4), 569–587 (2000)
17. Vaidyanathan, P.P.: Multirate Systems and Filter Banks. Prentice-Hall, Englewood Hills, NJ (1993)
18. Walter, G.G.: A sampling theorem for wavelet subspaces. IEEE Trans. Inform. Theory **38**, 881–884 (1992)
19. Zhou, X., Sun, W.: On the Sampling Theorem for Wavelet Subspaces. J. Fourier Anal. Appl. **5**(4), 347–354 (1999)

# Chapter 4
# A Review on Kramer Sampling Theorem

The Kramer sampling theorem (announced by Weiss in Ref. [46]) has been the cornerstone for a significant mathematical literature on the topic of sampling theorems associated with differential or difference problems which has flourished for the past few years. As a small but significant sample of examples see, for instance, Refs. [8, 15, 20, 50] and references therein. Here we just take a sample of this approach in Sect. 4.2.

The main goal in this chapter is to give a common framework gathering most of the sampling formulas coming from the Kramer sampling result. The starting point is an abstract version of the Kramer sampling theorem. To this end, we work in the reproducing kernel Hilbert space of functions introduced by Saitoh in Ref. [41] (see also Ref. [50]) as follows:

Let $(\mathcal{H}, \langle\cdot,\cdot\rangle_{\mathcal{H}})$ be a separable Hilbert space with orthonormal basis $\{e_n\}_{n=1}^\infty$. Suppose that $K:\Omega\longrightarrow\mathcal{H}$ is a $\mathcal{H}$-valued function defined on $\Omega$ a subset of $\mathbb{R}$ (or $\mathbb{C}$). For each $x\in\mathcal{H}$, define the function $f_x(t)=\langle x, K(t)\rangle_{\mathcal{H}}$ in $\Omega$, and let $\mathcal{H}_K$ denote the collection of all such functions $f_x$.

In this setting, an abstract version of the Kramer sampling theorem is obtained in $\mathcal{H}_K$ assuming the existence of two sequences, $\{t_n\}_{n=1}^\infty$ in $\Omega$, and $\{a_n\}_{n=1}^\infty$ in $\mathbb{C}\setminus\{0\}$, such that $K(t_n)=a_n\,e_n$ for each $n\in\mathbb{N}$. This is a slight modification of a sampling result derived by Higgins in Ref. [30].

The non-orthogonal case is also included by considering a Riesz basis $\{x_n\}_{n=1}^\infty$ instead of the orthonormal basis $\{e_n\}_{n=1}^\infty$.

Depending on the nature of the Hilbert space $\mathcal{H}$, an $L^2$-space, an $\ell^2$-space, a Sobolev space, a Hilbert space of distributions, etc., we exhibit different settings where the Kramer sampling theorem applies. This chapter is broadly in line with Ref. [23].

 
A. García García, *The Use of Frames in Sampling Theory*,
RSME Springer Series 14, https://doi.org/10.1007/978-3-031-63242-6_4

## 4.1 Sampling in RKHSs: Kramer Sampling Theorem

We start by considering the classical Kramer sampling theorem (see Ref. [34]):

### 4.1.1 The Classical Kramer Sampling Theorem

The classical Kramer sampling theorem provides a method for obtaining orthogonal sampling theorems; see Refs. [9, 15, 29, 34, 50]. The statement of this general result is as follows.

**Theorem 4.1 (Kramer's Sampling Theorem)** *Let $K$ be a complex function defined on $I \times D$, where $I \subset \mathbb{R}$ is an interval and $D$ is an open subset of $\mathbb{R}$, such that for every $t \in D$ the sections $K(\cdot, t)$ are in $L^2(I)$. Assume that there exists a sequence of distinct real numbers $\{t_n\}_{n\in\mathcal{I}} \subset D$, indexed by a subset $\mathcal{I}$ of $\mathbb{Z}$, such that the set $\{K(x, t_n)\}_{n\in\mathcal{I}}$ is a complete orthogonal sequence of functions for $L^2(I)$. Then for any $f$ of the form*

$$f(t) = \int_I F(x)K(x,t)\,dx\,, \quad t \in D\,, \tag{4.1}$$

*where $F \in L^2(I)$, we have*

$$f(t) = \sum_{n\in\mathcal{I}} f(t_n)S_n(t)\,, \quad t \in D\,, \tag{4.2}$$

*with*

$$S_n(t) := \frac{\int_I K(x,t)\overline{K(x,t_n)}\,dx}{\int_I |K(x,t_n)|^2\,dx}\,, \quad t \in D\,.$$

*The series in* (4.2) *converges absolutely and uniformly on subsets of $D$ where the function $t \mapsto \|K(\cdot,t)\|_{L^2(I)}$ is bounded.*

Its proof is an immediate consequence of Theorem 4.2 below.

Perhaps, the most important example of the Kramer sampling theorem is the Whittaker-Shannon-Kotel'nikov sampling result:

**Example 4.1 (Shannon Sampling Theorem Revisited)** For $I = [-\pi, \pi]$, $D = \mathbb{R}$, $K(x,t) = \mathrm{e}^{itx}/\sqrt{2\pi}$ and the sequence $\{t_n = n\}_{n\in\mathbb{Z}}$, we get the WSK sampling formula

$$f(t) = \sum_{n=-\infty}^{\infty} f(n)\frac{\sin \pi(t-n)}{\pi(t-n)}\,, \quad t \in \mathbb{R}\,,$$

for functions $f \in L^2(\mathbb{R})$ whose Fourier transform $\widehat{f}$ has support in the interval $[-\pi, \pi]$, i.e., $f(t) = \frac{1}{\sqrt{2\pi}} \int_{-\pi}^{\pi} \widehat{f}(w)\, e^{itw}\, dw,\ t \in \mathbb{R}$. The series converges absolutely and uniformly on $\mathbb{R}$.

Reconstructing integral transforms other than the Fourier one from some sampled values occurs frequently in some physical applications. One such integral transform is the *Bessel-Hankel transform*. Sampling associated with this transform was first introduced by Kramer in his seminal paper Ref. [34] (see also Ref. [29, p. 83]):

**Example 4.2 (Sampling Associated to the Bessel-Hankel Transform)** The Fourier-Bessel set $\{\sqrt{x}\, J_\nu(xt_n)\}_{n=1}^{\infty}$ is known to be an orthogonal basis in $L^2(0, 1)$, where $t_n$ is the $n$-th positive zero of the Bessel function $J_\nu(t)$, $\nu > -1$. The Bessel function of order $\nu$ is given by

$$J_\nu(t) = \frac{t^\nu}{2^\nu \Gamma(\nu+1)} \left[ 1 + \sum_{n=1}^{\infty} \frac{(-1)^n}{n!(1+\nu)_n} \left(\frac{t}{2}\right)^{2n} \right], \quad t \in \mathbb{R}^+,$$

where $(1+\nu)_n := (1+\nu)\cdots(n+\nu)$ denotes the Pochhammer symbol. Using special function formulas (see [1, 11.3.29]), for a fixed $t > 0$, we get the expansion

$$\sqrt{xt}\, J_\nu(xt) = \sum_{n=1}^{\infty} \frac{2\sqrt{t_n t}\, J_\nu(t)}{J_\nu'(t_n)(t^2 - t_n^2)} \sqrt{x}\, J_\nu(xt_n) \quad \text{in } L^2(0, 1)\,.$$

Thus, if we take $I = [0, 1]$, $D = \mathbb{R}$, $K(x, t) = \sqrt{xt}\, J_\nu(xt)$ and $\{t_n\}_{n=1}^{\infty}$ the sequence of the positive zeros of $J_\nu(t)$ we obtain:

Any function $f$ of the form $f(t) = \int_0^1 F(x)\sqrt{xt}\, J_\nu(xt)\, dx$, $t \in \mathbb{R}^+$, with $F \in L^2(0, 1)$, can be recovered from its samples at the positive zeros $\{t_n\}_{n=1}^{\infty}$ of the Bessel function $J_\nu$ of $\nu$-th order with $\nu > -1$, through the sampling formula

$$f(t) = \sum_{n=1}^{\infty} f(t_n)\, \frac{2\sqrt{t_n t}\, J_\nu(t)}{J_\nu'(t_n)(t^2 - t_n^2)}, \quad t \in \mathbb{R}^+\,.$$

### 4.1.2 An Abstract Version of the Kramer Sampling Theorem

Since the functions $f$ for which Kramer sampling Theorem 4.1 applies are images of an integral transform with kernel $K$, the reproducing kernel Hilbert spaces considered by Saitoh in Ref. [41] are the suitable spaces where a generalization of the classical Kramer sampling theorem also works (see also Refs. [47, 50]):

Let $\mathcal{H}$ be a separable Hilbert space, and $\Omega$ a fixed subset of $\mathbb{R}$. Given an $\mathcal{H}$-valued continuous function $K : \Omega \to \mathcal{H}$, for any $x \in \mathcal{H}$ we define the function

$$\begin{aligned} f_x : \Omega &\longrightarrow \mathbb{C} \\ t &\longmapsto f_x(t) := \langle x, K(t)\rangle_{\mathcal{H}} . \end{aligned}$$

We denote by $\mathcal{H}_K$ the set of functions obtained in this way and by $\mathcal{T}_K$ the linear transform

$$\mathcal{T}_K : \mathcal{H} \ni x \longmapsto f_x \in \mathcal{H}_K . \tag{4.3}$$

In other words, $\mathcal{H}_K := \mathcal{T}_K(\mathcal{H})$. Hereafter we refer the function $K$ as the kernel of the linear transform $\mathcal{T}_K$ and henceforth we omit the subscript $x$ for denoting the elements in $\mathcal{H}_K$.

We take into account two facts about the vector space $\mathcal{H}_K$:

(a) *The vector space $\mathcal{H}_K$ as a reproducing kernel Hilbert space of continuous functions on $\Omega$.*

Notice that the continuity of the kernel $K$ implies that the functions in $\mathcal{H}_K$ are continuous in $\Omega$. Next, we endow $\mathcal{H}_K$ with a Hilbert space structure. To this end, consider in $\mathcal{H}$ the set $M := \{x \in \mathcal{H} \; : \; \mathcal{T}_K(x) = 0\}$. It is easy to check that $M$ is a closed subspace of $\mathcal{H}$. We endow $\mathcal{H}_K$ with the norm of the quotient space $\widetilde{\mathcal{H}} = \mathcal{H}/M$: Namely, for $f \in \mathcal{H}_K$ we set

$$\|f\|_{\mathcal{H}_K} := \|[x]\|_{\widetilde{\mathcal{H}}} = \inf_{m\in M} \big\{\|x+m\|_{\mathcal{H}}\big\} = \inf\big\{\|x\|_{\mathcal{H}} \; : \; f = \mathcal{T}_K(x)\big\},$$

where $[x]$ denotes the coset of $x$, i.e., $[x] = \{x + m \; : \; m \in M\}$.

In fact, the infimum is actually reached: There exists $\widetilde{x} \in \mathcal{H}$ such that $f = \mathcal{T}_K(\widetilde{x})$ and

$$\|[x]\|_{\widetilde{\mathcal{H}}} = \|\widetilde{x}\|_{\mathcal{H}} = \|f\|_{\mathcal{H}_K} .$$

Indeed, given $x, x' \in [x]$ we can write

$$x = P_M(x) + P_{M^\perp}(x) ; \quad x' = P_M(x') + P_{M^\perp}(x') ,$$

where $P_M$ and $P_{M^\perp}$ denote the *orthogonal projection* onto $M$ and $M^\perp$, respectively. Since $x - x' \in M$, it implies that $P_{M^\perp}(x) = P_{M^\perp}(x') = \widetilde{x}$ for all elements in the coset $[x]$, and $\widetilde{x} \in [x]$.

Therefore, for any $x \in [x]$ we have $\|x\|^2 = \|P_M(x)\|^2 + \|\widetilde{x}\|^2$ and, as a consequence,

$$\|[x]\|_{\widetilde{\mathcal{H}}} = \inf_{x \in [x]} \|x\|_{\mathcal{H}} = \|\widetilde{x}\|_{\mathcal{H}}.$$

By using the *polarization identity* in a complex Hilbert space $\mathcal{H}$, i.e.,

$$\langle x, y \rangle = \frac{1}{4}\Big(\|x+y\|^2 - \|x-y\|^2 + i\big(\|x+iy\|^2 - \|x-iy\|^2\big)\Big), \quad x, y \in \mathcal{H},$$

it is easy to prove that

$$\langle f, g \rangle_{\mathcal{H}_K} = \langle \widetilde{x}, \widetilde{y} \rangle_{\mathcal{H}},$$

whenever $\|f\|_{\mathcal{H}_K} = \|\widetilde{x}\|_{\mathcal{H}}$ and $\|g\|_{\mathcal{H}_K} = \|\widetilde{y}\|_{\mathcal{H}}$.

Consequently, the linear mapping

$$\widetilde{\mathcal{T}}_K : M^\perp \ni \widetilde{x} \longmapsto f \in \mathcal{H}_K \text{ such that } f(t) := \big\langle \widetilde{x}, K(t) \big\rangle_{\mathcal{H}} \text{ for } t \in \Omega, \tag{4.4}$$

is a bijective isometry.

The function space $\mathcal{H}_K$ is a RKHS. Indeed, for fixed $t \in \Omega$, for each $f \in \mathcal{H}_K$ we have that $f(t) = \langle \widetilde{x}, K(t) \rangle_{\mathcal{H}}$. By using the Cauchy-Schwarz inequality we obtain

$$|f(t)| \le \|K(t)\|_{\mathcal{H}} \|\widetilde{x}\|_{\mathcal{H}} = C_t \|f\|_{\mathcal{H}_K},$$

that is, for each $t \in \Omega$ the evaluation functional $E_t(f) = f(t)$, $f \in \mathcal{H}_K$ is bounded.

As a consequence, convergence in the norm $\|\cdot\|_{\mathcal{H}_K}$ implies pointwise convergence which will be uniform on subsets of $\Omega$ where the function $t \mapsto \|K(t)\|_{\mathcal{H}}$ is bounded.

Its reproducing kernel is given by $k(t, s) = \langle K(s), K(t) \rangle_{\mathcal{H}}$, $t, s \in \Omega$. Indeed,

$$\begin{aligned} \langle f, k(\cdot, s) \rangle_{\mathcal{H}_K} &= \big\langle \mathcal{T}_K(x), \mathcal{T}_K(K(s)) \big\rangle_{\mathcal{H}} = \big\langle P_{M^\perp}(x), P_{M^\perp}(K(s)) \big\rangle_{\mathcal{H}} \\ &= \big\langle \widetilde{x}, P_{M^\perp}(K(s)) \big\rangle_{\mathcal{H}} = \langle \widetilde{x}, K(s) \rangle_{\mathcal{H}} = f(s), \qquad s \in \Omega. \end{aligned} \tag{4.5}$$

*(b) The operator $\mathcal{T}_K$ as a unitary operator between $\mathcal{H}$ and $\mathcal{H}_K$.*

Having in mind that the operator $\widetilde{\mathcal{T}}_K$ in (4.4) is a bijective isometry, hence the operator $\mathcal{T}_K : \mathcal{H} \longrightarrow \mathcal{H}_K$ is injective if and only if the linear mapping $\mathcal{T}_K$ is an isometry, or equivalently, if and only if the set $\{K(t)\}_{t \in \Omega}$ is complete in $\mathcal{H}$.

In particular, if there exists a sequence $\{t_n\}_{n=1}^\infty$ in $\Omega$ such that $\{K(t_n)\}_{n=1}^\infty$ is an orthogonal basis for $\mathcal{H}$, then the operator $\mathcal{T}_K$ is a linear isometry from $\mathcal{H}$ onto $\mathcal{H}_K$, i.e., a unitary operator.

A generalization of the Kramer sampling result consists of stating the Kramer condition in terms of Riesz bases. According to the Riesz basis Theorem 1.2, any Riesz basis $\{x_n\}_{n=1}^{\infty}$ has a unique biorthogonal (dual) Riesz basis $\{y_n\}_{n=1}^{\infty}$, i.e., $\langle x_n, y_m\rangle_{\mathcal{H}} = \delta_{n,m}$, such that, for every $x \in \mathcal{H}$, the expansions

$$x = \sum_{n=1}^{\infty} \langle x, y_n\rangle_{\mathcal{H}}\, x_n = \sum_{n=1}^{\infty} \langle x, x_n\rangle_{\mathcal{H}}\, y_n \quad \text{in } \mathcal{H}$$

hold.

**Definition 4.1 (Abstract Kramer Kernel)** An $\mathcal{H}$-valued function $K : \Omega \longrightarrow \mathcal{H}$ is said to be an abstract Kramer kernel with respect to the data $\{t_n\}_{n=1}^{\infty} \subset \Omega$ and $\{a_n\}_{n=1}^{\infty} \subset \mathbb{C} \setminus \{0\}$ if it satisfies $K(t_n) = \overline{a}_n\, x_n$, $n \in \mathbb{N}$, for some Riesz basis $\{x_n\}_{n=1}^{\infty}$ for $\mathcal{H}$.

For each fixed $t \in \Omega$, $K(t)$ can be written as $K(t) = \sum_{n=1}^{\infty} \overline{S_n(t)}\, x_n$, where the functions

$$S_n(t) := \langle y_n, K(t)\rangle_{\mathcal{H}}, \quad t \in \Omega, \quad n \in \mathbb{N}, \tag{4.6}$$

belong to $\mathcal{H}_K$; here $\{y_n\}_{n=1}^{\infty}$ denotes the dual Riesz basis of $\{x_n\}_{n=1}^{\infty}$ in $\mathcal{H}$.

The *Kramer condition*, $K(t_n) = \overline{a}_n\, x_n$ for every $n \in \mathbb{N}$, is equivalent to the interpolation property

$$S_n(t_m) = a_n \delta_{n,m}\,, \quad n, m \in \mathbb{N}\,.$$

The Kramer condition also implies that the linear transform $\mathcal{T}_K$ in (4.3) is a unitary operator between the Hilbert spaces $\mathcal{H}$ and $\mathcal{H}_K$. As a consequence, we obtain the following sampling theorem for functions in $\mathcal{H}_K$:

**Theorem 4.2 (Abstract Kramer Sampling Theorem)** *Assume that $K$ is an abstract Kramer kernel for the data $\{t_n\}_{n=1}^{\infty} \subset \Omega$, $\{a_n\}_{n=1}^{\infty} \subset \mathbb{C} \setminus \{0\}$, and the Riesz basis $\{x_n\}_{n=1}^{\infty}$ for $\mathcal{H}$. Then, the sequence of functions $\{S_n\}_{n=1}^{\infty}$ defined in (4.6) is a Riesz basis for $\mathcal{H}_K$. Expanding any $f \in \mathcal{H}_K$ in this Riesz basis we obtain the nonorthogonal sampling expansion*

$$f(t) = \sum_{n=1}^{\infty} f(t_n)\, \frac{S_n(t)}{a_n}\,, \quad t \in \Omega\,. \tag{4.7}$$

*The series converges in the $\mathcal{H}_K$-norm sense and also absolutely and uniformly on subsets of $\Omega$ where the function $t \mapsto \|K(t)\|_{\mathcal{H}}$ is bounded.*

***Proof*** From (4.6) we have that $\mathcal{T}_K(y_m) = S_m$ for each $m \in \mathbb{N}$; since the operator $\mathcal{T}_K$ is unitary we obtain that the sequence $\{S_m\}_{m=1}^{\infty}$ is a Riesz basis for $\mathcal{H}_K$ whose biorthogonal basis $\{T_m\}_{m=1}^{\infty}$ is given by $T_m := \mathcal{T}_K(x_m)$, $m \in \mathbb{N}$.

Expanding any $f \in \mathcal{H}_K$ in this Riesz basis, we have $f = \sum_{n=1}^{\infty} \langle f, T_n \rangle_{\mathcal{H}_K} S_n$, in the $\mathcal{H}_K$-norm sense and, consequently, pointwise in $\Omega$. Moreover, having in mind that $\mathcal{T}_K$ is a unitary operator, and that $K(t_n) = \overline{a}_n\, x_n$, $n \in \mathbb{N}$, we obtain

$$\begin{aligned}\langle f, T_n \rangle_{\mathcal{H}_K} &= \langle \mathcal{T}_K(x), \mathcal{T}_K(x_n) \rangle_{\mathcal{H}_K} = \langle x, x_n \rangle_{\mathcal{H}} \\ &= \Big\langle x, \frac{K(t_n)}{\overline{a}_n} \Big\rangle_{\mathcal{H}} = \frac{f(t_n)}{a_n}, \quad n \in \mathbb{N},\end{aligned}$$

and hence, the sampling expansion in (4.7).

Since a Riesz basis is an unconditional basis (notice that any orthonormal basis is an unconditional basis by using the Parseval equality), the sampling series (4.7) is pointwise unconditionally convergent for each $t \in \Omega$, and hence pointwise absolutely convergent. The uniform convergence of the series in (4.7) is a standard result in RKHSs theory. □

In the particular case when $\{x_n\}_{n=1}^{\infty}$ is an orthonormal basis for $\mathcal{H}$, it is self-dual and we have the following result:

**Corollary 4.1** *Whenever the sequence $\{x_n\}_{n=1}^{\infty}$ in Definition 4.1 is an orthonormal basis for $\mathcal{H}$, the sequence $\{S_n\}_{n=1}^{\infty}$ is an orthonormal basis for $\mathcal{H}_K$ and the sampling expansion* (4.7) *is an orthonormal expansion in $\mathcal{H}_K$ having the same pointwise convergence properties.*

Thus, Theorem 4.1 is a consequence of the above corollary. Next, we exhibit some examples illustrating the Kramer sampling theorem:

**Example 4.3 (Sampling Formula for Bandlimited Functions with respect to the Fourier Cosine Transform)** For $\mathcal{H} := L^2[0, \pi]$ we consider the kernel $K_c : \mathbb{R} \longrightarrow L^2[0, \pi]$ defined, for each $t \in \mathbb{R}$, by $\big[K_c(t)\big](x) = \cos tx$, $x \in [0, \pi]$.

Since the sequence $\{\cos nx\}_{n=0}^{\infty}$ is an orthogonal basis for $L^2[0, \pi]$, the kernel $K_c$ satisfies the Kramer condition for $n \in \mathbb{N}_0 := \mathbb{N} \cup \{0\}$.

Thus, any function

$$f(t) = \big\langle F, \cos tx \big\rangle_{L^2[0,\pi]} = \int_0^{\pi} F(x)\, \cos tx\, dx\,, \quad t \in \mathbb{R},$$

where $F \in L^2[0, \pi]$, can be expanded as

$$f(t) = f(0)\frac{\sin \pi t}{\pi t} + \frac{2}{\pi} \sum_{n=1}^{\infty} f(n) \frac{(-1)^n t \sin \pi t}{t^2 - n^2}, \quad t \in \mathbb{R}.$$

The series above converges absolutely and uniformly on $\mathbb{R}$.

**Example 4.4 (Sampling Formula for Bandlimited Functions with respect to the Fourier Sine Transform)** Analogously, consider the kernel $K_s : \mathbb{R} \longrightarrow L^2[0,\pi]$ defined, for each $t \in \mathbb{R}$, by $\big[K_s(t)\big](x) = \sin tx$, $x \in [0,\pi]$. The sequence $\{\sin nx\}_{n=1}^{\infty}$ is an orthogonal basis for $L^2[0,\pi]$, the kernel $K_s$ satisfies the Kramer condition for $n \in \mathbb{N}$. Then, any function

$$f(t) = \big\langle F, \sin tx \big\rangle_{L^2[0,\pi]} = \int_0^{\pi} F(x)\,\sin tx\,dx\,, \quad t \in \mathbb{R}\,,$$

where $F \in L^2[0,\pi]$, can be expanded as

$$f(t) = \frac{2}{\pi}\sum_{n=1}^{\infty} f(n)\frac{(-1)^n n \sin \pi t}{t^2 - n^2}\,, \quad t \in \mathbb{R}\,.$$

The series above converges absolutely and uniformly on $\mathbb{R}$.

**Example 4.5 (Sampling Formula for Bandlimited Functions with respect to the Fractional Fourier Transform)** Consider the kernel $K$ mapping $\mathbb{R} \ni t \longmapsto K(t) \in L^2[-\pi,\pi]$, such that

$$K(t)(x) := \frac{1}{\sqrt{2\pi}}\mathrm{e}^{i(t^2+x^2-xt)}\,, \quad x \in [-\pi,\pi]\,.$$

For the sequence of integers $\{t_n = n\}_{n\in\mathbb{Z}}$, the sequence $\{K(n)\}_{n\in\mathbb{Z}}$ is the orthonormal basis $\big\{\frac{1}{\sqrt{2\pi}}\mathrm{e}^{i(n^2+x^2-nx)}\big\}_{n\in\mathbb{Z}}$ of $L^2[-\pi,\pi]$.

Hence, any function $f$ defined by

$$f(t) = \big\langle F, K(t)\big\rangle_{L^2[-\pi,\pi]} = \frac{1}{\sqrt{2\pi}}\int_{-\pi}^{\pi} F(x)\,\mathrm{e}^{-i(t^2+x^2-xt)}\,dx\,, \quad t \in \mathbb{R}\,,$$

where $F \in L^2[-\pi,\pi]$, can be expanded as

$$f(t) = \sum_{n=-\infty}^{\infty} f(n)\,\mathrm{e}^{-i(t^2-n^2)}\frac{\sin \pi(t-n)}{\pi(t-n)}\,, \quad t \in \mathbb{R}\,. \tag{4.8}$$

Thus, we have obtained a sampling formula that looks like the sampling formula for bandlimited functions in the sense of the fractional Fourier transform.

The fractional Fourier transform with angle $\alpha \notin \{0,\pi\}$ of a function $f$ is defined as

$$\mathcal{F}_\alpha[f](x) := \int_{-\infty}^{\infty} f(t) K_\alpha(x,t)\,dt\,,$$

where, up to a normalization constant, the integral kernel $K_\alpha(x,t)$ is given by

$$e^{i\frac{\cot\alpha}{2}(t^2+x^2)-i\frac{xt}{\sin\alpha}}. \tag{4.9}$$

For $\alpha = 0$ the fractional Fourier transform is defined as $\mathcal{F}_0[f](x) := f(x)$, and for $\alpha = \pi$ as $\mathcal{F}_\pi[f](x) := f(-x)$. Whenever $\alpha = \pi/2$, the kernel (4.9) coincides with the Fourier kernel.

The kernel (4.9) can be also written as

$$e^{ia(\alpha)[t^2+x^2-2b(\alpha)xt]},$$

where $a(\alpha) = (\cot\alpha)/2$ and $b(\alpha) = \sec\alpha$. The inversion formula for the fractional Fourier transform involves, up to a normalization constant, the kernel $K_{-\alpha}(x,t)$.

In the above sampling example we have taken the constants $a = 1$ and $b = 1/2$ which do not correspond to any angle $\alpha$. However, if $f$ is a signal bandlimited to $[-\delta,\delta]$ in the fractional Fourier transform sense, i.e., the support of $F_\alpha(u)$ is contained in the interval $[-\delta,\delta]$, then the sampling expansion

$$f(t) = \sum_{n=-\infty}^{\infty} e^{-ia(\alpha)\left(t^2-t_n^2\right)} f(t_n)\frac{\sin\left[\frac{\delta}{\sin\alpha}(t-t_n)\right]}{\frac{\delta}{\sin\alpha}(t-t_n)}, \quad t \in \mathbb{R},$$

holds, where the sampling points are $t_n = n\pi\sin\alpha/\delta$, $n \in \mathbb{Z}$. Observe that this formula looks like the formula (4.8). See, for instance, Ref. [55].

### 4.1.3 *Kramer Sampling Theorem and Indeterminate Moment Problems*

A source of discrete Kramer kernels $K$, where $\mathcal{H} := \ell^2(\mathbb{N}_0)$, is provided from the *indeterminate Hamburger moment problem*. The interested reader can be find the details on this theory in Refs. [2, 42]. Next, we go into a quick summary of this theory.

Let $s = \{s_n\}_{n=0}^{\infty} \subset \mathbb{R}$ be an indeterminate Hamburger moment sequence, that is, there exists a non-unique positive Borel measure $\mu$ on $\mathbb{R}$ such that

$$s_n = \int_{-\infty}^{\infty} t^n d\mu(t), \quad n \in \mathbb{N}_0.$$

Let $V_s$ be the set of positive Borel measures $\mu$ on $\mathbb{R}$ satisfying the above moment problem. The functional $\mathcal{L}$ defined on the vector space $\mathbb{C}[x]$ of polynomials $p(t) = \sum_{k=0}^{n} p_k t^k$ by

$$\mathcal{L}(p) := \sum_{k=0}^{n} p_k s_k = \int_{-\infty}^{\infty} p(t) d\mu(t)$$

is independent of $\mu \in V_s$. Let $\{P_n\}_{n=0}^{\infty}$ be the corresponding orthonormal polynomials satisfying

$$\int_{-\infty}^{\infty} P_n(t) P_m(t) d\mu(t) = \delta_{nm}\,, \quad \text{for each } \mu \in V_s\,.$$

We assume that $P_n$ is of degree $n$ with positive leading coefficient. Recall that the sequence $\{P_n(t)\}_{n=0}^{\infty}$ satisfies a three-term recurrence relation

$$t P_n(t) = a_n P_{n+1}(t) + b_n P_n(t) + a_{n-1} P_{n-1}(t)\,, \quad n \geq 0\,,$$

where $P_{-1}(t) = 0$ and $P_0(t) = 1$.

The two sequences $\{b_n\}_{n=0}^{\infty}$ and $\{a_n\}_{n=0}^{\infty}$ of real and positive numbers, respectively, form the *semi-infinite Jacobi matrix* associated with the indeterminate Hamburger moment problem (see, for instance, Ref. [42]):

$$\mathcal{J} = \begin{pmatrix} b_0 & a_0 & 0 & 0 & \cdots \\ a_0 & b_1 & a_1 & 0 & \cdots \\ 0 & a_1 & b_2 & a_2 & \cdots \\ 0 & 0 & a_2 & b_3 & \cdots \\ \vdots & \vdots & \vdots & \vdots & \ddots \end{pmatrix}$$

Since we are dealing with an indeterminate Hamburger moment problem it is known that $\sum_{n=0}^{\infty} |P_n(t)|^2 < \infty$ for each $t \in \mathbb{R}$; in fact, uniformly on compact subsets of $\mathbb{C}$ (see Refs. [2, 42]).

Thus, we can consider the kernel $K$ defined as

$$\mathbb{R} \ni t \overset{K}{\longmapsto} K(t) := \big\{P_n(t)\big\}_{n=0}^{\infty} \in \ell^2(\mathbb{N}_0)\,,$$

and its associated $\mathcal{H}_K$ space.

Moreover, there exist sequences $\{t_m\}_{m=0}^{\infty}$ in $\mathbb{R}$ such that the sequence

$$\big\{\{P_0(t_m), P_1(t_m), P_2(t_m), \ldots, P_n(t_m), \ldots\}\big\}_{m=0}^{\infty}$$

is an orthogonal basis for $\ell^2(\mathbb{N}_0)$ (see Ref. [19]).

More can be said about these sequences $\{t_m\}_{m=0}^{\infty}$ by using the well-known Hamburger moment problem theory. An $N$-extremal measure $\mu$ is a solution of the Hamburger moment problem for which the polynomials $\{P_n\}_{n=0}^{\infty}$ are dense in $L^2(\mu)$ (see Ref. [42]).

The $N$-extremal measures $\mu_x$ can be parametrized with $x \in \mathbb{R} \cup \{\infty\}$ and we have that (see [42, p. 126]):

$$\frac{A(z) + xC(z)}{B(z) + xD(z)} = \int_{-\infty}^{\infty} \frac{d\mu_x(y)}{z - y}, \quad z \in \mathbb{C} \setminus \mathbb{R},$$

where $A(z)$, $B(z)$, $C(z)$ and $D(z)$ are the components of the so-called *Nevalinna matrix* of the Hamburger indeterminate moment problem (see Ref. [42, p. 124]).

The sequence of zeros of the entire function, $B(z) + xD(z)$ if $x \in \mathbb{R}$ or the zeros of $D(z)$ if $x = \infty$, are precisely the sampling points $\{t_m\}_{m=0}^{\infty}$ (see Ref. [42, p. 127]).

As a consequence, associated with any indeterminate Hamburger moment problem we obtain a discrete Kramer kernel $K$. Next we show a specific example:

**Example 4.6 (Sampling Associated with the $q^{-1}$-Hermite Polynomials where** $(0 < q < 1)$) We consider the particular case of the so-called $q^{-1}$-Hermite polynomials where $0 < q < 1$. These polynomials have the explicit representation (see Ref. [31])

$$h_n(t|q) = \sum_{k=0}^{n} \frac{(q;q)_n}{(q;q)_k (q;q)_{n-k}} (-1)^k q^{k(k-n)} (t + \sqrt{t^2+1})^{n-2k}, \quad t \in \mathbb{R},$$

where we use the $q$-shifted factorial notation

$$(c_1, c_2, \cdots, c_p; q)_n = \prod_{k=1}^{n} \prod_{j=1}^{p} (1 - c_j q^{k-1}), \quad \text{for } n = 0, 1, \ldots, \infty.$$

The moment problem associated with $\{h_n(t|q)\}_{n=0}^{\infty}$ is indeterminate, and the norms of the polynomials are given by $\|h_n\| = \sqrt{\frac{(q;q)_n}{q^{n(n+1)/2}}}$ (see Ref. [31]). In this case the kernel $K_q$ is given by

$$\begin{aligned} K_q : \mathbb{R} &\longrightarrow \ell^2(\mathbb{N}_0) \\ t &\longmapsto K_q(t) := \{P_n(t)\}_{n=0}^{\infty}, \end{aligned}$$

where $\{P_n := h_n/\|h_n\|\}_{n=0}^{\infty}$ is the sequence of orthonormal polynomials.

Consider, for example, the points $\pm t_m := \pm\frac{1}{2}(q^{-m-1/2} - q^{m+1/2})$, $m \in \mathbb{N}_0$, which are the zeros of the corresponding $D$ function, given by (see Ref. [31])

$$D(t) = -\frac{(qe^{2\xi}, qe^{-2\xi}; q^2)_\infty}{(q;q^2)_\infty^2}, \quad \text{where } t = \sinh \xi.$$

We obtain that

$$\Big\{\{P_0(t_m), P_1(t_m), P_2(t_m), \cdots\}\Big\}_{m\in\mathbb{N}_0} \cup \Big\{\{P_0(-t_m), P_1(-t_m), P_2(-t_m), \cdots\}\Big\}_{m\in\mathbb{N}_0}$$

is an orthogonal basis for $\ell^2(\mathbb{N}_0)$. Following Refs. [6, 20], for each $m \in \mathbb{N}_0$ we have

$$S_m^{\pm}(t) := \big\langle \{P_n(\pm t_m)\}_{n\in\mathbb{N}_0}, K_q(t)\big\rangle_{\ell^2(\mathbb{N}_0)} = \frac{D(t)}{(t \mp t_m)D'(\pm t_m)}, \quad t \in \mathbb{R}.$$

Thus, in this case Theorem 4.2 reads:

Any function $f$ given by

$$f(t) = \big\langle \{c_n\}, \{P_n(t)\}\big\rangle_{\ell^2(\mathbb{N}_0)} = \sum_{n=0}^{\infty} c_n P_n(t), \quad t \in \mathbb{R},$$

where $\{c_n\} \in \ell^2(\mathbb{N}_0)$, can be expanded as

$$f(t) = \sum_{m=0}^{\infty} f(-t_m)\frac{D(t)}{(t+t_m)D'(-t_m)} + \sum_{m=0}^{\infty} f(t_m)\frac{D(t)}{(t-t_m)D'(t_m)}, \quad t \in \mathbb{R}.$$

The series converges absolutely and uniformly on compact subsets of $\mathbb{R}$.

There is another way to obtain the RKHS $\mathcal{H}_K$ where the Kramer sampling theorem applies. It is the so-called *Moore-Aronszajn procedure* (see Ref. [5]) which uses the concept of *positive matrix* (see below). In our case this positive matrix would be the function defined as

$$k(t,s) := \big\langle K(s), K(t)\big\rangle_{\mathcal{H}}, \quad t, s \in \Omega. \tag{4.10}$$

In next section we go into the details. Besides, the Moore-Aronszajn procedure allows to prove some necessary results in the RKHS theory. Next we include a summary about the alluded procedure.

### *4.1.4 Positive Matrices and RKHSs*

**Definition 4.2** A function $k : \Omega \times \Omega \longrightarrow \mathbb{C}$ is said to be a positive matrix[1] if for all $N \in \mathbb{N}$, $t_1, t_2, \ldots, t_N \in \Omega$ and $\xi_1, \xi_2, \ldots, \xi_N \in \mathbb{C}$ we have that

$$\sum_{i,j=1}^{N} k(t_i, t_j)\,\overline{\xi_i}\xi_j \geq 0. \tag{4.11}$$

[1] A positive matrix is also called a positive definite function.

As an example, the function $k$ defined in (4.10) is a positive matrix. Indeed,

$$\sum_{i,j=1}^{N} \langle K(t_j), K(t_i)\rangle\, \overline{\xi_i}\xi_j = \Big\langle \sum_{j=1}^{N} \xi_j K(t_j), \sum_{i=1}^{N} \xi_i K(t_i)\Big\rangle = \Big\| \sum_{i=1}^{N} \xi_i K(t_i)\Big\|^2 \geq 0\,.$$

A positive definite function $k$ has the following properties:

**Proposition 4.1** *Let $k : \Omega \times \Omega \longrightarrow \mathbb{C}$ be a positive matrix. The following properties hold:*

*(i)* $k(t,t) \geq 0$ *for each* $t \in \Omega$.
*(ii)* $k(t,s) = \overline{k(s,t)}$ *for* $t, s \in \Omega$.
*(iii)* $|k(t,s)|^2 \leq k(t,t)\,k(s,s)$ *for* $t, s \in \Omega$.

***Proof*** To prove $(i)$, take $N = 1$, $t_1 = t$ and $\xi_1 = 1$ in (4.11) to get $k(t,t) \geq 0$.

For proving $(ii)$, by taking $N = 2$, $t_1 = t$, $t_2 = s$, $\xi = 1$ and $\eta \in \mathbb{C}$ in (4.11) we get

$$k(t,t) + k(s,s)|\eta|^2 + k(t,s)\overline{\eta} + k(s,t)\eta \geq 0\,.$$

The first two summands are $\geq 0$ which implies that $k(t,s)\overline{\eta} + k(s,t)\eta \in \mathbb{R}$ for all $\eta \in \mathbb{C}$. Taking $\eta = 1$ we obtain that $k(t,s) + k(s,t) \in \mathbb{R}$, which implies that

$$\operatorname{Im} k(t,s) = -\operatorname{Im} k(s,t)\,.$$

Taking $\eta = i$ we obtain that $i(-k(t,s) + k(s,t)) \in \mathbb{R}$ which implies that

$$\operatorname{Re} k(t,s) = \operatorname{Re} k(s,t)\,.$$

Hence, $k(t,s) = \overline{k(s,t)}$.

Finally, we prove $(iii)$. Taking $N = 2$, $t_1 = t$, $t_2 = s$ and $\xi, \eta \in \mathbb{C}$ in (4.11) we get

$$k(t,t)|\xi|^2 + k(t,s)\xi\,\overline{\eta} + k(s,t)\overline{\xi}\,\eta + k(s,s)|\eta|^2 \geq 0\,.$$

Take $\xi \in \mathbb{C}$, necessarily with $|\xi| = 1$, such that $\xi\, k(t,s) = -|k(t,s)|$. Then,

$$k(t,t) + (-|k(t,s)|)\overline{\eta} + (-|k(t,s)|)\eta + k(s,s)|\eta|^2 \geq 0\,,$$

which implies $k(t,t) - 2|k(t,s)|\eta + k(s,s)\eta^2 \geq 0$ for all $\eta \in \mathbb{R}$. That is,

$$4|k(t,s)|^2 - 4k(t,t)k(s,s) \leq 0 \quad \text{or} \quad |k(t,s)|^2 \leq k(t,t)\,k(s,s)\,.$$

□

In a RKHS $\mathcal{H}$, its reproducing kernel $k$ defines a positive matrix. Indeed,

$$\sum_{i,j=1}^{N} k(t_i,t_j)\,\overline{\xi_i}\xi_j = \Big\langle \sum_{j=1}^{N} \xi_j\, k(\cdot,t_j), \sum_{i=1}^{N} \xi_i\, k(\cdot,t_i) \Big\rangle = \Big\| \sum_{i=1}^{N} \xi_i\, k(\cdot,t_i) \Big\|^2 \geq 0\,.$$

The Moore-Aronszajn theorem assures that given a positive matrix $k$ there exists a unique RKHS $\mathcal{H}_k$ of functions defined on $\Omega$ such that $k$ is its reproducing kernel.

**Theorem 4.3 (Moore-Aronszajn's Theorem)** *For any positive matrix $k : \Omega \times \Omega \longrightarrow \mathbb{C}$ there exists a unique Hilbert space $\mathcal{H}_k$ of functions defined on $\Omega$ such that $k$ is its reproducing kernel.*

***Proof*** To this end, consider the vector space

$$\mathcal{H}_0 := \text{span}\,\big\{k(\cdot,t)\,:\, t\in\Omega\big\} = \Big\{\sum_{i=1}^{N} \xi_i\, k(s,t_i)\,:\, N\in\mathbb{N},\ t_i\in\Omega,\ \xi_i\in\mathbb{C}\Big\} \tag{4.12}$$

endowed with the inner product

$$\langle f,g\rangle_{\mathcal{H}_0} = \Big\langle \sum_i \eta_i\, k(s,t_i), \sum_j \xi_j\, k(s,t_j)\Big\rangle_{\mathcal{H}_0} := \sum_{i,j} \eta_i\,\overline{\xi}_j\, k(t_j,t_i)\,,$$

for $f = \sum_i \eta_i\, k(s,t_i)$ and $g = \sum_j \xi_j\, k(s,t_j)$ in $\mathcal{H}_0$.

The definition of $\langle f,g\rangle_{\mathcal{H}_0}$ is independent of the representations of $f$ and $g$ chosen in $\mathcal{H}_0$. Indeed, suppose that

$$f(t) = \sum_i k(t,t_i)\xi_i \quad \text{and} \quad g(t) = \sum_j k(t,s_j)\eta_j\,.$$

Then,

$$\langle f,g\rangle_{\mathcal{H}_0} = \sum_j f(s_j)\,\overline{\eta}_j \quad \text{and} \quad \langle f,g\rangle_{\mathcal{H}_0} = \sum_i \overline{g(t_i)}\,\xi_i\,.$$

The first equality shows that $\langle f,g\rangle_{\mathcal{H}_0}$ is independent of the representation of $f$, and the second one that $\langle f,g\rangle_{\mathcal{H}_0}$ is independent of the representation of $g$.

Observe that, for any $f\in\mathcal{H}_0$, the reproducing property

$$f(t) = \langle f, k(\cdot,t)\rangle_{\mathcal{H}_0}\,, \quad t\in\Omega\,,$$

holds.

Next, the pre-Hilbert space $\mathcal{H}_0$ satisfies the necessary and sufficient condition for the existence of a *proper functional completion* (see Ref. [25, p. 185]):

*If for each Cauchy sequence* $\{f_n\}_{n=1}^\infty$ in $\mathcal{H}_0$, $\lim_{n\to\infty} f_n(s) = 0$ for all $s \in \Omega$ implies $\lim_{n\to\infty} \|f_n\| = 0$, then, the pre-Hilbert space $\mathcal{H}_0$ admits a completion $\mathcal{H}_k$ which is a RKHS with k as reproducing kernel.

In our case, the above condition on the existence of a proper functional completion for $\mathcal{H}_0$ is obtained from the inequality

$$\big|\|f_n\|^2 - \langle f_m, f_n\rangle_{\mathcal{H}_0}\big| \le \|f_n - f_m\|\,\|f_n\|\,,$$

having in mind that $\lim_{n,m\to\infty} \|f_n - f_m\| = 0$ since $\{f_n\}_{n=1}^\infty$ is a Cauchy sequence in $\mathcal{H}_0$, and that, for any fixed $n \in \mathbb{N}$, we have $\lim_{m\to\infty}\langle f_m, f_n\rangle_{\mathcal{H}_0} = 0$. Therefore, $\lim_{n\to\infty} \|f_n\| = 0$.

Observe that

$$\langle f_m, f_n\rangle_{\mathcal{H}_0} = \Big\langle f_m, \sum_i \xi_i^{(n)}\, k(\cdot, t_i^{(n)})\Big\rangle = \sum_i \overline{\xi_i^{(n)}}\, f_m(t_i^{(n)}) \to 0 \quad \text{as } m \to \infty$$

since we are assuming that $\lim_{n\to\infty} f_n(s) = 0$ for all $s \in \Omega$.

The completion space can be described as

$$\mathcal{H}_k = \big\{f : \Omega \longrightarrow \mathbb{C} \,:\, f \text{ is the pointwise limit of a Cauchy sequence in } \mathcal{H}_0\big\}\,.$$

Observe that, given a Cauchy sequence $\{f_n\}$ in $\mathcal{H}_0$ there exists the pointwise limit $f(t) := \lim_{n\to\infty} f_n(t)$, $t \in \Omega$. Indeed, it comes from the inequality

$$\big|f_n(t) - f_m(t)\big| = \big|\langle f_n - f_m, k(\cdot, t)\rangle_{\mathcal{H}_0}\big| \le \|f_n - f_m\|\, k(t,t)\,, \quad t \in \Omega\,.$$

Any function $f \in \mathcal{H}_k$ has norm $\|f\|_{\mathcal{H}_k} = \lim_{n\to\infty} \|f_n\|$, where $\{f_n\}_{n=1}^\infty$ is a Cauchy sequence in $\mathcal{H}_0$ converging to $f$ pointwise. This norm is independent of the choice of the Cauchy sequence $\{f_n\}_{n=1}^\infty$ in $\mathcal{H}_0$. Indeed, let $\{g_n\}_{n=1}^\infty$ be another Cauchy sequence in $\mathcal{H}_0$ converging to $f$ pointwise. Then, the Cauchy sequence $\{f_n - g_n\}_{n=1}^\infty$ satisfies $\lim_{n\to\infty}\big(f_n(t) - g_n(t)\big) = 0$ for any $t \in \Omega$. Hence, $\lim_{n\to\infty} \|f_n - g_n\| = 0$, and since

$$0 \le \big|\|f_n\| - \|g_n\|\big| \le \|f_n - g_n\| \to 0 \quad \text{as } n \to \infty\,,$$

we obtain that $\lim_{n\to\infty} \|f_n\| = \lim_{n\to\infty} \|g_n\|$.

The space $\mathcal{H}_k$ carries the natural Hilbert space structure induced by $\mathcal{H}_0$. Concerning the reproducing property, for any $f \in \mathcal{H}_k$ with $\|f\|_{\mathcal{H}_k} = \lim_{n\to\infty} \|f_n\|$ we have that $f_n \to f$ in $\mathcal{H}_k$ as $n \to \infty$. Thus we obtain

$$f(t) = \lim_{n\to\infty} f_n(t) = \lim_{n\to\infty} \langle f_n, k(\cdot, t)\rangle_{\mathcal{H}_k} = \langle f, k(\cdot, t)\rangle_{\mathcal{H}_k}\,, \quad t \in \Omega\,.$$

See the details, for instance, in Refs. [25, 41]. □

In a RKHS $\mathcal{H}$ of functions defined on $\Omega$ and reproducing kernel $k$, for any $f \in \mathcal{H}$ the following inequality holds:

$$\Big|\sum_{i=1}^{N} \xi_i f(t_i)\Big|^2 \leq \|f\|^2 \sum_{i,j=1}^{N} k(t_i, t_j)\, \xi_i \overline{\xi}_j\,, \tag{4.13}$$

with $N \in \mathbb{N}$, $\xi_1, \xi_2, \ldots, \xi_N \in \mathbb{C}$ and $t_1, t_2, \ldots, t_N \in \Omega$.

Indeed, denoting $k_{t_i} = k(\cdot, t_i)$, the reproducing property and the Cauchy-Schwarz inequality give

$$\Big|\sum_{i=1}^{N} \xi_i f(t_i)\Big|^2 = \Big|\Big\langle f, \sum_{i=1}^{N} \overline{\xi}_i k_{t_i}\Big\rangle\Big|^2 \leq \|f\|^2 \,\Big\|\sum_{i=1}^{N} \overline{\xi}_i k_{t_i}\Big\|^2$$
$$= \|f\|^2 \sum_{i,j=1}^{N} k(t_i, t_j)\, \xi_i \overline{\xi}_j\,.$$

The above property can be used as a test for checking if a given function belongs to a RKHS. Next we include a result in this direction borrowed from Ref. [43]:

**Proposition 4.2 (RKHS Test)** *Let $\mathcal{H}$ be a RKHS with reproducing kernel $k$ on a set $\Omega$. A function $f$ belongs to $\mathcal{H}$ if and only if there is a constant $C > 0$ such that*

$$\Big|\sum_{i=1}^{M} f(s_i)\, \overline{\xi}_i\Big|^2 \leq C^2 \sum_{i,j=1}^{M} k(s_i, s_j)\, \overline{\xi}_i\, \xi_j\,, \tag{4.14}$$

*for all $M \in \mathbb{N}$, where $\xi_1, \xi_2, \cdots, \xi_M \in \mathbb{C}$ and $s_1, s_2, \cdots, s_M \in \Omega$.*

*Moreover, in this case $\|f\| = \inf\{C\}$, where the infimum is taken over all the constants $C$ satisfying* (4.14).

***Proof*** The necessary condition has been proved in (4.13). Next we prove the sufficient condition. For a fixed $f \in \mathcal{H}$, in the dense subspace $\mathcal{H}_0$ of $\mathcal{H}$ given in (4.12) we define a linear functional as

$$\Phi\Big(\sum_i \overline{\xi}_i k_{t_i}\Big) := \sum_i \overline{\xi}_i \overline{f(t_i)}\,, \quad t_i \in \Omega\,,\ \xi_i \in \mathbb{C}\,,$$

where the index $i$ runs over a finite set of $\mathbb{N}$. Using (4.14) we obtain that $\Phi$ is bounded and $\|\Phi\| \leq C$. Then, extending the functional $\Phi$ to $\mathcal{H}$, from the Riesz representation theorem there exists a unique $g \in \mathcal{H}$ such that $\Phi(h) = \langle h, g\rangle$ for each $h \in \mathcal{H}$ and $\|\Phi\| = \|g\|$. Hence we obtain

$$\overline{f(t)} = \Phi(k_t) = \langle k_t, g\rangle = \overline{g(t)} \quad \text{for all } t \in \Omega\,,$$

that is, $f = g$ and $f$ belongs to $\mathcal{H}$. Obviously, by using (4.13) the statement for $\|f\|$ holds. □

This leads to the following result on two positive matrices $k_1$ and $k_2$ where $k_1 \ll k_2$ means that the function $k_2 - k_1$ is a positive matrix too.

**Corollary 4.2** *Let $k_1$ and $k_2$ be two positive matrices on $\Omega$ and $\|\cdot\|_{k_1}$ and $\|\cdot\|_{k_2}$ be the norms in their associated RKHSs. Then*

$$c^2 k_1 \ll k_2 \quad \textit{if and only if} \quad c\,\|\cdot\|_{k_2} \le \|\cdot\|_{k_1}\,,$$

*with some $c > 0$.*

***Proof*** Suppose $c^2 k_1 \ll k_2$. Then

$$\Big|\sum_{i=1}^{M} f(s_i)\overline{\xi}_i\Big|^2 \le \|f\|_{k_1}^2 \sum_{i,j=1}^{M} k_1(s_i, s_j)\overline{\xi}_i\,\xi_j \le \|f\|_{k_1}^2 \sum_{i,j=1}^{M} c^{-2}k_2(s_i, s_j)\overline{\xi}_i\,\xi_j\,,$$

and the RKHS test gives us $c\,\|f\|_{k_2} \le \|f\|_{k_1}$.

Suppose the second condition holds. Then $B_{k_1}(1) \subset B_{k_2}(1/c)$ where $B$ stands for a ball of center at 0 in a corresponding space. Thus, denoting $k_{1,t}(\cdot) = k_1(\cdot, t)$ and $k_{2,t}(\cdot) = k_2(\cdot, t)$, we have

$$\begin{aligned}
\sum_{i,j=1}^{M} \overline{\xi}_j\,\xi_i\,k_1(t_j, t_i) = \Big\|\sum_{i=1}^{M} \xi_i\,k_{1,t_i}\Big\|_{k_1}^2 &= \sup_{f\in B_{k_1}(1)} \Big\{\Big|\Big\langle f, \sum_{i=1}^{M} \xi_i\,k_{1,t_i}\Big\rangle_{k_1}\Big|^2\Big\} \\
&= \sup_{f\in B_{k_1}(1)} \Big\{\Big|\sum_{i=1}^{M} \overline{\xi}_i\,f(t_i)\Big|^2\Big\} = \sup_{f\in B_{k_1}(1)} \Big\{\Big|\sum_{i=1}^{M} \overline{\xi}_i\,\langle f, k_{2,t_i}\rangle_{k_2}\Big|^2\Big\} \\
&\le \sup_{f\in B_{k_2}(1/c)} \Big\{\Big|\sum_{i=1}^{M} \overline{\xi}_i\,\langle f, k_{2,t_i}\rangle_{k_2}\Big|^2\Big\} = c^{-2}\Big\|\sum_{i=1}^{M} \xi_i\,k_{2,t_i}\Big\|_{k_2}^2 \\
&= c^{-2} \sum_{i,j=1}^{M} \overline{\xi}_j\,\xi_i\,k_2(t_j, t_i)\,,
\end{aligned}$$

and $c^2 k_1 \ll k_2$. □

### 4.1.5 A Converse of the Kramer Theorem

The method proposed in Sect. 4.1.2 is limited to those linear transforms $\mathcal{T}_K$ whose Kramer kernel $K$, evaluated at each $t \in \Omega$, can be written as

$$K(t) = \sum_{n=1}^{\infty} \overline{S_n(t)}\, y_n\,, \tag{4.15}$$

where $S_n(t) = \langle x_n, K(t)\rangle$, with $\{x_n\}_{n=1}^{\infty}$ and $\{y_n\}_{n=1}^{\infty}$ a pair of dual Riesz bases for $\mathcal{H}$, and the interpolation property

$$S_n(t_m) = a_n \delta_{n,m}\,, \quad \{a_n\}_{n=1}^{\infty} \subset \mathbb{C} \setminus \{0\}\,, \tag{4.16}$$

is fulfilled.

However, we prove that, under plausible hypotheses, the kernel $K$ adopts the required form. Namely, consider a linear transform $\mathcal{T}_K$ as in (4.3), and let $\mathcal{H}_K$ be its range space, a RKHS whose reproducing kernel is given by (4.10), i.e., $k(t,s) = \langle K(s), K(t)\rangle_{\mathcal{H}}$ for $t, s \in \Omega$.

Assume also that $\mathcal{T}_K$ is one-to-one, although as we will point out at the end of the section, this assumption can be dropped so as to get a similar result.

Under these circumstances we prove the following converse result which was originally proved in Ref. [22] as a converse result of the classical Kramer sampling theorem.

**Theorem 4.4** *Let $\mathcal{H}_K$ be the range of a linear transform $\mathcal{T}_K$ as in* (4.3) *considered as a RKHS with reproducing kernel* (4.10). *Let $\{S_n\}_{n=1}^{\infty}$ be a sequence in $\mathcal{H}_K$ such that $\big\{S_n(t)\big\}_{n=1}^{\infty}$ belongs to $\ell^2(\mathbb{N})$ for each $t \in \Omega$. Suppose that the following conditions are fulfilled:*

*(i) $\sum_{n=1}^{\infty} \alpha_n S_n(t) = 0$ for all $t \in \Omega$ and $\{\alpha_n\}_{n=1}^{\infty}$ in $\ell^2(\mathbb{N})$ implies $\alpha_n = 0$ for all $n$.*
*(ii) There exist sequences $\{t_n\}_{n=1}^{\infty}$ in $\Omega$ and $\{a_n\}_{n=1}^{\infty}$ in $\mathbb{C} \setminus \{0\}$ such that*

$$\left\{\frac{f(t_n)}{a_n}\right\}_{n=1}^{\infty} \in \ell^2(\mathbb{N}) \quad \text{and} \quad f(t) = \sum_{n=1}^{\infty} f(t_n)\frac{S_n(t)}{a_n} \quad \text{for any } f \in \mathcal{H}_K,$$

*where the sampling series is pointwise convergent in $\Omega$.*

*Then, the sequence* $\{S_n\}_{n=1}^{\infty}$ *is a Riesz basis for* $\mathcal{H}_K$ *and the kernel* $K$ *of the linear transform* $\mathcal{T}_K$ *evaluated at* $t \in \Omega$ *can be expressed as*

$$K(t) = \sum_{n=1}^{\infty} \overline{S_n(t)}\, y_n\,,$$

*where* $\{y_n\}_{n=1}^{\infty}$ *is the dual basis of the Riesz basis* $\{x_n = \mathcal{T}_K^{-1}(S_n)\}_{n=1}^{\infty}$ *in* $\mathcal{H}$.

***Proof*** By defining

$$\widetilde{k}(t,s) := \sum_{n=1}^{\infty} S_n(t)\,\overline{S_n(s)}\,, \quad t, s \in \Omega\,,$$

we obtain a positive matrix. By using the Moore-Aronszajn procedure outlined in Sect. 4.1.4, the positive matrix $\widetilde{k}$ defines a RKHS $\widetilde{\mathcal{H}}$ such that $\widetilde{\mathcal{H}} \subseteq \mathcal{H}_K$. □

Firstly, by using the RKHS test in Proposition 4.2, we prove some results concerning the RKHS $\widetilde{\mathcal{H}}$ adapted from Ref. [44]:

**Lemma 4.1** *In the RKHS* $\widetilde{\mathcal{H}}$, *the following statements hold:*

- *For each sequence* $\mathbf{a} = \{a_n\}_{n=1}^{\infty}$ *in* $\ell^2(\mathbb{N})$, *the function* $f_{\mathbf{a}}(t) = \sum_{n=1}^{\infty} a_n\, S_n(t)$, $t \in \Omega$, *belongs to* $\widetilde{\mathcal{H}}$. *In particular, for each* $n \in \mathbb{N}$, *the function* $S_n$ *belongs to* $\widetilde{\mathcal{H}}$.
- *For each sequence* $\mathbf{a} = \{a_n\}_{n=1}^{\infty}$ *in* $\ell^2(\mathbb{N})$, *the series* $\sum_{n=1}^{\infty} a_n\, S_n$ *converges to* $f_{\mathbf{a}}$ *in* $\widetilde{\mathcal{H}}$*-norm, and* $\{S_n\}_{n=1}^{\infty}$ *is a Bessel sequence for* $\widetilde{\mathcal{H}}$. *In particular, denoting* $\widetilde{k}_t(s) = \widetilde{k}(s,t)$ *for each fixed* $t \in \Omega$, *we obtain that*

$$\widetilde{k}_t = \sum_{n=1}^{\infty} \overline{S_n(t)}\, S_n \quad \text{in } \widetilde{\mathcal{H}}\text{-norm}\,. \tag{4.17}$$

***Proof*** For $M \in \mathbb{N}$, we take $M$ scalars $\xi_1, \xi_2, \cdots, \xi_M$ in $\mathbb{C}$ and $M$ points $s_1, s_2, \cdots, s_M$ in $\Omega$. We get

$$\Big|\sum_{i=1}^{M}\Big(\sum_{n=1}^{\infty} a_n\, S_n(s_i)\Big)\overline{\xi_i}\Big|^2 = \Big|\sum_{n=1}^{\infty} a_n \Big(\sum_{i=1}^{M} S_n(s_i)\,\overline{\xi_i}\Big)\Big|^2$$
$$\le \sum_{n=1}^{\infty} |a_n|^2\, \Big\|\sum_{i=1}^{M} S_n(s_i)\,\overline{\xi_i}\Big\|^2_{\ell^2(\mathbb{N})} = \|\mathbf{a}\|^2_{\ell^2(\mathbb{N})} \sum_{i,j=1}^{M} \overline{\xi_i}\,\xi_j\,\widetilde{k}(s_i, s_j)\,,$$

where we have used the Cauchy-Schwarz inequality and the definition of $\widetilde{k}$. By using Proposition 4.2 the function $f_{\mathbf{a}}$ belongs to $\widetilde{\mathcal{H}}$; in particular, any function $S_n \in \widetilde{\mathcal{H}}$.

For the second statement, we define $f_{\mathbf{a},N} = f_{\mathbf{a}} - \sum_{n=1}^{N} a_n\, S_n$. As before, we get

$$\Big|\sum_{i=1}^{M}\overline{\xi}_i\, f_{\mathbf{a},N}(s_i)\Big|^2 = \Big|\sum_{n=N+1}^{\infty} a_n \sum_{i=1}^{M}\overline{\xi}_i\, S_n(s_i)\Big|^2$$
$$\leq\Big(\sum_{n=N+1}^{\infty}|a_n|^2\Big)\Big(\sum_{n=1}^{\infty}\Big|\sum_{i=1}^{M}\overline{\xi}_i\, S_n(s_i)\Big|^2\Big)$$
$$=\Big(\sum_{n=N+1}^{\infty}|a_n|^2\Big)\Big(\sum_{i,j=1}^{M}\overline{\xi}_i\,\xi_j\,\widetilde{k}(s_i,s_j)\Big).$$

Using again Proposition 4.2 we obtain

$$\|f_{\mathbf{a},N}\|^2_{\widetilde{\mathcal{H}}} \leq \sum_{n=N+1}^{\infty}|a_n|^2 \to 0 \quad \text{as } N\to\infty\,.$$

From Proposition 1.2, the sequence $\{S_n\}_{n=1}^{\infty}$ is a Bessel sequence for $\widetilde{\mathcal{H}}$.

In particular, the above proof shows that $\widetilde{k}_t = \sum_{n=1}^{\infty}\overline{S_n(t)}\, S_n$ in $\widetilde{\mathcal{H}}$-norm. □

Coming back to our theorem, we prove that condition $(i)$ implies that the sequence $\{S_n\}_{n=1}^{\infty}$ is an orthonormal basis for $\widetilde{\mathcal{H}}$. In fact, both are equivalent conditions. Using (4.17) and the reproducing property in $\widetilde{\mathcal{H}}$ we obtain

$$\overline{S_m(t)} = \langle \widetilde{k}_t, S_m\rangle_{\widetilde{\mathcal{H}}} = \sum_{n=1}^{\infty}\overline{S_n(t)}\,\langle S_n, S_m\rangle_{\widetilde{\mathcal{H}}}\,.$$

For any fixed $m$, the sequence $\big\{\langle S_n, S_m\rangle_{\widetilde{\mathcal{H}}}\big\}_{n\in\mathbb{N}}$ belongs to $\ell^2(\mathbb{N})$ since $\{S_n\}_{n=1}^{\infty}$ is a Bessel sequence for $\widetilde{\mathcal{H}}$. Thus, condition in $(i)$ implies that $\langle S_n, S_m\rangle_{\widetilde{\mathcal{H}}} = \delta_{n,m}$.

For the completeness of the sequence $\{S_n\}_{n=1}^{\infty}$, suppose that $\langle S_n, f\rangle_{\widetilde{\mathcal{H}}} = 0$ for all $n\in\mathbb{N}$. Hence, $0 = \sum_{n=1}^{\infty}\overline{S_n(t)}\,\langle S_n, f\rangle_{\widetilde{\mathcal{H}}} = \langle \widetilde{k}_t, f\rangle_{\widetilde{\mathcal{H}}}$ for each $t\in\Omega$. By using the reproducing property in $\widetilde{\mathcal{H}}$ we obtain $f = 0$ in $\Omega$.

Now we prove that $\widetilde{\mathcal{H}} = \mathcal{H}_K$ and that the identity mapping $\widetilde{\mathcal{H}} \hookrightarrow \mathcal{H}_K$ is continuous. Take $f\in\mathcal{H}_K$, by condition $(ii)$ the sequence $\big\{f(t_n)a_n^{-1}\big\}_{n=1}^{\infty}$ is in $\ell^2(\mathbb{N})$. As a consequence, the series $\sum_{n=1}^{\infty} f(t_n)\, a_n^{-1} S_n$ converges in the norm of $\widetilde{\mathcal{H}}$.

By the reproducing kernel property, we have that the series $\sum_{n=1}^{\infty} f(t_n)\, a_n^{-1} S_n$ is pointwise convergent. Comparing this with what we get from the sampling formula for $f$ we deduce that

$$f = \sum_{n=1}^{\infty} f(t_n)\, a_n^{-1} S_n\,,$$

where the convergence is in $\widetilde{\mathcal{H}}$ and, consequently, $f \in \widetilde{\mathcal{H}}$.

Next we show the continuity of the identity mapping by application of the closed graph theorem.

Indeed, let $\{f_n\}_{n=1}^{\infty}$ be a sequence such that $f_n \to f$ in $\widetilde{\mathcal{H}}$ and $f_n \to g$ in $\mathcal{H}_K$ as $n \to \infty$. Using the reproducing property in both $\mathcal{H}_K$ and $\widetilde{\mathcal{H}}$, for $t \in \Omega$ we have

$$|f_n(t) - f(t)| \le \|f_n - f\|_{\widetilde{\mathcal{H}}} \sqrt{\widetilde{k}(t,t)}$$

$$|f_n(t) - g(t)| \le \|f_n - g\|_{\mathcal{H}_K} \sqrt{k(t,t)}\,,$$

and therefore, $\lim_{n\to\infty} f_n(t) = f(t) = g(t)$ for each $t \in \Omega$, and hence $f = g$.

Since it is also surjective, we infer that the norms $\|\cdot\|_{\mathcal{H}_K}$ and $\|\cdot\|_{\widetilde{\mathcal{H}}}$ are equivalent from the open mapping theorem. As a consequence, the orthonormal basis $\{S_n\}_{n=1}^{\infty}$ in $\widetilde{\mathcal{H}}$ is a Riesz basis for $\mathcal{H}_K$.

Finally, consider $\{x_n = \mathcal{T}_K^{-1}(S_n)\}_{n=1}^{\infty}$, a Riesz basis in $\mathcal{H}$, and denote by $\{y_n\}_{n=1}^{\infty}$ its biorthonormal basis. Expanding $K(t)$ with respect to $\{y_n\}_{n=1}^{\infty}$, for each fixed $t \in \Omega$, we obtain

$$K(t) = \sum_{n=1}^{\infty} \langle K(t), x_n\rangle_{\mathcal{H}}\, y_n = \sum_{n=1}^{\infty} \overline{\langle \mathcal{T}_K^{-1}(S_n), K(t)\rangle}\, y_n$$

$$= \sum_{n=1}^{\infty} \overline{S_n(t)}\, y_n\,,$$

i.e., the required expansion for $K(t)$.

Observe that the interpolation property (4.16) comes out of a direct application of condition $(ii)$ to $S_n$, followed by condition $(i)$. □

Notice that the equivalence of the norms $\|\cdot\|_{\mathcal{H}_K}$ and $\|\cdot\|_{\widetilde{\mathcal{H}}}$ implies the existence of two constants $0 < a \le b$ such that $a\|f\|_{\widetilde{\mathcal{H}}} \le \|f\|_{\mathcal{H}_K} \le b\|f\|_{\widetilde{\mathcal{H}}}$ for all $f \in \widetilde{\mathcal{H}} = \mathcal{H}_K$. Having in mind Corollary 4.2, this fact is equivalent to

$$a^2\, k \ll \widetilde{k} \ll b^2\, k\,.$$

If $a = b$, then $a^2 k = \widetilde{k}$ and the transform $\mathcal{T}_K$ is an isometry (up to the positive factor $a$) between $\mathcal{H}$ and $\widetilde{\mathcal{H}}$. As a consequence, the sequence $\{x_n = \mathcal{T}_K^{-1}(S_n)\}_{n=1}^{\infty}$ is an orthogonal basis for $\mathcal{H}$.

As to the case when, a priori, $\mathcal{T}_K$ is not known to be one-to-one, let $\{x_n\}_{n=1}^{\infty}$ be a sequence in $\mathcal{H}$ with $P(x_n) \neq 0$ for all $n$, where $P$ denotes the orthogonal projection onto the closed subspace $(\operatorname{Ker} \mathcal{T}_K)^{\perp}$. Consider $S_n = \mathcal{T}_K(x_n) \in \mathcal{H}_K$, and suppose that these functions satisfy the hypotheses in Theorem 4.4. In this case, $\{S_n\}_{n=1}^{\infty}$ is a Riesz basis in $\mathcal{H}_K$. Consequently, since $S_n = \mathcal{T}_K[P(x_n)]$ and $\mathcal{T}_K|_{P(\operatorname{Ker} \mathcal{T}_K)} = 0$, we obtain that $\{P(x_n)\}_{n=1}^{\infty}$ is a Riesz basis in $P(\mathcal{H}) = (\operatorname{Ker} \mathcal{T}_K)^{\perp}$. The result comes out taking into account the orthogonal sum $\mathcal{H} = (\operatorname{Ker} \mathcal{T}_K)^{\perp} \oplus (\operatorname{Ker} \mathcal{T}_K)$.

### 4.1.6 Sampling Through Samples of a Related Function

Throughout this section the sequences $\{x_n\}_{n=1}^{\infty} \cup \{y_n\}_{n=1}^{\infty}$ and $\{x_n^*\}_{n=1}^{\infty} \cup \{y_n^*\}_{n=1}^{\infty}$ denote a pair of dual Riesz bases for a separable Hilbert space $\mathcal{H}$. Consider $\mathcal{H}_K$, the range space of a linear transform $\mathcal{T}_K$ as in (4.3) with kernel $K$.

Assume that, for each fixed $t \in \Omega$, we have the following expansion for $K(t)$

$$K(t) = \sum_{n=1}^{\infty} \overline{S_n(t)}\, x_n^* + \sum_{n=1}^{\infty} \overline{T_n(t)}\, y_n^* ,$$

where the functions $S_n, T_n : \Omega \mapsto \mathbb{C}$ satisfy that there exists a sequence $\{t_n\}_{n=1}^{\infty}$ in $\Omega$ such that

$$S_n(t_m) = a_n\, \delta_{n,m}\, ; \quad T_n(t_m) = b_n\, \delta_{n,m} \quad \text{with } \{a_n\}_{n=1}^{\infty}, \{b_n\}_{n=1}^{\infty} \subset \mathbb{C}. \tag{4.18}$$

Assume also that there exists another $\mathcal{H}$-valued kernel $\widetilde{K}$ defining the linear transform $\mathcal{T}_{\widetilde{K}}$ in $\mathcal{H}$ such that for each $x \in \mathcal{H}$, $\mathcal{T}_{\widetilde{K}}(x) = \widetilde{f}$, where $\widetilde{f}(t) := \langle x, \widetilde{K}(t)\rangle_{\mathcal{H}}$ for $t \in \Omega$.

Now, for each fixed $t \in \Omega$, assume the following expansion for $\widetilde{K}(t)$

$$\widetilde{K}(t) = \sum_{n=1}^{\infty} \overline{U_n(t)}\, x_n^* + \sum_{n=1}^{\infty} \overline{V_n(t)}\, y_n^* ,$$

where the functions $U_n, V_n : \Omega \mapsto \mathbb{C}$ satisfy, for the sequence $\{t_n\}_{n=1}^{\infty}$ in $\Omega$, that

$$U_n(t_m) = c_n\, \delta_{n,m}\, ; \quad V_n(t_m) = d_n\, \delta_{n,m} \quad \text{with } \{c_n\}_{n=1}^{\infty}, \{d_n\}_{n=1}^{\infty} \subset \mathbb{C}. \tag{4.19}$$

In addition, in (4.18) and (4.19) we also assume that

$$\Delta_n := a_n\, d_n - b_n\, c_n \neq 0 \quad \text{for all } n \in \mathbb{N}. \tag{4.20}$$

Finally, suppose that the linear transforms $\mathcal{T}_K$ and $\mathcal{T}_{\widetilde{K}}$ are related in the sense that their null spaces satisfy

$$\operatorname{Ker}\mathcal{T}_K \subseteq \operatorname{Ker}\mathcal{T}_{\widetilde{K}}\,.$$

The above condition implies that there exists a unique function $\widetilde{f}$ associated with each $f \in \mathcal{H}_K$.

Indeed, suppose that $\mathcal{T}_{\widetilde{K}}(x_1) = \widetilde{f}_1$ and $\mathcal{T}_{\widetilde{K}}(x_2) = \widetilde{f}_2$ with $\mathcal{T}_K(x_1) = \mathcal{T}_K(x_2) = f$. Then, $x_1 - x_2 \in \operatorname{Ker}\mathcal{T}_K \subseteq \operatorname{Ker}\mathcal{T}_{\widetilde{K}}$ which implies that $\mathcal{T}_{\widetilde{K}}(x_1 - x_2) = \widetilde{f}_1 - \widetilde{f}_2 = 0$, i.e., $\widetilde{f}_1 = \widetilde{f}_2$.

Under the above hypotheses we prove a sampling result for $\mathcal{H}_K$ allowing to recover any function $f$ in $\mathcal{H}_K$ from the sequence of samples $\{f(t_n)\}_{n=1}^\infty \cup \{\widetilde{f}(t_n)\}_{n=1}^\infty$.

**Proposition 4.3** *Suppose that the linear transforms $\mathcal{T}_K$ and $\mathcal{T}_{\widetilde{K}}$ defined on $\mathcal{H}$ satisfy* $\operatorname{Ker}\mathcal{T}_K \subseteq \operatorname{Ker}\mathcal{T}_{\widetilde{K}}$, *and their respective kernels the properties* (4.18) *and* (4.19) *together with condition* (4.20).

*Then, $\mathcal{T}_K$ is a bijective isometry between $\mathcal{H}$ and the RKHS $\mathcal{H}_K$, and consequently, the sequence $\{S_n\}_{n=1}^\infty \cup \{T_n\}_{n=1}^\infty$ is a Riesz basis for $\mathcal{H}_K$.*

*Moreover, any function $f = \mathcal{T}_K(x)$ in $\mathcal{H}_K$, with $x \in \mathcal{H}$, can be recovered from the samples $\{f(t_n)\}_{n=1}^\infty$ and the samples $\{\widetilde{f}(t_n)\}_{n=1}^\infty$ of its related function $\widetilde{f} = \mathcal{T}_{\widetilde{K}}(x)$, by means of the sampling formula*

$$f(t) = \sum_{n=1}^{\infty}\Big[f(t_n)\frac{d_n\,S_n(t) - c_n\,T_n(t)}{\Delta_n} + \widetilde{f}(t_n)\frac{a_n\,T_n(t) - b_n\,S_n(t)}{\Delta_n}\Big], \quad t \in \Omega\,.$$

*The convergence of the series is absolute and uniform on subsets of $\Omega$ where the function $t \mapsto k(t,t)$ is bounded.*

***Proof*** Regarding that $\mathcal{T}_K$ is a bijective isometry, we prove that $\mathcal{T}_K$ is one-to-one. Indeed, suppose that $f_x = f_y$, where we are using the notation $f_x := \mathcal{T}_K(x)$. Then $f_x - f_y = f_{x-y} = 0$ and also $\widetilde{f}_{x-y} = 0$. In particular,

$$f_{x-y}(t_n) = \langle x-y, K(t_n)\rangle_{\mathcal{H}} = a_n\,\langle x-y, x_n^*\rangle_{\mathcal{H}} + b_n\,\langle x-y, y_n^*\rangle_{\mathcal{H}} = 0$$
$$\widetilde{f}_{x-y}(t_n) = \langle x-y, \widetilde{K}(t_n)\rangle_{\mathcal{H}} = c_n\,\langle x-y, x_n^*\rangle_{\mathcal{H}} + d_n\,\langle x-y, y_n^*\rangle_{\mathcal{H}} = 0$$

for all $n \in \mathbb{N}$.

Condition (4.20) and the completeness of the Riesz basis $\{x_n^*\}_{n=1}^\infty \cup \{y_n^*\}_{n=1}^\infty$ imply $x = y$. Consequently, $\mathcal{T}_K : \mathcal{H} \mapsto \mathcal{H}_K$ is a bijective isometry and

$$\big\{S_n = \mathcal{T}_K(x_n)\big\}_{n=1}^\infty \cup \big\{T_n = \mathcal{T}_K(y_n)\big\}_{n=1}^\infty$$

is a Riesz basis for $\mathcal{H}$, whose biorthonormal Riesz basis is given by the sequence

$$\{S_n^* = \mathcal{T}_K(x_n^*)\}_{n=1}^{\infty} \cup \{T_n^* = \mathcal{T}_K(y_n^*)\}_{n=1}^{\infty}.$$

Expanding $f = \mathcal{T}_K(x)$ in $\mathcal{H}$ with respect to the above Riesz basis we obtain

$$f = \sum_{n=1}^{\infty} \big[\langle f, S_n^*\rangle_{\mathcal{H}_K} S_n + \langle f, T_n^*\rangle_{\mathcal{H}_K} T_n\big]. \tag{4.21}$$

From

$$f(t_n) = \langle x, K(t_n)\rangle_{\mathcal{H}} = a_n \langle x, x_n^*\rangle_{\mathcal{H}} + b_n \langle x, y_n^*\rangle_{\mathcal{H}},$$
$$\widetilde{f}(t_n) = \langle x, \widetilde{K}(t_n)\rangle_{\mathcal{H}} = c_n \langle x, x_n^*\rangle_{\mathcal{H}} + d_n \langle x, y_n^*\rangle_{\mathcal{H}},$$

we get

$$\langle x, x_n^*\rangle_{\mathcal{H}} = \frac{d_n f(t_n) - b_n \widetilde{f}(t_n)}{\Delta_n} \quad \text{and} \quad \langle x, y_n^*\rangle_{\mathcal{H}} = \frac{a_n \widetilde{f}(t_n) - c_n f(t_n)}{\Delta_n}.$$

Since $\mathcal{T}_K$ is a unitary operator,

$$\langle x, x_n^*\rangle_{\mathcal{H}} = \langle f, S_n^*\rangle_{\mathcal{H}_K} \quad \text{and} \quad \langle x, y_n^*\rangle_{\mathcal{H}} = \langle f, T_n^*\rangle_{\mathcal{H}_K}.$$

Substituting in (4.21) and after some calculations we obtain the desired formula in Proposition 4.3. The convergence properties come out of the RKHS setting. □

Sampling formulas for $\mathcal{H}_K$ by using samples of several related functions can be obtained in a similar manner. Given $f \in \mathcal{H}_K$, a candidate to play the role of $\widetilde{f}$ is its derivative $f'$.

Next we deal with the case of the derivative $f'$. Indeed, assume that the kernel $K : \Omega \longmapsto \mathcal{H}$ has a derivative $K'$ in $\Omega$. In this case, any function $f$ in $\mathcal{H}_K$ is differentiable in $\Omega$ and its derivative is given by

$$f'(t) = \langle x, K'(t)\rangle_{\mathcal{H}}, \quad t \in \Omega.$$

If the kernel $K(t)$ admits the expansion

$$K(t) = \sum_{n=1}^{\infty} \overline{S_n(t)}\, x_n^* + \sum_{n=1}^{\infty} \overline{T_n(t)}\, y_n^*,$$

then

$$K'(t) = \sum_{n=1}^{\infty} \overline{S_n'(t)}\, x_n^* + \sum_{n=1}^{\infty} \overline{T_n'(t)}\, y_n^*.$$

Indeed, the coefficients in the expansion of $K'(t)$ with respect to the Riesz basis $\{x_n^*\}_{n=1}^\infty \cup \{y_n^*\}_{n=1}^\infty$ are $\langle K'(t), x_n\rangle_{\mathcal{H}} = \overline{S_n'(t)}$ and $\langle K'(t), y_n\rangle_{\mathcal{H}} = \overline{T_n'(t)}$ since $S_n = \mathcal{T}_K(x_n)$ and $T_n = \mathcal{T}_K(y_n)$, respectively.

Assume also that there exists a sequence $\{t_n\}_{n=1}^\infty$ in $\Omega$ such that

$$S_n(t_m) = a_n\,\delta_{n,m}\,;\quad T_n(t_m) = b_n\,\delta_{n,m}\quad \text{with}\quad \{a_n\}_{n=1}^\infty, \{b_n\}_{n=1}^\infty \subset \mathbb{C},$$

$$S_n'(t_m) = c_n\,\delta_{n,m}\,;\quad T_n'(t_m) = d_n\,\delta_{n,m}\quad \text{with}\quad \{c_n\}_{n=1}^\infty, \{d_n\}_{n=1}^\infty \subset \mathbb{C},$$

hold with $\Delta_n \neq 0$ for all $n \in \mathbb{N}$. Then, as a consequence of Proposition 4.3, we obtain the following sampling formula with derivatives for functions in $\mathcal{H}_K$.

**Corollary 4.3** *Under the above hypotheses on the kernels $K$ and $K'$, any function $f$ in the RKHS $\mathcal{H}_K$ can be recovered from its samples $\{f(t_n)\}_{n=1}^\infty$ and those of its derivative $\{f'(t_n)\}_{n=1}^\infty$ by means of the sampling formula*

$$f(t) = \sum_{n=1}^\infty \Big[ f(t_n)\frac{d_n\,S_n(t) - c_n\,T_n(t)}{\Delta_n} + f'(t_n)\frac{a_n\,T_n(t) - b_n\,S_n(t)}{\Delta_n}\Big], \quad t \in \Omega.$$

*The convergence of the series is absolute and uniform on subsets of $\Omega$ where the function $t \mapsto k(t,t)$ is bounded.*

**Example 4.7 (Sampling in $PW_\pi$ by using the Derivative Revisited)** Now we put to work this result in order to obtain, as a straightforward consequence, the sampling formula with derivatives in the Paley-Wiener space $PW_\pi$. To this end, note that the Fourier kernel

$$\begin{aligned} K : \mathbb{R} &\longrightarrow L^2[-\pi,\pi] \\ t &\longmapsto K(t)(x) = \frac{\mathrm{e}^{-itx}}{\sqrt{2\pi}}, \quad x \in [-\pi,\pi], \end{aligned}$$

is differentiable in $\mathbb{R}$ and its derivative is given by

$$\begin{aligned} K' : \mathbb{R} &\longrightarrow L^2[-\pi,\pi] \\ t &\longmapsto K'(t)(x) = -ix\frac{\mathrm{e}^{-itx}}{\sqrt{2\pi}}, \quad x \in [-\pi,\pi]. \end{aligned}$$

Now we consider the following Riesz basis in $L^2[-\pi,\pi]$ (see Chap. 2)

$$\Big\{x_n = \frac{1}{\sqrt{\pi}}\,\mathrm{e}^{-2inx}\Big\}_{n=-\infty}^{\infty} \cup \Big\{y_n = \frac{1}{\sqrt{\pi}}\,ix\,\mathrm{e}^{-2inx}\Big\}_{n=-\infty}^{\infty} \tag{4.22}$$

whose biorthonormal basis is given by

$$\left\{x_n^* = \frac{1}{\sqrt{\pi}}\left(1 - \frac{|x|}{\pi}\right)\mathrm{e}^{-2inx}\right\}_{n=-\infty}^{\infty} \cup \left\{y_n^* = \frac{-i\,\mathrm{sgn}\,x}{\pi\sqrt{\pi}}\,\mathrm{e}^{-2inx}\right\}_{n=-\infty}^{\infty}.$$

Expanding the function $K(t)$ in $L^2[-\pi, \pi]$ with respect to the first basis and using special Fourier transforms, we obtain

$$\begin{aligned}
K(t) &= \sum_{n=-\infty}^{\infty} \langle K(t), x_n^*\rangle\, x_n + \sum_{n=-\infty}^{\infty} \langle K(t), y_n^*\rangle\, y_n \\
&= \sum_{n=-\infty}^{\infty} \left\langle \frac{\mathrm{e}^{-itx}}{\sqrt{2\pi}}, \left(1 - \frac{|x|}{\pi}\right)\frac{\mathrm{e}^{-2inx}}{\sqrt{\pi}}\right\rangle x_n + \sum_{n=-\infty}^{\infty} \left\langle \frac{\mathrm{e}^{-itx}}{\sqrt{2\pi}}, \frac{-i\,\mathrm{sgn}\,x}{\pi\sqrt{\pi}}\,\mathrm{e}^{-2inx}\right\rangle y_n \\
&= \sum_{n=-\infty}^{\infty} \frac{\sqrt{2}}{2}\left(\frac{\sin\frac{\pi}{2}(t-2n)}{\frac{\pi}{2}(t-2n)}\right)^2 x_n - \sum_{n=-\infty}^{\infty} \frac{\sqrt{2}}{\pi}\,\frac{\sin^2\frac{\pi}{2}(t-2n)}{\frac{\pi}{2}(t-2n)}\, y_n\,.
\end{aligned}$$

Taking the functions

$$S_n(t) = \frac{\sqrt{2}}{2}\left(\frac{\sin\frac{\pi}{2}(t-2n)}{\frac{\pi}{2}(t-2n)}\right)^2 \quad \text{and} \quad T_n(t) = -\frac{\sqrt{2}}{\pi}\,\frac{\sin^2\frac{\pi}{2}(t-2n)}{\frac{\pi}{2}(t-2n)},$$

and the sequence $\{t_n = 2n\}_{n\in\mathbb{Z}}$, it is easy to check that

$$S_n(2m) = \frac{\sqrt{2}}{2}\,\delta_{n,m}\,; \quad T_n(2m) = 0\,,$$

$$S_n'(2m) = 0\,; \quad T_n'(2m) = -\frac{\sqrt{2}}{2}\,\delta_{n,m}\,.$$

Hence, from Corollary 4.3 we obtain again the sampling formula (2.15) in $PW_\pi$:

$$\begin{aligned}
f(t) &= \sum_{n=-\infty}^{\infty} f(2n)\left[\frac{\sin\frac{\pi}{2}(t-2n)}{\frac{\pi}{2}(t-2n)}\right]^2 + \sum_{n=-\infty}^{\infty} f'(2n)\frac{2}{\pi}\,\frac{\sin^2\frac{\pi}{2}(t-2n)}{\frac{\pi}{2}(t-2n)} \\
&= \sum_{n=-\infty}^{\infty} \left[f(2n) + (t-2n)f'(2n)\right]\left[\frac{\sin\frac{\pi}{2}(t-2n)}{\frac{\pi}{2}(t-2n)}\right]^2, \qquad t\in\mathbb{R}\,.
\end{aligned}$$

**Example 4.8 (Lagrange and Hermite-Type Interpolation Series)** Let $\{t_n\}_{n=1}^{\infty}$ be a sequence of distinct real numbers such that $\sum_{n=1}^{\infty} |t_n|^{-2} < \infty$. There exists an entire function $P(t)$ with simple zeros at the sequence $\{t_n\}_{n=1}^{\infty}$. Specifically, the function $P(t)$ is given by the *canonical product*

$$P(t) = \begin{cases} \prod_{n=1}^{\infty}(1 - \frac{t}{t_n})\exp(t/t_n) & \text{if } \sum_{n=1}^{\infty}|t_n|^{-1} = \infty\,, \\ \prod_{n=1}^{\infty}(1 - \frac{t}{t_n}) & \text{if } \sum_{n=1}^{\infty}|t_n|^{-1} < \infty\,, \end{cases}$$

whenever $t_n \neq 0$ for all $n \in \mathbb{N}$, or by

$$P(t) = \begin{cases} t\prod_{n=2}^{\infty}(1 - \frac{t}{t_n})\exp(t/t_n) & \text{if } \sum_{n=2}^{\infty}|t_n|^{-1} = \infty \\ t\prod_{n=2}^{\infty}(1 - \frac{t}{t_n}) & \text{if } \sum_{n=2}^{\infty}|t_n|^{-1} < \infty\,, \end{cases}$$

when, for instance, $t_1 = 0$ (see, for instance, Ref. [48, p. 55] for the details).

It is easy to construct a kernel $K$ such that the functions in the corresponding RKHS $\mathcal{H}_K$ can be recovered by means of an interpolation process at the fixed sequence $\{t_n\}_{n=1}^{\infty}$.

To this end, consider $\{x_n\}_{n=1}^{\infty}$ and $\{x_n^*\}_{n=1}^{\infty}$, a pair of dual Riesz bases for a separable Hilbert space $\mathcal{H}$. For $t \in \mathbb{R}$, define $K(t)$ as

$$K(t) = \sum_{n=1}^{\infty}\overline{S_n(t)}\,x_n^*\,,\quad t \in \mathbb{R}\,,\qquad \text{where}\quad S_n(t) = \frac{P(t)}{t - t_n}\,.$$

Since $S_n(t_m) = P'(t_n)\delta_{n,m}$, according to the abstract Kramer theorem, any function $f$ defined by $f(t) = \langle x, K(t)\rangle_{\mathcal{H}}$ for $t \in \mathbb{R}$, where $x \in \mathcal{H}$, can be expanded as the *Lagrange-type interpolation series*

$$f(t) = \sum_{n=1}^{\infty} f(t_n)\frac{P(t)}{(t - t_n)P'(t_n)}\,,\qquad t \in \mathbb{R}\,.$$

For an account of Lagrange interpolation and sampling results, see Ref. [53].

Next, we show that we can obtain a RKHS $\mathcal{H}_K$ such that any function $f$ in $\mathcal{H}_K$ can be recovered from the samples $\{f(t_n)\}_{n=1}^{\infty}$ and $\{f'(t_n)\}_{n=1}^{\infty}$, taken at the above sequence $\{t_n\}_{n=1}^{\infty}$ in $\mathbb{R}$, by means of a *Hermite-type interpolation series.*

Indeed, consider the function $Q(t) = P^2(t)$, which has double zeros at $\{t_n\}_{n=1}^{\infty}$, and take, for $n \in \mathbb{N}$, the functions

$$S_n(t) = \frac{Q(t)}{(t - t_n)^2}\quad \text{and}\quad T_n(t) = \frac{Q(t)}{t - t_n}\,,\qquad t \in \mathbb{R}\,.$$

Define the kernel $K$ as

$$K(t)=\sum_{n=1}^{\infty}\Big[\frac{Q(t)}{(t-t_n)^2}\,x_n^*+\frac{Q(t)}{t-t_n}\,y_n^*\Big],\qquad t\in\mathbb{R},$$

where $\{x_n^*\}_{n=1}^{\infty}\cup\{y_n^*\}_{n=1}^{\infty}$ denotes the biorthogonal Riesz basis of a Riesz basis $\{x_n\}_{n=1}^{\infty}\cup\{y_n\}_{n=1}^{\infty}$ for a separable Hilbert space $\mathcal{H}$ (this basis might be obtained by partitioning a given Riesz basis into two arbitrary sequences).

It is easy to check the interpolation properties:

$$S_n(t_m)=\frac{Q''(t_n)}{2}\,\delta_{n,m}\,;\qquad T_n(t_m)=0\,,$$

$$S_n'(t_m)=\frac{Q'''(t_n)}{6}\,\delta_{n,m}\,;\qquad T_n'(t_m)=\frac{Q''(t_n)}{2}\,\delta_{n,m}\,.$$

Taking into account that $Q''(t_n)\neq 0$ for all $n\in\mathbb{N}$, any function $f\in\mathcal{H}_K$ can be expanded, following Proposition 4.3, as the *Hermite-type interpolation series*

$$f(t)=\sum_{n=1}^{\infty}\Big\{f(t_n)\Big[1-\frac{Q'''(t_n)}{3Q''(t_n)}(t-t_n)\Big]+f'(t_n)(t-t_n)\Big\}\frac{2Q(t)}{Q''(t_n)(t-t_n)^2}\,,\qquad t\in\mathbb{R}\,.$$

Taking the sampling points $\{t_n=2n\}_{n\in\mathbb{Z}}$ and the Riesz basis (4.22) in $L^2[-\pi,\pi]$, the above construction leads to the derivative sampling in $PW_\pi$ stated in Example 4.7.

## 4.2 Analytic Version of the Kramer Sampling Theorem

Suppose that $\Omega=\mathbb{C}$ and let $K$ be an analytic $\mathcal{H}$-valued function defined on $\mathbb{C}$ where $\mathcal{H}$ denotes a separable complex Hilbert space. Set $f_x(z):=\langle K(z),x\rangle_{\mathcal{H}}$, $z\in\mathbb{C}$, and denote by $\mathcal{H}_K$ the collection of all such functions $f_x$, $x\in\mathcal{H}$. We place $K$ in the first argument of the inner product in order to transfer the analyticity of $K$ to the functions of $\mathcal{H}_K$.

In fact, the kernel $K$ is an analytic $\mathcal{H}$-valued function defined on $\mathbb{C}$ if and only if every $f_x\in\mathcal{H}_K$ is an entire function (see Ref. [45, p. 266]).

As in Sect. 4.1.2, the space $\mathcal{H}_K$ is a reproducing kernel Hilbert space. In this case, the reproducing kernel is

$$k(z,w)=\big\langle K(z),K(w)\big\rangle_{\mathcal{H}},\qquad z,w\in\mathbb{C}\,.$$

Notice that the mapping $\mathcal{T}_K$ given by

$$\mathcal{H} \ni x \overset{\mathcal{T}_K}{\longmapsto} f_x \in \mathcal{H}_K \tag{4.23}$$

is an anti-linear mapping from $\mathcal{H}$ onto $\mathcal{H}_K$. Thus, the space $\mathcal{H}_K$ is an RKHS of entire functions.

Another characterization of the analyticity of the functions in $\mathcal{H}_K$ is given in terms of Riesz bases by using the *Montel theorem* in complex analysis. Indeed, suppose that a Riesz basis $\{x_n\}_{n=1}^\infty$ for $\mathcal{H}$ is given and let $\{y_n\}_{n=1}^\infty$ be its dual Riesz basis; expanding $K(z)$, for each fixed $z \in \mathbb{C}$, with respect to the basis $\{x_n\}_{n=1}^\infty$ we obtain

$$K(z) = \sum_{n=1}^\infty \langle K(z), y_n\rangle_{\mathcal{H}}\, x_n\,, \quad z \in \mathbb{C}\,,$$

where the coefficients $\langle K(z), y_n\rangle_{\mathcal{H}}$, as functions in $z$, belong to $\mathcal{H}_K$.

**Lemma 4.2** *The space $\mathcal{H}_K$ is an RKHS of entire functions if and only if for each $n \in \mathbb{N}$ the function*

$$S_n(z) := \langle K(z), y_n\rangle_{\mathcal{H}}, \quad z \in \mathbb{C}\,, \tag{4.24}$$

*is entire, and the function $z \mapsto \|K(z)\|_{\mathcal{H}}$ is bounded on compact subsets of $\mathbb{C}$.*

***Proof*** The necessary condition is obvious. The sufficient condition is a straightforward consequence of Montel's theorem. Indeed, $K$ will be analytic in $\mathbb{C}$ if and only if the function $f(z) := \langle K(z), x\rangle_{\mathcal{H}}$ is analytic in $\mathbb{C}$ for each $x \in \mathcal{H}$. By using the continuity of the inner product we get

$$f(z) = \langle K(z), x\rangle_{\mathcal{H}} = \langle \sum_{n=1}^\infty S_n(z)\, x_n, x\rangle_{\mathcal{H}} = \sum_{n=1}^\infty S_n(z)\, \langle x_n, x\rangle_{\mathcal{H}}\,. \tag{4.25}$$

Applying the Cauchy-Schwarz inequality we get

$$\begin{aligned}\Big|\sum_{n=1}^N S_n(z)\, \langle x_n, x\rangle_{\mathcal{H}}\Big|^2 &\le \Big(\sum_{n=1}^N |S_n(z)|^2\Big)\Big(\sum_{n=1}^N |\langle x_n, x\rangle_{\mathcal{H}}|^2\Big)\\ &\le C \sum_{n=1}^\infty |S_n(z)|^2\, \|x\|^2\,,\end{aligned}$$

since $\{x_n\}_{n=1}^{\infty}$ is, in particular, a Bessel sequence. The Riesz basis condition for $\{y_n\}_{n=1}^{\infty}$ implies that there exist two constants $0 < A \le B$ such that

$$A\|K(z)\|_{\mathcal{H}}^2 \le \sum_{n=1}^{\infty} |S_n(z)|^2 \le B\|K(z)\|_{\mathcal{H}}^2 .$$

As a consequence, the partial sums in the series of (4.25) are uniformly bounded on compact sets of $\mathbb{C}$. Consequently, Montel's theorem assures the existence of a subsequence which will converge to an entire function which necessarily coincides with $f$. Hence, $f$ is an entire function. □

### 4.2.1 *Analytic Kramer Kernels*

According to Definition 4.1 the analytic kernel $K : \mathbb{C} \longrightarrow \mathcal{H}$ is said to be an *analytic Kramer kernel* with respect to the data $\{z_n\}_{n=1}^{\infty} \in \mathbb{C}$ and $\{a_n\}_{n=1}^{\infty} \in \mathbb{C} \setminus \{0\}$ if it satisfies $K(z_n) = a_n\, x_n$, $n \in \mathbb{N}$, for some Riesz basis $\{x_n\}_{n=1}^{\infty}$ for $\mathcal{H}$.

An analytic kernel $K$ is an analytic Kramer one if and only if the sequence of functions $\{S_n\}_{n=1}^{\infty}$ in $\mathcal{H}_K$ given by (4.24), where $\{y_n\}_{n=1}^{\infty}$ is the dual Riesz basis of $\{x_n\}_{n=1}^{\infty}$, satisfies the interpolation property $S_n(z_m) = a_n\, \delta_{n,m}$, $m, n \in \mathbb{N}$.

As a consequence, $\lim_{n\to\infty} |z_n| = +\infty$; otherwise the sequence $\{z_n\}_{n=1}^{\infty}$ contains a bounded subsequence and hence, according to the *isolated zeros principle*, the entire function $S_n \equiv 0$ for all $n \in \mathbb{N}$ which contradicts that $S_n(z_n) = a_n$ for each $n \in \mathbb{N}$.

Under the notation introduced so far, a slight modification of Theorem 4.2 holds:

**Theorem 4.5 (Analytic Kramer Sampling Theorem)** *Let $K : \mathbb{C} \longrightarrow \mathcal{H}$ be an analytic Kramer kernel with respect to the data $\{z_n\}_{n=1}^{\infty} \subset \mathbb{C}$ and $\{a_n\}_{n=1}^{\infty} \subset \mathbb{C} \setminus \{0\}$. Let $\mathcal{H}_K$ be the corresponding RKHS of entire functions. Then any $f \in \mathcal{H}_K$ can be recovered from its samples $\{f(z_n)\}_{n=1}^{\infty}$ by means of the sampling series*

$$f(z) = \sum_{n=1}^{\infty} f(z_n)\, \frac{S_n(z)}{a_n}, \quad z \in \mathbb{C}, \tag{4.26}$$

*where the reconstruction functions $S_n$ are given in* (4.24). *The series converges absolutely and uniformly on compact subsets of* $\mathbb{C}$.

***Proof*** The anti-linear mapping $\mathcal{T}_K$ given by (4.23) is a bijective isometry between $\mathcal{H}$ and $\mathcal{H}_K$. Due to the anti-linearity, we have that $\langle f, g\rangle_{\mathcal{H}_K} = \overline{\langle x, y\rangle_{\mathcal{H}}}$ in case $f = \mathcal{T}_K(x)$ and $g = \mathcal{T}_K(y)$.

As a consequence, the functions $\{S_n = \mathcal{T}_K(y_n)\}_{n=1}^{\infty}$ form a Riesz basis for $\mathcal{H}_K$; let $\{T_n\}_{n=1}^{\infty}$ be its dual Riesz basis. Expanding any $f \in \mathcal{H}_K$ in this basis we obtain $f = \sum_{n=1}^{\infty}\langle f, T_n\rangle_{\mathcal{H}_K} S_n$ in $\mathcal{H}_K$. Moreover,

$$\langle f, T_n\rangle_{\mathcal{H}_K} = \overline{\langle x, x_n\rangle_{\mathcal{H}}} = \Big\langle \frac{K(z_n)}{a_n}, x\Big\rangle_{\mathcal{H}} = \frac{f(z_n)}{a_n}, \quad n \in \mathbb{N}.$$

Since a Riesz basis is an unconditional basis, the sampling series will be pointwise unconditionally convergent and, hence, absolutely convergent. The uniform convergence comes from the RKHS theory since the function $z \mapsto \|K(z)\|_{\mathcal{H}}$ is bounded on compact subsets of $\mathbb{C}$. □

Concerning the existence of analytic Kramer kernels we have the following result:

**Proposition 4.4** *Given an arbitrary sequence of complex numbers $\{z_n\}_{n=1}^{\infty}$ such that $\lim_{n\to\infty}|z_n| = +\infty$, there exists an analytic Kramer kernel $K$ having the sequence $\{z_n\}_{n=1}^{\infty}$ as the data sampling points.*

***Proof*** Given the sequence $\{z_n\}_{n=1}^{\infty}$, consider a sequence $\{a_n\}_{n=1}^{\infty}$ in $\mathbb{C}\setminus\{0\}$ such that $\sum_{z_n\neq 0}|a_n/z_n|^2 < \infty$ (set $a_k = 1$ in the case $z_k = 0$). Now, let $P$ be an entire function having only simple zeros at $\{z_n\}_{n=1}^{\infty}$. This is allowed by the *Weierstrass theorem*; see Ref. [48, p. 54].

Now, let $\mathcal{H}$ be a separable Hilbert space with dual Riesz bases $\{x_n\}_{n=1}^{\infty}$ and $\{y_n\}_{n=1}^{\infty}$, and define the mapping $K : \mathbb{C} \longrightarrow \mathcal{H}$ as:

$$K(z) := \sum_{n=1}^{\infty}\frac{a_n P(z)}{z - z_n}\, x_n \in \mathcal{H}.$$

Then, the kernel $K$ defines an analytic Kramer kernel.

To this end, for each $n \in \mathbb{N}$ the entire function $S_n(z) = \dfrac{a_n P(z)}{z - z_n} = \langle K(z), y_n\rangle_{\mathcal{H}}$, $z \in \mathbb{C}$ satisfies the interpolation property $S_n(z_m) = a_m P'(z_m)\delta_{n,m}$.

In addition, the function $z \mapsto \|K(z)\|_{\mathcal{H}}$ is uniformly bounded on compact subsets of $\mathbb{C}$. The Riesz basis condition on $\{y_n\}_{n=1}^{\infty}$ gives the existence of a constant $A > 0$ such that

$$\|K(z)\|_{\mathcal{H}}^2 \le A\sum_{n=1}^{\infty}\left|\frac{a_n P(z)}{z - z_n}\right|^2, \quad z \in \mathbb{C},$$

and the series is uniformly bounded on compact subsets. Indeed, given a compact $K$ in $\mathbb{C}$ there exists a closed disk $\Delta_R$ centered at the origin with radius $R > 0$ such that $K \subseteq \Delta_R$. Apart from a possible finite number of points $\{z_k\}$, $k \in \mathbb{I}_R$, we have that $|z - z_n| \geq \big||z| - |z_n|\big| \geq |z_n| - R$ for all $z \in K$ and $n \in \mathbb{N} \setminus \mathbb{I}_R$. Thus,

$$\sum_{n=1}^{\infty} \left| \frac{a_n P(z)}{z - z_n} \right|^2 \leq \sum_{n \in \mathbb{I}_R} \left| \frac{a_n P(z)}{z - z_n} \right|^2 + \sum_{n \in \mathbb{N} \setminus \mathbb{I}_R} \frac{|a_n|^2 |P(z)|^2}{(|z_n| - R)^2},$$

and both summands are bounded on the compact $K$.

Thus, by using Lemma 4.2, the kernel $K$ defines an analytic Kramer kernel.

As a consequence, Theorem 4.5 assures that any function $f \in \mathcal{H}_K$ can be recovered from its samples $\{f(z_n)\}_{n=1}^{\infty}$ by means of the Lagrange-type interpolation formula

$$f(z) = \sum_{n=1}^{\infty} f(z_n) \frac{a_n P(z)/(z - z_n)}{a_n\, P'(z_n)} = \sum_{n=1}^{\infty} f(z_n) \frac{P(z)}{(z - z_n) P'(z_n)}, \quad z \in \mathbb{C}.$$

The convergence of the series is absolute and uniform on compact subsets of $\mathbb{C}$. □

A sequence $\{P_n(z)\}_{n=0}^{\infty}$ of orthonormal polynomials associated with an indeterminate Hamburger moment problem provides an example of a discrete analytic Kramer kernel. Indeed, let $\{P_n(z)\}_{n=0}^{\infty}$ be a sequence of orthonormal polynomials associated with an indeterminate Hamburger moment problem. It is known (see Refs. [2, 42]) that the series $\sum_{n=0}^{\infty} |P_n(z)|^2 < \infty$, uniformly on compact subsets of $\mathbb{C}$.

Thus, we can define a discrete analytic kernel $K$ as

$$\mathbb{C} \ni z \overset{K}{\longmapsto} K(z) := \big\{P_n(z)\big\}_{n=0}^{\infty} \in \ell^2(\mathbb{N}_0).$$

Let $\mathcal{H}_K$ be its corresponding RKHS of entire functions; any $f \in \mathcal{H}_K$ has the form

$$f(z) = \big\langle \{P_n(z)\}, \{c_n\} \big\rangle_{\ell^2(\mathbb{N}_0)}, \quad z \in \mathbb{C},$$

where $\{c_n\} \in \ell^2(\mathbb{N}_0)$.

For the sequences $\{t_m\}_{m=0}^{\infty}$ associated with indeterminate moment problems considered in the section about the Kramer sampling theorem and indeterminate moment problems, the kernel $K$ also becomes an analytic Kramer kernel. Next we exhibit an example taken from polynomials associated with birth and death processes (see the classical Ref. [33]):

**Example 4.9 (Sampling Associated with Birth and Death Orthogonal Polynomials)** Consider birth and death polynomials $\{P_n\}_{n=0}^{\infty}$ with quartic rates in Ref. [7]. These polynomials satisfy the three–term recurrence relation

$$x P_n(x) = a_n P_{n+1}(x) + b_n P_n(x) + a_{n-1} P_{n-1}(x), \quad n \geq 0,$$
$$P_{-1}(x) = 0; \quad P_0(x) = 1,$$

where, for $n \geq 0$, the coefficients are given by

$$a_n = \sqrt{\lambda_n \mu_{n+1}}\,, \qquad b_n = \lambda_n + \mu_n\,,$$
$$\lambda_n = (4n+1)(4n+2)^2(4n+3)\,,\ \mu_n = (4n-1)(4n)^2(4n+1)\,.$$

By Favard's theorem these polynomials are orthonormal polynomials with respect to a normalized Hamburger moment sequence $s$ which is indeterminate; see Ref. [7]. The entire function $D$ in Nevanlinna parametrization is given by

$$D(z) = \frac{4}{\pi}\sqrt{z}\sin\Big(\frac{\sqrt[4]{z}}{2}K_0\Big)\sinh\Big(\frac{\sqrt[4]{z}}{2}K_0\Big), \quad z \in \mathbb{C},$$

whose zeros are $z_m = (2\pi m/K_0)^4$, $m \in \mathbb{N}_0$, and $K_0$ denotes the constant $\dfrac{\Gamma^2(1/4)}{4\sqrt{\pi}}$. For each $m \in \mathbb{N}_0$ we have (see Refs. [7, 20]):

$$S_m(z) := \big\langle K(z), \{P_n(z_m)\}_{n=0}^{\infty}\big\rangle_{\ell^2(\mathbb{N}_0)} = \frac{D(z)}{(z-z_m)D'(z_m)}, \quad z \in \mathbb{C}.$$

Hence, any entire function of the form

$$f(z) = \langle K(z), \{c_n\}\rangle_{\ell^2(\mathbb{N}_0)} = \sum_{n=0}^{\infty} \overline{c_n}\, P_n(z)\,, \quad z \in \mathbb{C},$$

where $\{c_n\} \in \ell^2(\mathbb{N}_0)$ can be expanded as

$$f(z) = f(0)\,\frac{4\sqrt{z}\sin\Big(\frac{\sqrt[4]{z}}{2}K_0\Big)\sinh\Big(\frac{\sqrt[4]{z}}{2}K_0\Big)}{zK_0^2} +$$
$$+\frac{16\pi}{K_0^2}\sum_{m=1}^{\infty} f\Big[\Big(\frac{2\pi m}{K_0}\Big)^4\Big]\frac{(-1)^m m}{\sinh(m\pi)}\,\frac{\sqrt{z}\sin\Big(\frac{\sqrt[4]{z}}{2}K_0\Big)\sinh\Big(\frac{\sqrt[4]{z}}{2}K_0\Big)}{\Big(z-\big(\frac{2\pi m}{K_0}\big)^4\Big)}, \quad z \in \mathbb{C}.$$

The series converges absolutely and uniformly on compact subsets of $\mathbb{C}$.

We have used that

$$D'(0) = \frac{K_0^2}{\pi} \quad \text{and} \quad D'\Big[\Big(\frac{2\pi m}{K_0}\Big)^4\Big] = \frac{K_0^2}{4m\pi^2}(-1)^m \sinh(m\pi)\,.$$

**Example 4.10 (An Example where $\mathcal{H}$ is the Sobolev Hilbert Space $H^1(-\pi,\pi)$)** Let $H^1(-\pi,\pi)$ be the Sobolev Hilbert space with its usual inner product

$$\langle f, g\rangle_1 := \int_{-\pi}^{\pi} f(x)\,\overline{g(x)}\,dx + \int_{-\pi}^{\pi} f'(x)\,\overline{g'(x)}\,dx\,, \quad f, g \in H^1(-\pi,\pi)\,.$$

The system $\{e^{inx}\}_{n\in\mathbb{Z}} \cup \{\sinh x\}$ forms an orthogonal basis for $H^1(-\pi, \pi)$. It is easy to prove that the orthogonal complement of $\{e^{inx}\}_{n\in\mathbb{Z}}$ in $H^1(-\pi, \pi)$ is one-dimensional and $\sinh x$ belongs to it.

Given $a \in \mathbb{C} \setminus \mathbb{Z}$ we define a kernel $K_a : \mathbb{C} \longrightarrow H^1(-\pi, \pi)$ by setting

$$[K_a(z)](x) = (z-a)\,e^{izx} + \sin\pi z\,\sinh x\,, \quad x \in (-\pi, \pi)\,.$$

Expanding $K_a(z) \in H^1(-\pi, \pi)$ with respect to the above orthogonal basis we obtain

$$K_a(z) = [1 - i(z-a)]\,\sin\pi z\,\sinh x + (z-a)\sum_{n=-\infty}^{\infty} \frac{1+zn}{1+n^2}\,\mathrm{sinc}(z-n)\,e^{inx}$$

in $H^1(-\pi, \pi)$. As a consequence, Theorem 4.5 reads:

Any entire function $f$ defined by

$$\begin{aligned} f(z) &= \int_{-\pi}^{\pi} F(x)\,[K_a(z)](x)\,dx + \int_{-\pi}^{\pi} F'(x)\,[K_a(z)]'(x)\,dx \\ &= \langle K_a(z), \overline{F}\rangle_1\,, \quad z \in \mathbb{C}\,, \end{aligned}$$

where $F \in H^1(-\pi, \pi)$, can be recovered from its samples $\{f(a)\} \cup \{f(n)\}_{n\in\mathbb{Z}}$ by means of the sampling formula

$$f(z) = [1-i(z-a)]\,\frac{\sin\pi z}{\sin\pi a}\,f(a) + \sum_{n=-\infty}^{\infty} f(n)\,\frac{z-a}{n-a}\,\frac{1+zn}{1+n^2}\,\mathrm{sinc}(z-n)\,, \quad z \in \mathbb{C}\,.$$

The series converges absolutely and uniformly on compact subsets of $\mathbb{C}$.

### 4.2.2 *Analytic Kramer Kernels and Lagrange-Type Interpolation Series*

A more difficult question concerns whether the sampling expansion (4.26) can be written, in general, as a Lagrange-type interpolation series. Remember that, for any $f \in PW_\pi$, the corresponding sampling formula can be written as:

$$f(z) = \sum_{n=-\infty}^{\infty} f(n)\,\frac{\sin\pi(z-n)}{\pi(z-n)} = \sum_{n=-\infty}^{\infty} f(n)\,\frac{P(z)}{P'(n)(z-n)}\,, \quad z \in \mathbb{C}\,,$$

where $P(z) = \sin \pi z$ is an entire function having simple zeros at the integers set $\mathbb{Z}$. A complete answer to this question involves the following algebraic property:

**Definition 4.3 (ZR Property)** A space $\mathcal{H}$ of entire functions has the zero-removing property (ZR property hereafter) if for any $g \in \mathcal{H}$ and any zero $w$ of $g$ the function $g(z)/(z-w)$ belongs to $\mathcal{H}$.

The ZR property is ubiquitous in mathematics. For instance:

1. The set $\mathcal{P}_N(\mathbb{C})$ of polynomials with complex coefficients of degree less than or equal to $N$ has the ZR property.
2. The classical Paley-Wiener space $PW_\pi$ satisfies the ZR property. It follows from its characterization by using the classical Paley-Wiener theorem (Theorem 2.3), i.e.,

$$PW_\pi = \Big\{ f \in H(\mathbb{C}) \,:\, |f(z)| \leq A\mathrm{e}^{\pi|z|}, \quad f|_{\mathbb{R}} \in L^2(\mathbb{R}) \Big\}.$$

   If $w \in \mathbb{C}$ is a zero of $f \in PW_\pi$, it is easy to check that the function $f(z)/(z-w)$ satisfies the above three conditions defining $PW_\pi$.
3. A more involved example sharing this property corresponds to the entire functions in the *Pólya class*. Functions in this class satisfy the ZR property Ref. [13, p. 15].

An entire function $E(z)$ is said to be of Pólya class if it has no zeros in the upper half-plane, $|E(x-iy)| \leq |E(x+iy)|$ for $y > 0$, and $|E(x+iy)|$ is a nondecreasing function of $y > 0$ for each fixed $x$.

Examples of functions belonging to this class are: $\sin z$, $\cos\sqrt{z}$, $\mathrm{e}^{-z^2}$ or $1/\Gamma(z)$.

On the other hand, it is easy to construct analytic Kramer kernels $K$ such that the corresponding spaces $\mathcal{H}_K$ do not satisfy the ZR property. Indeed, let $K : \mathbb{C} \to L^2[-\pi, \pi]$ be the kernel defined by $[K(z)](x) := \frac{1}{\sqrt{2\pi}} \mathrm{e}^{iz^2 x}$, $x \in [-\pi, \pi]$. It is an analytic Kramer kernel and its Taylor series around $z = 0$ is given by

$$[K(z)](x) = \sum_{k=0}^{\infty} \frac{(ix)^k z^{2k}}{k!}.$$

Hence, the Taylor series of any function $f(z) = \langle K(z), F \rangle_{L^2[-\pi,\pi]}$ in $\mathcal{H}_K$, where $F \in L^2[-\pi, \pi]$, is of the form

$$f(z) = \sum_{k=0}^{\infty} \frac{c_k}{k!} z^{2k}, \quad z \in \mathbb{C},$$

where $c_k = \langle (ix)^k, F \rangle$, $k \in \mathbb{N}_0$. Thus, any function $f \in \mathcal{H}_K$ is an even function.

Now let $G$ be a nonzero function in $L^2[-\pi,\pi]$ such that $\int_{-\pi}^{\pi} G = 0$. Then, the function $G$ is orthogonal to $K(0)$ and, consequently, the entire function $g(z) = \langle K(z), G\rangle_{L^2[-\pi,\pi]}$, $z \in \mathbb{C}$, satisfies $g(0) = 0$. We have,

$$\frac{g(z)}{z} = \sum_{k=1}^{\infty} \frac{c_k}{k!} z^{2k-1} \quad \text{where } c_k = \big\langle (ix)^k, G\big\rangle,\ k \in \mathbb{N}.$$

Clearly, the entire function $g(z)/z$ does not belong to $\mathcal{H}_K$ and, therefore, the space $\mathcal{H}_K$ does not satisfy the ZR property.

In case of a polynomial kernel $K$ with coefficients in $\mathcal{H}$ we have the following result:

**Proposition 4.5** *The Hilbert space $\mathcal{H}_K$ associated with a polynomial kernel $K(z) := \sum_{n=0}^{N} p_n z^n$, where $p_n \in \mathcal{H}$ and $p_N \neq 0$, has the ZR property if and only if the set of coefficients $\{p_0, p_1, \ldots, p_N\}$ is linearly independent in $\mathcal{H}$.*

***Proof*** Consider $f(z) = a_N z^N + \cdots + a_1 z + a_0 \in \mathcal{H}_K$ with $a_N \neq 0$; there exists $x \in \mathcal{H}$ such that $f(z) = \langle K(z), x\rangle$ and, consequently, $a_n = \langle p_n, x\rangle$ for $n = 0, 1, \ldots, N$.

If the space $\mathcal{H}_K$ has the ZR property and $\alpha_1, \alpha_2, \ldots, \alpha_N$ are the roots of the polynomial $f$ then the constant $a_N$ and the polynomials

$$a_N(z-\alpha_N);\ a_N(z-\alpha_N)(z-\alpha_{N-1});\ \ldots;\ a_N(z-\alpha_N)(z-\alpha_{N-1})\cdots(z-\alpha_1)$$

belong to $\mathcal{H}_K$. Let $b_0, b_1, \ldots, b_N \in \mathbb{C}$ such that

$$b_N p_N + b_{N-1} p_{N-1} + \cdots + b_0 p_0 = 0 \quad \text{in } \mathcal{H}. \tag{4.27}$$

The vector $(b_N, \ldots, b_0)^\top$ is orthogonal in $\mathbb{C}^{N+1}$ to any vector $(c_N, \ldots, c_0)^\top \in \mathbb{C}^{N+1}$ with $c_N z^N + \cdots + c_1 z + c_0 \in \mathcal{H}_K$. Since $a_N \neq 0$, $b_0 a_N = 0$ implies that $b_0 = 0$. Analogously, since $a_N(z-\alpha_N)$ belongs to $\mathcal{H}_K$ we have that $a_N b_1 - (a_N \alpha_N) b_0 = 0$ and consequently $b_1 = 0$. Proceeding iteratively it is straightforward to obtain that $b_2 = \cdots = b_{N-1} = 0$; finally, from (4.27) we conclude that $b_N = 0$. Hence, the set $\{p_0, p_1, \ldots, p_N\}$ is linearly independent in $\mathcal{H}$.

Conversely, suppose that the set $\{p_0, p_1, \ldots, p_N\}$ is linearly independent in $\mathcal{H}$. In this case, the mapping

$$\begin{aligned} \Phi : \mathcal{H} &\longrightarrow \mathbb{C}^{N+1} \\ x &\longmapsto \Phi(x) = \big(\langle p_0, x\rangle, \langle p_1, x\rangle, \ldots, \langle p_N, x\rangle\big)^\top \end{aligned}$$

is surjective. As a consequence, any complex polynomial of degree less than or equal to $N$ belongs to $\mathcal{H}_K$. Let $f(z) = a_N z^N + \cdots + a_1 z + a_0 \in \mathcal{H}_K$ and let $w \in \mathbb{C}$ be a root of $f$. Hence $f(z)/(z-w) = c_0 + c_1 z + \cdots + c_{N-1} z^{N-1}$ is a polynomial of degree less than or equal to $N-1$.

Since $\Phi$ is onto there exists $x \in \mathcal{H}$ such that $\Phi(x) = (c_0, c_1, \ldots, c_{N-1}, 0)$. From the definition of $\Phi$ we conclude that $f(z)/(z-w) = \langle K(z), x\rangle$, that is, the function $f(z)/(z-w) \in \mathcal{H}_K$. □

There exists a necessary and sufficient condition to ensure when the sampling formula (4.26) associated with an analytic Kramer kernel can be expressed as a Lagrange-type interpolation series. This result was proved for the first time in Ref. [14] for the orthogonal case; see also Ref. [17].

**Theorem 4.6 (Analytic Kramer Sampling and Lagrange-Type Interpolation Series)** *Let $\mathcal{H}_K$ be an RKHS of entire functions obtained from an analytic Kramer kernel $K$ with respect to the sequence $\{z_n\}_{n=1}^{\infty} \subset \mathbb{C}$. Then, the sampling formula* (4.26) *for $\mathcal{H}_K$ can be written as a Lagrange-type interpolation series*

$$f(z) = \sum_{n=1}^{\infty} f(z_n) \frac{P(z)}{(z-z_n)P'(z_n)}, \quad z \in \mathbb{C}, \tag{4.28}$$

*where $P$ denotes an entire function having simple zeros at $\{z_n\}_{n=1}^{\infty}$ if and only if the space $\mathcal{H}_K$ satisfies the ZR property.*

***Proof***

(*a*) For the sufficient condition we have to prove that sampling formula (4.26) can be written as a Lagrange-type interpolation series (4.28) for some entire function $P$ having simple zeros at $\{z_n\}_{n=1}^{\infty}$ in case the ZR property holds.

(1) First, we prove that the only zeros of the sampling function $S_n$ are given by $\{z_r\}_{r \neq n}$. Suppose that $S_n(w) = 0$, then by hypothesis the function $S_n(z)/(z-w)$ is in $\mathcal{H}_K$. Hence, the function

$$\frac{z-z_n}{z-w} S_n(z) = S_n(z) + \frac{w-z_n}{z-w} S_n(z)$$

also belongs to $\mathcal{H}_K$. If $w \notin \{z_r\}_{r \neq n}$, then the function $\frac{z-z_n}{z-w} S_n(z)$ in $\mathcal{H}_K$ vanishes at the sequence $\{z_r\}_{r=1}^{\infty}$ which implies that $S_n \equiv 0$, to give a contradiction.

(2) In addition, the zeros of $S_n$ are simple; indeed, suppose that $z_m$ is a multiple zero of $S_n$. Proceeding as above, the function $\frac{z-z_n}{z-z_m} S_n(z)$ belongs to $\mathcal{H}_K$ and vanishes at $\{z_r\}_{r=1}^{\infty}$ which again implies that $S_n \equiv 0$.

Consequently, choosing an entire function $Q$ having only simple zeros at $\{z_n\}_{n=1}^{\infty}$, for each $n \in \mathbb{N}$ there exists an entire function $A_n$ without zeros such that

$$(z-z_n)S_n(z) = Q(z)A_n(z), \quad z \in \mathbb{C}.$$

Next, we prove that there exists an entire function $A$ without zeros and a sequence $\{\sigma_n\}_{n=1}^{\infty}$ in $\mathbb{C}\setminus\{0\}$ such that $A_n(z)=\sigma_n A(z)$ for all $z\in\mathbb{C}$.

For $m\neq n$ the function $\frac{z-z_n}{z-z_m}S_n(z)$ in $\mathcal{H}_K$ has its zeros at $\{z_r\}_{r\neq m}$. Thus the sampling formula (4.26) gives

$$\frac{z-z_n}{z-z_m}S_n(z)=[(z_m-z_n)S_n'(z_m)]\frac{S_m(z)}{a_m}\,,\qquad z\in\mathbb{C}\,.$$

Fixing $m=1$, we conclude that $A_n(z)=\sigma_n A(z)$, where $A=A_1$ and

$$\sigma_n=\frac{1}{a_1}(z_1-z_n)S_n'(z_1)\neq 0\quad \text{for } n\in\mathbb{N}\setminus\{1\} \text{ and } \sigma_1=1\,.$$

Hence, $S_n(z)=\frac{\sigma_n Q(z)A(z)}{z-z_n}$ for $z\neq z_n$ and $S_n(z_n)=a_n=\sigma_n Q'(z_n)A(z_n)$. Substituting in Equation (4.26) we obtain the Lagrange-type interpolation series (4.28), where $P(z)=A(z)Q(z)$.

$(b)$ For the necessary condition, assume that the sampling formula in $\mathcal{H}_K$ takes the form of a Lagrange-type interpolation series (4.28). Given $g\in\mathcal{H}_K$, there exists $x\in\mathcal{H}$ such that $g(z)=\langle K(z),x\rangle$, $z\in\mathbb{C}$. Assuming that $g(w)=0$, we have to prove that the function $g(z)/(z-w)$ belongs to $\mathcal{H}_K$.

The sampling expansion for $g$ at $w$ gives

$$\sum_{n=1}^{\infty} g(z_n)\frac{P(w)}{(w-z_n)P'(z_n)}=0\,. \tag{4.29}$$

We distinguish two cases:

$(i)$ The zero $w\in\mathbb{C}\setminus\{z_n\}_{n=1}^{\infty}$.

As $P(w)\neq 0$, from (4.29) we obtain

$$\sum_{n=1}^{\infty} g(z_n)\frac{1}{(w-z_n)P'(z_n)}=0\,.$$

Thus,

$$\begin{aligned} g(z)&=\sum_{n=1}^{\infty} g(z_n)\frac{P(z)}{(z-z_n)P'(z_n)}-\sum_{n=1}^{\infty} g(z_n)\frac{P(z)}{(w-z_n)P'(z_n)}\\ &=(z-w)\sum_{n=1}^{\infty} g(z_n)\frac{P(z)}{P'(z_n)}\frac{1}{(z-z_n)(z_n-w)}\,. \end{aligned}$$

Therefore, the entire function $G(z) := g(z)/(z-w)$ can be recovered from its samples at $\{z_n\}_{n=1}^{\infty}$ through the formula

$$G(z) = \sum_{n=1}^{\infty} G(z_n) \frac{P(z)}{(z-z_n)P'(z_n)}, \quad z \in \mathbb{C}. \tag{4.30}$$

Moreover, the function $G$ is in $\mathcal{H}_K$ because $G(z) = \langle K(z), y\rangle_{\mathcal{H}}$, where $y \in \mathcal{H}$ has the expansion $y = \sum_{n=1}^{\infty} \langle y, x_n\rangle\, y_n$ with respect to the dual Riesz basis $\{y_n\}_{n=1}^{\infty}$ of $\{x_n\}_{n=1}^{\infty}$ and the coefficients are given by

$$\left\{ \langle y, x_n\rangle := \frac{1}{\overline{z_n} - \overline{w}} \langle x, x_n\rangle \right\}_{n=1}^{\infty} \in \ell^2(\mathbb{N}).$$

Indeed, sampling formula (4.28) for $S_n$ gives

$$S_n(z) = a_n \frac{P(z)}{(z-z_n)P'(z_n)}.$$

Hence, by using the biorthogonality $\langle x_n, y_n\rangle = \delta_{n,m}$ , we obtain

$$\langle K(z), y\rangle = \sum_{n=1}^{\infty} \frac{S_n(z)\,\overline{\langle x, x_n\rangle}}{z_n - w} = G(z), \quad z \in \mathbb{C},$$

where we have used (4.30), and the result that $\overline{\langle x, x_n\rangle} = g(z_n)/a_n$, $n \in \mathbb{N}$. Therefore, the function $g(z)/(z-w)$ belongs to $\mathcal{H}_K$.

*(ii)* The zero $w$ satisfies $w = z_m$ for some $m \in \mathbb{N}$.

As $g(z_m) = 0$, the sampling expansion for $g$ reads

$$g(z) = \sum_{\substack{n=1\\ n\neq m}}^{\infty} g(z_n) \frac{P(z)}{(z-z_n)P'(z_n)}, \quad z \in \mathbb{C}.$$

Setting $P(z) = (z-z_m)Q_m(z)$ we have $P'(z) = Q_m(z) + (z-z_m)Q'_m(z)$ and consequently

$$P'(z_k) = \begin{cases} (z_k - z_m)Q'_m(z_k) & \text{if } k \neq m\,, \\ Q_m(z_m) & \text{if } k = m\,, \end{cases}$$

Hence,

$$\frac{g(z)}{z-z_m} = \sum_{\substack{n=1\\ n\neq m}}^{\infty} \frac{g(z_n)}{z_n - z_m} \frac{Q_m(z)}{(z-z_n)Q'_m(z_n)}, \quad z \in \mathbb{C}. \tag{4.31}$$

Using the uniform convergence of the series in Eq. (4.31) we deduce that this series defines a continuous function. As a consequence, taking the limit as $z \to z_m$ we obtain

$$g'(z_m) = \sum_{\substack{n=1 \\ n\neq m}}^{\infty} \frac{g(z_n)}{z_n - z_m} \frac{Q_m(z_m)}{(z_m - z_n)Q'_m(z_n)} \,. \tag{4.32}$$

Now we prove that

$$\frac{g(z)}{z - z_m} = \sum_{\substack{n=1 \\ n\neq m}}^{\infty} \frac{g(z_n)}{z_n - z_m} \frac{P(z)}{(z - z_n)P'(z_n)} + g'(z_m)\frac{P(z)}{(z - z_m)P'(z_m)} \,. \tag{4.33}$$

Indeed, substituting Eq. (4.32) into the right-hand of Eq. (4.33) we obtain

$$\sum_{\substack{n=1 \\ n\neq m}}^{\infty} \Big[\frac{g(z_n)}{z_n - z_m} \frac{P(z)}{(z - z_n)P'(z_n)} + \frac{g(z_n)}{z_n - z_m} \frac{Q_m(z)}{(z_m - z_n)Q'_m(z_n)}\Big]$$

$$= \sum_{\substack{n=1 \\ n\neq m}}^{\infty} \frac{g(z_n)}{z_n - z_m} \frac{Q_m(z)}{Q'_m(z_n)} \Big[\frac{z - z_m}{(z_n - z_m)(z - z_n)} - \frac{1}{z_n - z_m}\Big]$$

$$= \sum_{\substack{n=1 \\ n\neq m}}^{\infty} \frac{g(z_n)}{z_n - z_m} \frac{Q_m(z)}{(z - z_n)Q'_m(z_n)} = \frac{g(z)}{z - z_m} \,.$$

Thus, defining $y \in \mathcal{H}$ by the expansion $y = \sum_{n=1}^{\infty} \langle y, x_n \rangle\, y_n$, where the coefficients $\{\langle y, x_n \rangle\}_{n=1}^{\infty}$ in $\ell^2(\mathbb{N})$ are given by

$$\langle y, x_n \rangle := \begin{cases} \frac{\langle x, x_n \rangle}{\overline{z_n} - \overline{z_m}} & \text{if } n \neq m \,, \\[2ex] \frac{\overline{g'(z_m)}}{a_m} & \text{if } n = m \,, \end{cases}$$

and proceeding as in case (i), it may be checked that

$$\frac{g(z)}{z - z_m} = \langle K(z), y \rangle \,, \quad z \in \mathbb{C} \,,$$

which proves that the function $g(z)/(z - z_m)$ belongs to $\mathcal{H}_K$.

This concludes the proof of the theorem. □

**Example 4.11 (Paley-Wiener-Levinson Theorem Revisited)** Let $\{z_n\}_{n\in\mathbb{Z}}$ be a sequence in $\mathbb{C}$ satisfying

$$\sup_n |\operatorname{Re} z_n - n| < 1/4 \quad \text{and} \quad \sup_n |\operatorname{Im} z_n| < \infty\,.$$

It is known that the sequence $\{e^{iz_n w}/\sqrt{2\pi}\}_{n\in\mathbb{Z}}$ is a Riesz basis for $L^2[-\pi,\pi]$ (see Ref. [48, p. 196]).

The Fourier kernel $\mathbb{C} \ni z \mapsto K(z) \in L^2[-\pi,\pi]$, where $[K(z)](x) = e^{izx}/\sqrt{2\pi}$, $x \in [-\pi,\pi]$, is an analytic Kramer kernel for the data $\{z_n\}_{n\in\mathbb{Z}}$ and $a_n = 1$.

Thus, for any function $f \in PW_\pi$ we have the sampling expansion

$$f(z) = \sum_{n=-\infty}^{\infty} f(z_n)\, S_n(z) = \sum_{n=-\infty}^{\infty} f(z_n)\, \frac{P(z)}{(z-z_n)P'(z_n)}\,, \quad z \in \mathbb{C}\,,$$

where, for each $n \in \mathbb{Z}$, the sampling function $S_n$ is given by

$$S_n(z) = \big\langle K(z), h_n \big\rangle_{L^2[-\pi,\pi]}\,, \quad z \in \mathbb{C}\,,$$

being $\{h_n(w)\}_{n\in\mathbb{Z}}$ the dual Riesz basis of $\{e^{iz_n w}/\sqrt{2\pi}\}_{n\in\mathbb{Z}}$ in $L^2[-\pi,\pi]$, and $P$ is an entire function having only simple zeros at $\{z_n\}_{n\in\mathbb{Z}}$.

Now Theorem 2.4 assures that functions in $PW_\pi$ are completely determined by their zeros. Thus we derive that, up to a constant factor, the entire function $P(z)$ coincides with the (convergent) infinite product

$$(z - z_0) \prod_{n=1}^{\infty} \Big(1 - \frac{z}{z_n}\Big)\Big(1 - \frac{z}{z_{-n}}\Big)\,, \quad z \in \mathbb{C}\,.$$

**Example 4.12 (Example 4.10 Revisited)** The space $\mathcal{H}_{K_a}$ in Example 4.10 does not satisfy the ZR property. Indeed, the function $(z-a)\operatorname{sinc} z$ belongs to $\mathcal{H}_{K_a}$ since

$$(z-a)\operatorname{sinc} z = \big\langle K_a(z), 1/2\pi \big\rangle_1\,, \quad z \in \mathbb{C}\,.$$

However, by using the sampling formula for $\mathcal{H}_{K_a}$ it is straightforward to check that the function $\operatorname{sinc} z$ does not belong to $\mathcal{H}_{K_a}$. As a consequence, the sampling formula in Example 4.10 cannot be expressed as a Lagrange-type interpolation series.

Let $K : \mathbb{C} \to \mathcal{H}$ be an analytic kernel such that $K(z_0) = 0$ for some $z_0 \in \mathbb{C}$. Then every function in the associated space $\mathcal{H}_K$ has a zero at $z_0$ and the ZR property does not hold in $\mathcal{H}_K$.

Indeed, let $f$ be a nonzero entire function in $\mathcal{H}_K$ and let $r$ denote the multiplicity of its zero at $z_0$. The function $f(z)/(z-z_0)^r$ is not in $\mathcal{H}_K$ since it does not vanish at $z_0$.

**Example 4.13 (Example 4.4 Revisited)** Consider the sine kernel

$$\begin{aligned} K_s : \mathbb{C} &\longrightarrow L^2[0,\pi] \\ z &\longmapsto K_s(z) \qquad \text{where} \quad [K_s(z)](x) = \sin zx\,, \quad x \in [0,\pi]\,. \end{aligned}$$

It defines an analytic Kramer kernel, and for any function $f$ defined by

$$f(z) = \big\langle \sin zx\,, F \big\rangle_{L^2[0,\pi]}\,, \quad z \in \mathbb{C}\,,$$

where $F \in L^2[0,\pi]$, the sampling expansion

$$f(z) = \frac{2}{\pi}\sum_{n=1}^{\infty} f(n)\frac{(-1)^n n \sin \pi z}{z^2 - n^2}\,, \qquad z \in \mathbb{C}\,,$$

holds. Since $K_s(0) = 0$, the ZR property does not hold in $\mathcal{H}_{K_s}$, and the above sampling formula cannot be written as a Lagrange-type interpolation series.

### 4.2.3 *Sampling Related to the One-Dimensional Dirac Operator*

As announced at the beginning of this chapter, the connection between sampling theorems and boundary-values problems has been the focus of many research papers. In Ref. [56], sampling theorems associated with regular Sturm-Liouville problems were investigated. It was shown, among others things, that for such boundary-value problems Kramer's sampling series is nothing more than a Lagrange-type interpolation formula. These results were refined and extended to singular Sturm-Liouville boundary-value problems in Ref. [49]. Sampling theorems associated with differential equations and more general types of boundary-value problems have been investigated, for instance, in Refs. [16, 37, 57], and in Ref. [51], where the Green's function was used to generate the interpolation functions. Sampling theorems associated with integral equations have also been reported (see, for instance, Ref. [4]).

In this section, following Ref. [54], we include a sampling theorem associated with a boundary-value problem involving the *one-dimensional Dirac operator*. We show that the sampling series in this case, as in the Sturm-Liouville case, is nothing but a Lagrange-type interpolation series.

## Some Preliminaries on the One-Dimensional Dirac Operator

Consider the one-dimensional Dirac system of differential equations

$$\begin{aligned} y_2' + p_{11}(x)y_1 + p_{12}(x)y_2 &= \lambda y_1 \\ -y_1' + p_{21}(x)y_1 + p_{22}(x)y_2 &= \lambda y_2, \end{aligned}$$

where $p_{ij}(x)$, $i, j = 1, 2$, are real-valued functions defined on $[0, \pi]$ and $\lambda$ is a parameter. Using matrix notation, we can write this system in the form

$$\mathbf{B}\frac{d\mathbf{y}}{dx} + \mathbf{P}(x)\mathbf{y} = \lambda \mathbf{y} \,,$$

where

$$\mathbf{B} = \begin{pmatrix} 0 & 1 \\ -1 & 0 \end{pmatrix}, \quad \mathbf{P}(x) = \begin{pmatrix} p_{11}(x) & p_{12}(x) \\ p_{21}(x) & p_{22}(x) \end{pmatrix}, \quad \mathbf{y} = \begin{pmatrix} y_1(x) \\ y_2(x) \end{pmatrix}.$$

By using a smooth and orthogonal transformation of the form

$$\mathbf{H}(x) = \begin{pmatrix} \cos\phi(x) & -\sin\phi(x) \\ \sin\phi(x) & \cos\phi(x) \end{pmatrix},$$

with

$$\phi(x) = \frac{1}{2}\tan^{-1}\frac{2p_{12}(x)}{p_{11}(x) - p_{22}(x)},$$

and setting $\mathbf{y} = \mathbf{H}(x)\mathbf{z}$, we can transform the system into the canonical form:

$$\mathbf{B}\frac{d\mathbf{z}}{dx} + \mathbf{Q}(x)\mathbf{z} = \lambda \mathbf{z}\,, \tag{4.34}$$

where

$$\mathbf{Q}(x) = \begin{pmatrix} p(x) & 0 \\ 0 & r(x) \end{pmatrix}$$

for some appropriate functions $p(x)$ and $r(x)$ (see Ref. [35, p. 186]). Therefore, from now on, we shall confine our attention to Dirac systems in their canonical form (4.34).

Let us consider the following boundary-value problem for Eq. (4.34):

$$y_2' + p(x)y_1 = \lambda y_1\,,\ \ y_1' - r(x)y_2 = -\lambda y_2\,, \tag{4.35}$$

$$y_1(0)\sin\alpha + y_2(0)\cos\alpha = 0\,, \tag{4.36}$$

$$y_1(\pi)\sin\beta + y_2(\pi)\cos\beta = 0\,, \tag{4.37}$$

where $p(x)$ and $r(x)$ are continuous on $[0, \pi]$.

It is known (see Ref. [35]) that the problem (4.35)–(4.37) has a countable number of eigenvalues $\{\lambda_n\}_{n=-\infty}^{\infty}$, which are all real and simple, and to every eigenvalue $\lambda_n$, there corresponds a vector-valued eigenfunction

$$\mathbf{y}_n^{\top}(x, \lambda_n) = (y_{n,1}(x, \lambda_n), y_{n,2}(x, \lambda_n))\,,$$

where $\top$ stands for the transpose. Moreover, vector-valued eigenfunctions belonging to different eigenvalues are orthogonal, i.e.,

$$\int_0^{\pi} \mathbf{y}_\mathbf{n}^{T}(x, \lambda_n)\mathbf{y}_\mathbf{m}(x, \lambda_m)dx$$
$$= \int_0^{\pi} [y_{n,1}(x, \lambda_n)y_{m,1}(x, \lambda_m) + y_{n,2}(x, \lambda_n)y_{m,2}(x, \lambda_m)]dx = 0\,.$$

Let $\mathbf{y}(x, \lambda)$ and $\mathbf{z}(x, \lambda')$ be solutions of (4.35); hence

$$y_2' - p(x)y_1 = \lambda y_1\,,\ \ y_1' + r(x)y_2 = -\lambda y_2 \tag{4.38}$$

$$z_2' - p(x)z_1 = \lambda' z_1\,,\ \ z_1' + r(x)z_2 = -\lambda' z_2\,. \tag{4.39}$$

Multiplying (4.38) by $z_1$ and $-z_2$, and (4.39) by $-y_1$ and $y_2$ respectively, and adding them together, we obtain

$$\frac{d}{dx}\big\{z_1(x, \lambda')y_2(x, \lambda) - z_2(x, \lambda')y_1(x, \lambda)\big\}$$
$$= (\lambda - \lambda')\big\{y_1(x, \lambda)z_1(x, \lambda') + y_2(x, \lambda)z_2(x, \lambda')\big\}\,. \tag{4.40}$$

It is also known that if $\mathbf{f}^{\top}(x) = (f_1(x), f_2(x))$ is a vector-valued function which has a continuous derivative and satisfies the boundary conditions (4.36) and (4.37), then it can be expanded into an absolutely and uniformly convergent series of the vector-valued eigenfunctions of problem (4.35)–(4.37), viz.,

$$\mathbf{f}(x) = \sum_{n=-\infty}^{\infty} \widehat{f}_n \frac{\mathbf{\Phi}_n(x)}{\|\mathbf{\Phi}_n\|^2}\,, \tag{4.41}$$

where

$$\widehat{f}_n = \int_0^{\pi} \mathbf{f}^{\top}(x)\boldsymbol{\Phi}_n(x)dx = \int_0^{\pi} [f_1(x)\Phi_{n,1}(x) + f_2(x)\Phi_{n,2}(x)]dx,$$

$\boldsymbol{\Phi}_n^{\top}(x) = (\Phi_{n,1}(x), \Phi_{n,2}(x))$ is the vector-valued eigenfunction corresponding to the eigenvalue $\lambda_n$, and

$$\|\boldsymbol{\Phi}_\mathbf{n}\|^2 = \int_0^{\pi} \boldsymbol{\Phi}_\mathbf{n}^{\top}(x)\boldsymbol{\Phi}_n(x)dx = \int_0^{\pi} [\Phi_{n,1}^2(x) + \Phi_{n,2}^2(x)]dx.$$

If $\mathbf{f}$ is merely a square-integrable function, then the series (4.41) converges to $\mathbf{f}$ in the mean.

### The Associated Sampling Result

Now we derive a sampling theorem associated with the boundary-value problem (4.35)–(4.37). By using properties from Dirac's systems, we prove that the associated sampling formula is a Lagrange-type interpolation one.

As a consequence of Theorem 4.6, the associated functions $F$ with the analytic Kramer kernel $\boldsymbol{\Phi}(x, \lambda)$ (see Eq. (4.42) below) satisfy the ZR property.

**Theorem 4.7** *Let* $\mathbf{f}(\mathbf{x})$ *be a square-integrable vector-valued function on* $[0, \pi]$. *Set*

$$F(\lambda) = \int_0^{\pi} \mathbf{f}^{\top}(x)\boldsymbol{\Phi}(x, \lambda)dx\,, \quad \lambda \in \mathbb{C}, \tag{4.42}$$

*where* $\boldsymbol{\Phi}(\mathbf{x}, \boldsymbol{\lambda})$ *is a solution of the differential equation (4.35), together with* $\Phi_1(0, \lambda) = \cos\alpha$ *and* $\Phi_2(0, \lambda) = -\sin\alpha$.

*Then* $F(\lambda)$ *is an entire function of exponential type at most* $\pi$ *that can be reconstructed using its values at the eigenvalues sequence* $\{\lambda_n\}_{n\in\mathbb{Z}}$ *by means of the Lagrange-type interpolation formula*

$$F(\lambda) = \sum_{n=-\infty}^{\infty} F(\lambda_n)\frac{G(\lambda)}{(\lambda - \lambda_n)G'(\lambda_n)}, \quad \lambda \in \mathbb{C}, \tag{4.43}$$

*where the series converges uniformly on compact sets of* $\mathbb{C}$, *and* $G(\lambda)$ *is an entire function whose zeros are exactly the eigenvalues of the problem (4.35)–(4.37) and which can be written in the form*

$$G(\lambda) = (\lambda - \lambda_0)\prod_{n=1}^{\infty}\Big(1 - \frac{\lambda}{\lambda_n}\Big)\Big(1 - \frac{\lambda}{\lambda_{-n}}\Big), \quad \lambda \in \mathbb{C}. \tag{4.44}$$

***Proof*** First, the infinite product given by (4.44) defines an entire function because it is known that the eigenvalues of the problem (4.35)–(4.37) satisfy the asymptotic formula $\lambda_n = a_0 + n + a_1/n + O(1/n^2)$ as $|n| \longrightarrow \infty$. See Ref. [35, p. 193].

That $F(\lambda)$ is an entire function of exponential type at most $\pi$ follows from the relation

$$|F(\lambda)| \leq \int_0^\pi |f_1(x)||\Phi_1(x,\lambda)|dx + \int_0^\pi |f_2(x)||\Phi_2(x,\lambda)|dx$$

and the fact that $\Phi_1(x,\lambda)$ and $\Phi_2(x,\lambda)$ are entire functions of exponential type at most $\pi$ (see Ref. [35, Lemma 7.2.1]).

Since $\Phi(x,\lambda)$ is in $L^2(0,\pi)$ for any $\lambda$, we have

$$\mathbf{\Phi}(x,\lambda) = \sum_{n=-\infty}^{\infty} \widehat{\Phi}_n(\lambda) \frac{\mathbf{\Phi}_n(x)}{||\Phi_n||^2}\,, \tag{4.45}$$

where

$$\begin{aligned}\widehat{\Phi}_n(\lambda) &= \int_0^\pi \mathbf{\Phi}^\top(x,\lambda)\mathbf{\Phi}_n(x)dx \\ &= \int_0^\pi [\Phi_1(x,\lambda)\Phi_{n,1}(x) + \Phi_2(x,\lambda)\Phi_{n,2}(x)]dx\,,\end{aligned} \tag{4.46}$$

with

$$\mathbf{\Phi}^\top(x,\lambda) = (\Phi_1(x,\lambda), \Phi_2(x,\lambda)) \quad \text{and} \quad \mathbf{\Phi}_n^\top(x) = (\Phi_{n,1}(x), \Phi_{n,2}(x))\,.$$

Since $f$ is in $L^2[0,\pi]$, it has an expansion of the form (4.41). In view of Parseval's relation and definition (4.42), we obtain

$$F(\lambda) = \sum_{n=-\infty}^{\infty} \widehat{f_n} \frac{\widehat{\Phi}_n(\lambda)}{\|\mathbf{\Phi}_n\|^2}\,.$$

But

$$F(\lambda_n) = \int_0^\pi \mathbf{f}^\top(x)\mathbf{\Phi}(x,\lambda_n)dx = \int_0^\pi \mathbf{f}^\top(x)\mathbf{\Phi}_n(x)dx = \widehat{f_n}\,, \quad n \in \mathbb{Z}.$$

Thus

$$F(\lambda) = \sum_{n=-\infty}^{\infty} F(\lambda_n) \frac{\widehat{\Phi}_n(\lambda)}{\|\mathbf{\Phi}_n\|^2}\,. \tag{4.47}$$

From relation (4.40), we have

$$(\lambda - \lambda') \int_0^{\pi} \mathbf{y}^{\top}(x, \lambda)\mathbf{z}(x, \lambda')dx = \mathbf{W}(\mathbf{z}, \mathbf{y})|_{x=\pi} - \mathbf{W}(\mathbf{z}, \mathbf{y})|_{x=0}\,, \tag{4.48}$$

where

$$\mathbf{W}(\mathbf{y}, \mathbf{z}) = \begin{vmatrix} y_1(x, \lambda) & y_2(x, \lambda) \\ z_1(x, \lambda') & z_2(x, \lambda') \end{vmatrix}.$$

Let $\mathbf{\Psi}(x, \lambda)$ be the solution of the differential equation (4.35) with $\Psi_1(\pi, \lambda) = \cos\beta$, and $\Psi_2(\pi, \lambda) = -\sin\beta$. Set

$$\begin{aligned} G(\lambda) &= \mathbf{W}(\mathbf{\Phi}(\pi, \lambda), \mathbf{\Psi}(\pi, \lambda)) \\ &= \Phi_1(\pi, \lambda)\Psi_2(\pi, \lambda) - \Phi_2(\pi, \lambda)\Psi_1(\pi, \lambda) \\ &= -\sin\beta\Phi_1(\pi, \lambda) - \cos\beta\Phi_2(\pi, \lambda)\,. \end{aligned} \tag{4.49}$$

Similarly,

$$G(\lambda') = -\sin\beta\Phi_1(\pi, \lambda') - \cos\beta\Phi_2(\pi, \lambda')\,. \tag{4.50}$$

Multiplying (4.49) by $\Phi_1(\pi, \lambda')$ and (4.50) by $\Phi_1(\pi, \lambda)$ and subtracting the resulting equations, we have

$$\begin{aligned} &G(\lambda)\Phi_1(\pi, \lambda') - G(\lambda')\Phi_1(\pi, \lambda) \\ &\qquad = \cos\beta[\Phi_2(\pi, \lambda')\Phi_1(\pi, \lambda) - \Phi_1(\pi, \lambda')\Phi_2(\pi, \lambda)]\,. \end{aligned} \tag{4.51}$$

If we put $\mathbf{y}(x, \lambda) = \mathbf{\Phi}(x, \lambda)$ and $\mathbf{z}(x, \lambda') = \mathbf{\Phi}(x, \lambda')$ in (4.48), and observe that $\mathbf{W}(\mathbf{\Phi}(0, \lambda), \mathbf{\Phi}(0, \lambda')) = 0$ since both $\mathbf{\Phi}(0, \lambda)$ and $\mathbf{\Phi}(0, \lambda')$ satisfy the same boundary condition (4.36) at 0, then it follows that

$$(\lambda - \lambda') \int_0^{\pi} \mathbf{\Phi}^{\top}(x, \lambda)\mathbf{\Phi}(x, \lambda')dx = \Phi_1(\pi, \lambda')\Phi_2(\pi, \lambda) - \Phi_2(\pi, \lambda')\Phi_1(\pi, \lambda) \tag{4.52}$$

which, in view of (4.51), can be reduced to

$$(\lambda - \lambda') \int_0^{\pi} \mathbf{\Phi}^{\top}(x, \lambda)\mathbf{\Phi}(x, \lambda')dx = \frac{G(\lambda')\Phi_1(\pi, \lambda) - G(\lambda)\Phi_1(\pi, \lambda')}{\cos\beta}\,,$$

provided that $\cos\beta \neq 0$. By taking the limit as $\lambda' \to \lambda_n$, we obtain

$$(\lambda - \lambda_n) \int_0^{\pi} \mathbf{\Phi}^{\top}(x, \lambda)\mathbf{\Phi}_n(x)dx = \frac{-G(\lambda)\Phi_{n,1}(\pi)}{\cos\beta} \tag{4.53}$$

since $G(\lambda_n) = 0$ by (4.49) and (4.37). Similarly, we can show that

$$(\lambda - \lambda_n) \int_0^\pi \mathbf{\Phi}^\top(x, \lambda)\mathbf{\Phi}_n(x)dx = \frac{-G(\lambda)\Phi_{n,2}(\pi)}{\sin\beta}, \tag{4.54}$$

provided that $\sin\beta \neq 0$. Differentiating with respect to $\lambda$ and taking the limit as $\lambda \to \lambda_n$, we obtain

$$\|\mathbf{\Phi}_n\|^2 = \int_0^\pi \mathbf{\Phi}_n^\top(x)\mathbf{\Phi}_n(x)dx = \frac{-G'(\lambda_n)\Phi_{n,1}(\pi)}{\cos\beta} \tag{4.55}$$

$$= \frac{-G'(\lambda_n)\Phi_{n,2}(\pi)}{\sin\beta}. \tag{4.56}$$

From (4.46), (4.53) and (4.55), we have for $\beta \neq \pi/2$

$$\frac{\widehat{\Phi}_n(\lambda)}{\|\mathbf{\Phi}_n\|^2} = \frac{G(\lambda)}{(\lambda - \lambda_n)G'(\lambda_n)}, \tag{4.57}$$

and if $\beta = \pi/2$, we use (4.46), (4.54) and (4.56) to obtain the same result. Now the substitution of Eq. (4.57) into (4.47) yields (4.43).

To show the uniform convergence of the series (4.43) on a compact set $K$ of the complex plane, let

$$\mathbf{S}_N(x) = \sum_{n=-N}^{N} F(\lambda_n)\frac{\mathbf{\Phi_n}(x)}{\|\mathbf{\Phi}_n\|^2}.$$

For $\beta \neq \pi/2$, we have by the Cauchy-Schwarz inequality, Equations (4.47), (4.53) (or (4.54) if $\beta = \pi/2$) and (4.55),

$$\begin{aligned}\left|F(\lambda) - \sum_{n=-N}^{N} F(\lambda_n)\frac{G(\lambda)}{(\lambda - \lambda_n)G'(\lambda_n)}\right|^2 &= \left|F(\lambda) - \sum_{n=-N}^{N} F(\lambda_n)\frac{\widehat{\Phi}_n}{\|\mathbf{\Phi_n}\|^2}\right|^2 \\ &= \left|\int_0^\pi [\mathbf{f}(x) - \mathbf{S}_N(x)]^\top \mathbf{\Phi}(x, \lambda)dx\right|^2 \\ &\leq \left(\int_0^\pi \mathbf{\Phi}^\top(x, \lambda)\mathbf{\Phi}(x, \lambda)dx\right)\|\mathbf{f} - \mathbf{S}_N\|^2 \\ &\leq C_K\|\mathbf{f} - \mathbf{S}_N\|^2,\end{aligned}$$

but the last term goes to zero as $N \to \infty$, where

$$C_K = \max_{\lambda\in K}\left(\int_0^\pi \mathbf{\Phi}^\top(x, \lambda)\mathbf{\Phi}(x, \lambda)dx\right) < \infty.$$

To complete the proof, we must show that the function $G(\lambda)$ in (4.49) is essentially the infinite product in (4.44). Of course, the function defined in (4.49) is an entire function whose only zeros are the eigenvalues of the Dirac problem. If we denote the infinite product in (4.44) temporarily by $\widehat{G}(\lambda)$, we can write $G(\lambda) = g(\lambda)\widehat{G}(\lambda)$, where $g(\lambda)$ is an entire function with no zeros. Thus,

$$\frac{G(\lambda)}{G'(\lambda_n)} = \frac{g(\lambda)\widehat{G}(\lambda)}{g(\lambda_n)\widehat{G}'(\lambda_n)},$$

and the conclusion of the theorem remains valid for the function $F(\lambda)/g(\lambda)$. □

**Example 4.14 (Sampling Formula Associated with the Hartley Transform)** As a special case of Theorem 4.7 we obtain a sampling series for the Hartley transform of a bandlimited function. The *Hartley transform* of a function $f(t)$ is formally defined as

$$F(x) = \int_0^\infty f(t)\left\{\cos xt + \sin xt\right\} dt\,, \quad x \in \mathbb{R}.$$

It was introduced in Ref. [28] by the electric engineer Ralph V. L. Hartley as a way to overcome what he considered a drawback of the Fourier transform, i.e.,

$$F(x) = \int_{-\infty}^\infty f(t)\left\{\cos xt - i\sin xt\right\} dt\,, \quad x \in \mathbb{R},$$

namely, representing a real-valued signal $f(t)$ by a complex-valued one, $F(x)$. For more details, see Ref. [52, Chapter 13].

Consider the boundary-value problem (4.35)–(4.37) in which $p(x) = r(x) = 0$:

$$y_2' = \lambda y_1\,, \quad y_1' = -\lambda y_2\,, \tag{4.58}$$

$$y_1(0)\sin\alpha + y_2(0)\cos\alpha = 0\,, \tag{4.59}$$

$$y_1(\pi)\sin\alpha + y_2(\pi)\cos\alpha = 0\,. \tag{4.60}$$

It is easy to see that a solution of (4.58) and (4.59) is given by

$$\mathbf{\Phi}^\top(x,\lambda) = \big(\cos(\lambda x - \alpha), \sin(\lambda x - \alpha)\big)\,.$$

By substituting this solution in (4.60), we obtain

$$\cos(\lambda\pi - \alpha)\sin\alpha + \sin(\lambda\pi - \alpha)\cos\alpha = \sin(\lambda\pi) = 0\,.$$

Hence, the eigenvalues are $\lambda_n = n \in \mathbb{Z}$.

Therefore, the function $G(\lambda)$ defined in Eq. (4.44) is given by

$$G(\lambda) = \lambda \prod_{n=1}^{\infty} \left(1 - \frac{\lambda^2}{n^2}\right) = \frac{\sin \pi\lambda}{\pi}.$$

Therefore, Theorem 4.7 now takes on the form:
If

$$F(\lambda) = \int_0^{\pi} \{f_1(x)\cos(\lambda x - \alpha) + f_2(x)\sin(\lambda x - \alpha)\}dx\,, \quad \lambda \in \mathbb{C}\,, \tag{4.61}$$

for some $f_1$ and $f_2 \in L^2[0, \pi]$, then

$$F(\lambda) = \sum_{n=-\infty}^{\infty} F(n)\frac{\sin \pi(\lambda - n)}{\pi(\lambda - n)}, \quad \lambda \in \mathbb{C}.$$

Observe that the set of functions $F$ described in (4.61) coincides with the Paley-Wiener space $PW_\pi$.

In particular, if $f_1 = f_2 = f \in L^2[0, \pi]$, we have a generalization of the Hartley transform $F$ of a bandlimited function $f$ to the interval $[0, \pi]$

$$F(\lambda) = \int_0^{\pi} f(x)\big\{\cos(\lambda x - \alpha) + \sin(\lambda x - \alpha)\big\}\,dx\,,$$

which can be reconstructed via

$$F(\lambda) = \sum_{n=-\infty}^{\infty} F(n)\frac{\sin \pi(\lambda - n)}{\pi(\lambda - n)}, \quad \lambda \in \mathbb{C}.$$

When $\alpha = 0$ we obtain the Hartley transform

$$F(\lambda) = \int_0^{\pi} f(x)\big\{\cos \lambda x + \sin \lambda x\big\}\,dx\,.$$

The set of the above bandlimited functions to $[0, \pi]$ in the Hartley transform sense is included in $PW_\pi$.

## 4.3 Other Versions of the Kramer Sampling Theorem

We end this chapter by showing two more generalizations of the Kramer sampling theorem. The first one in a Banach space endowed with a semi-inner product, and the second one in the setting of Hilbert spaces of distributions having a reproducing distribution. These versions use some unfamiliar mathematical techniques and they can be omitted in a first reading.

## 4.3.1 The Semi-inner-Product Version in Banach Spaces

A *semi-inner-product* in a Banach space possesses some but not all properties of an inner product. Thus, some Hilbert space arguments and results become available in the presence of a semi-inner-product in a Banach space. The sampling results in this section are taken from Ref. [21].

### Reproducing Kernel Banach Spaces

**Definition 4.4** A reproducing kernel Banach space (RKBS hereafter) on $\Omega \subseteq \mathbb{R}$ (or $\mathbb{C}$) is a reflexive Banach space $\mathcal{B}$ of functions on $\Omega$ for which its dual space $\mathcal{B}^*$ is isometric to a Banach space $\widetilde{\mathcal{B}}$ of functions on $\Omega$, and the point evaluation functional at $t$ is continuous on both $\mathcal{B}$ and $\widetilde{\mathcal{B}}$ for each $t \in \Omega$.

It has been proved in Ref. [59] that there exists a reproducing kernel for an RKBS as defined above.

To this end, we introduce the bilinear form on $\mathcal{B} \times \mathcal{B}^*$ by setting

$$\big(u, v^*\big)_{\mathcal{B}} := v^*(u)\,, \quad u \in \mathcal{B},\ v^* \in \mathcal{B}^*\,.$$

As $\mathcal{B}$ is a *reflexive Banach space*, then for any bounded linear functional $T$ on $\mathcal{B}^*$ there exists a unique $u \in \mathcal{B}$ such that $T(v^*) = \big(u, v^*\big)_{\mathcal{B}}$ for each $v^* \in \mathcal{B}^*$. The following result, proved in [59, Theorem 2], holds:

**Theorem 4.8** *Suppose that $\mathcal{B}$ is an RKBS on $\Omega$. Then there exists a unique function $k : \Omega \times \Omega \longrightarrow \mathbb{C}$ such that the following statements hold:*

*(a) For every $t \in \Omega$, $k(\cdot, t) \in \mathcal{B}^*$ and $f(t) = \big(f, k(\cdot, t)\big)_{\mathcal{B}}$ for all $f \in \mathcal{B}$.*
*(b) For every $t \in \Omega$, $k(t, \cdot) \in \mathcal{B}$ and $f^*(t) = \big(k(t, \cdot), f^*\big)_{\mathcal{B}}$ for all $f^* \in \mathcal{B}^*$.*
*(c) The linear span of $\big\{k(t, \cdot) : t \in \Omega\big\}$ is dense in $\mathcal{B}$.*
*(d) The linear span of $\big\{k(\cdot, t) : t \in \Omega\big\}$ is dense in $\mathcal{B}^*$.*
*(e) For all $t, s \in \Omega$, $k(t, s) = \big(k(t, \cdot), k(\cdot, s)\big)_{\mathcal{B}}$.*

The function $k$ in Theorem 4.8 is the reproducing kernel for the RKBS $\mathcal{B}$. This reproducing kernel is unique. However, as showed in Ref. [59], different RKBSs may have the same reproducing kernel: For $1 < p < \infty$, the *Paley-Wiener classes*

$$\mathcal{B}_p := \Big\{f \in C(\mathbb{R})\ :\ \operatorname{supp} \widehat{f} \subset [-1/2, 1/2] \text{ and } \widehat{f} \in L^p[-1/2, 1/2]\Big\}$$

with norm $\|f\|_{\mathcal{B}_p} := \|\widehat{f}\|_{L^p[-1/2,1/2]}$ are RKBSs (not isomorphic), and they all have the function $k(t,s) = \operatorname{sinc}(t-s)$, $t, s \in \mathbb{R}$, as the reproducing kernel.

Here the inverse Fourier transform is given by $f(t) = \int_{-1/2}^{1/2} \widehat{f}(w)\, e^{2\pi i w t} dw$, $t \in \mathbb{R}$.

In other words, although we have at hand a reproducing kernel $k$, we cannot determine the norm on $\mathcal{B}$. Also, the reproducing kernel for a general RKBS may be an arbitrary function on $\Omega \times \Omega$ which, in particular, can be non-symmetric or non-positive definite (see Ref. [59, Proposition 5]).

In order to have the reproducing kernel of a RKBS the desired properties of a RKHS, we impose certain structures on RKBSs, which in some sense are substitutes of the inner product for RKHSs. For this purpose, we adopt the semi-inner-product introduced by Lumer in Ref. [36] (see also Refs. [24, 59]):

### Semi-inner-Product Reproducing Kernel Banach Space

Let $\mathcal{B}$ be a *Banach space*. A *semi-inner-product* on $\mathcal{B}$ is a function

$$[\cdot,\cdot] : \mathcal{B} \times \mathcal{B} \longrightarrow \mathbb{C},$$

such that, for all $x_1, x_2, x_3 \in \mathcal{B}$ and $\alpha \in \mathbb{C}$:

1. $[x_1 + x_2, x_3] = [x_1, x_3] + [x_2, x_3]$,
2. $[\alpha x_1, x_2] = \alpha[x_1, x_2]$ and $[x_1, \alpha x_2] = \overline{\alpha}\,[x_1, x_2]$,
3. $[x_1, x_1] > 0$ for all $x_1 \neq 0$,
4. $|[x_1, x_2]|^2 \leq [x_1, x_1][x_2, x_2]$.

The difference between a semi-inner-product and an inner product is the conjugate symmetry and, as a consequence, a semi-inner-product may not be additive in the second variable. Every normed vector space $\mathcal{B}$ has a semi-inner-product that induces its norm (see Refs. [24, 36]), that is,

$$\|x\|_{\mathcal{B}} = [x, x]^{1/2} \quad \text{for each } x \in \mathcal{B}.$$

In general, a semi-inner-product for a normed vector space may not be unique; however, if the space $\mathcal{B}$ is *uniformly Fréchet differentiable* we obtain the uniqueness of the semi-inner-product (see Ref. [59] for the details).

The space $\mathcal{B}$ is uniformly Fréchet differentiable if for all $x, y \in \mathcal{B}$ with $x \neq 0$, the limit

$$\lim_{\substack{t \in \mathbb{R} \\ t \to 0}} \frac{\|x + ty\|_{\mathcal{B}} - \|x\|_{\mathcal{B}}}{t}$$

exists and it is uniform on $\mathcal{S}(\mathcal{B}) \times \mathcal{S}(\mathcal{B})$ where $\mathcal{S}(\mathcal{B}) := \{x \in \mathcal{B} : \|x\|_{\mathcal{B}} = 1\}$.

Assuming also that the Banach space is *uniformly convex* we obtain a *Riesz representation theorem*: For each $f \in \mathcal{B}^*$ there exists a unique $x \in \mathcal{B}$ such that $f = x^*$. In other words,

$$f(y) = \big[y, x\big]_{\mathcal{B}} \quad \text{for all } y \in \mathcal{B}\,.$$

Moreover, $\|f\|_{\mathcal{B}^*} = \|x\|_{\mathcal{B}}$ (see, for instance, Ref. [24]).

The Banach space $\mathcal{B}$ is uniformly convex if for all $\varepsilon > 0$ there exists $\delta > 0$ such that $\|x + y\|_{\mathcal{B}} \leq 2 - \delta$ for all $x, y \in \mathcal{S}(\mathcal{B})$ with $\|x - y\|_{\mathcal{B}} \geq \varepsilon$.

Notice that if $\mathcal{B}$ is uniformly convex, then it is reflexive (see Ref. [40, p. 410]) and *strictly convex*, that is, for every $x, y \in \mathcal{B}$ with $x \neq y$ and $\|x\| = \|y\| = 1$, we have that $\|x + y\| < 2$.

For $1 < p < \infty$, the classical $L^p(I)$, where $I$ denotes any interval on $\mathbb{R}$, and $\ell^p(\mathbb{N})$ spaces are uniformly convex and uniformly Fréchet differentiable Banach spaces. Their semi-inner-product are given, respectively, by

$$[f, g]_p = \|g\|_p^{2-p} \int_I f(t)\,\overline{g(t)}\,|g(t)|^{p-2}\,dt$$

and

$$[x, y]_p = \|y\|_p^{2-p} \sum_{n=1}^{\infty} x_n\,\overline{y}_n\,|y_n|^{p-2}.$$

**Definition 4.5 (s.i.p. RKBS)** A semi-inner-product reproducing kernel Banach space (s.i.p. RKBS hereafter) on $\Omega$ is a uniformly convex and uniformly Fréchet differentiable reproducing kernel Banach space on $\Omega$.

An RKHS is an s.i.p. RKBS. Also, the dual of an s.i.p. RKBS remains an s.i.p. RKBS. An s.i.p. RKBS $\mathcal{B}$ has a unique semi-inner-product which represents the interaction between $\mathcal{B}$ and $\mathcal{B}^*$. This leads to a more specific representation of the reproducing kernel. Precisely, the following result proved in Ref. [59, Theorem 9] holds:

**Theorem 4.9** *Let $\mathcal{B}$ be an s.i.p. RKBS on $\Omega$ and $k$ its reproducing kernel. Then, there exists a unique function $G : \Omega \times \Omega \longrightarrow \mathbb{C}$ such that $\{G(t, \cdot) : t \in \Omega\} \subset \mathcal{B}$ and, for all $f \in \mathcal{B}$, we have the reproducing formula*

$$f(t) = [f, G(t, \cdot)]_{\mathcal{B}}\,, \quad t \in \Omega\,.$$

*Moreover,*

$$k(\cdot, t) = (G(t, \cdot))^*\,, \quad t \in \Omega$$

*and*

$$f^*(t) = [k(t, \cdot), f]_{\mathcal{B}} \quad \textit{for all } f \in \mathcal{B}, \quad t \in \Omega$$

We call the unique function $G$ in theorem above the *s.i.p. kernel* of the s.i.p. RKBS $\mathcal{B}$. It coincides with the reproducing kernel $k$ when $\mathcal{B}$ is an RKHS.

In general, whenever $G = k$ in Theorem 4.9, we call $G$ an *s.i.p. reproducing kernel*. Thus, an s.i.p. reproducing kernel $G$ satisfies that

$$G(t, s) = [G(t, \cdot), G(s, \cdot)]_{\mathcal{B}}, \quad t, s \in \Omega .$$

**The Analogous Kramer Sampling Result**

Consider a separable complex uniform Banach space $\mathcal{B}$ (i.e., both uniformly Fréchet differentiable and uniformly convex) and denote by $[\cdot, \cdot]_{\mathcal{B}}$ the unique compatible semi-inner product on $\mathcal{B}$; let $\mathcal{B}^*$ be its dual space. The counterpart to Riesz bases expansions in an s.i.p. RKBS is given by $X_d$-Riesz and $X_d^*$-Riesz bases.

Let $X_d$ be a *BK-space* on $\mathbb{N}$, that is, a Banach space of complex sequences $c = \{c_n\}_{n\in\mathbb{N}}$ such that, for each $m \in \mathbb{N}$, the linear functional

$$X_d \ni c = \{c_n\}_{n\in\mathbb{N}} \longmapsto c_m \in \mathbb{C}$$

is continuous on $X_d$.

It is known, see Ref. [32], that its dual space $X_d^*$ is also a BK-space such that the series $\sum_{n=1}^{\infty} c_n\, d_n$ converges for every $c \in X_d$ and $d \in X_d^*$. We suppose that if the series above converges for every $c \in X_d$, then $d \in X_d^*$ and if it converges for every $d \in X_d^*$, then $c \in X_d$.

We also assume that $X_d$ is reflexive, and that the sequence of the canonical unit vectors $\{\delta_n\}_{n=1}^{\infty}$ is a *Schauder basis* for both $X_d$ and $X_d^*$.

An example of such BK-spaces is $X_d = \ell^p(\mathbb{N})$ for $1 < p < \infty$; in this case, $X_d^* = \ell^q(\mathbb{N})$ with $1/p + 1/q = 1$.

Let $\{x_n^*\}_{n=1}^{\infty} \subset \mathcal{B}^*$ be an $X_d^*$-Riesz basis for $\mathcal{B}^*$. This means that

1. $\overline{\text{span}}\big\{x_n^* : n \in \mathbb{N}\big\} = \mathcal{B}^*$.
2. $\sum_{n=1}^{\infty} c_n x_n^*$ converges in $\mathcal{B}^*$ for all $c \in X_d^*$.
3. There exist $0 < A \le B < \infty$ such that

$$A\, \|c\|_{X_d^*} \le \Big\| \sum_{n=1}^{\infty} c_n x_n^* \Big\|_{\mathcal{B}^*} \le B\, \|c\|_{X_d^*} \quad \text{for all } c \in X_d^* . \tag{4.62}$$

From [58, Theorem 2.15], there exists a unique (dual) $X_d$-Riesz basis $\{y_n\}_{n=1}^{\infty}$ for $\mathcal{B}$ such that $[y_m, x_n]_{\mathcal{B}} = \delta_{m,n}$ for $m, n \in \mathbb{N}$, and satisfying the expansions:

$$x = \sum_{n=1}^{\infty} [x, x_n]_{\mathcal{B}}\, y_n \quad \text{for all } x \in \mathcal{B}$$

$$\text{and} \tag{4.63}$$

$$x^* = \sum_{n=1}^{\infty} [y_n, x]_{\mathcal{B}}\, x_n^* \quad \text{for all } x^* \in \mathcal{B}^*.$$

If the spaces $X_d$ and $X_d^*$ possess the additional property that for all $c \in X_d$ and $d \in X_d^*$ the series $\sum_{n=1}^{\infty} c_n d_n$ converges absolutely, then the expansions in (4.63) are unconditionally convergent, i.e., independent of the summation order (see Ref. [58, p. 7]). In particular, it is true for $\ell^p$-Riesz bases due to Hölder inequality.

Next, we obtain an s.i.p. RKBS with explicit s.i.p. reproducing kernel where a sampling theory holds. It is the Banach counterpart of the Kramer sampling theorem given in Sect. 4.1.2. Consider a $\mathcal{B}$-valued function $K : \Omega \subset \mathbb{C} \to \mathcal{B}$ and define, for each $x \in \mathcal{B}$, the function

$$\begin{aligned} f_x : \Omega &\longrightarrow \mathbb{C} \\ t &\longmapsto f_x(t) = [x, K(t)]_{\mathcal{B}}\,. \end{aligned}$$

Then, we have a linear transform $\mathcal{T}_K$ on $\mathcal{B}$ with values in $\mathbb{C}^{\Omega}$ such that $\mathcal{T}_K x = f_x$. Indeed, for $x, y \in \mathcal{B}$ and $\alpha, \beta \in \mathbb{C}$, we have

$$f_{\alpha x+\beta y}(t) = [\alpha x + \beta y, K(t)]_{\mathcal{B}} = \alpha[x, K(t)]_{\mathcal{B}} + \beta[y, K(t)]_{\mathcal{B}} = \alpha f_x(t) + \beta f_y(t)\,,$$

for all $t \in \Omega$.

Having in mind (4.63), we can write

$$\big[K(t)\big]^* = \sum_{n=1}^{\infty} [y_n, K(t)]_{\mathcal{B}}\, x_n^*\,, \qquad \text{for each } t \in \Omega\,.$$

We denote $S_n(t) := [y_n, K(t)]_{\mathcal{B}} = f_{y_n}(t)$, $t \in \Omega$. Suppose that there exists a sequence $\{t_n\}_{n=1}^{\infty}$ in $\Omega$ and $\{a_n\}_{n=1}^{\infty}$ in $\mathbb{C} \setminus \{0\}$ such that the interpolation property

$$S_n(t_m) = a_n\, \delta_{n,m}\,, \tag{4.64}$$

holds. As a consequence, we have that $\big[K(t_m)\big]^* = a_m\, x_m^*$, $m \in \mathbb{N}$, and the operator $\mathcal{T}_K$ is one-to-one.

Indeed, if $f_x(t) = 0$ for all $t \in \Omega$,

$$0 = f_x(t) = [x, K(t)]_{\mathcal{B}} = \Big[\sum_{n=1}^{\infty}[x, x_n]_{\mathcal{B}}\, y_n, K(t)\Big]$$

$$= \sum_{n=1}^{\infty}[x, x_n]_{\mathcal{B}}\,[y_n, K(t)]_{\mathcal{B}} = \sum_{n=1}^{\infty}[x, x_n]_{\mathcal{B}}\, S_n(t)\,,$$

where we have used that $x \mapsto [x, y]_{\mathcal{B}}$ is a continuous functional for any fixed $y \in \mathcal{B}$.

Evaluating at $t_m$, we have that $[x, x_m]_{\mathcal{B}} = 0$ for each $m \in \mathbb{N}$. This implies that $x = 0$ and then, the operator $\mathcal{T}_K$ is one-to-one.

Denote $\mathcal{B}_K := \mathcal{T}_K(\mathcal{B})$, the range of the operator $\mathcal{T}_K$. If we define $\|f_x\|_{\mathcal{B}_K} := \|x\|_{\mathcal{B}}$ we obtain that $\mathcal{B}_K$ is a Banach space of functions defined on $\Omega$ and valued on $\mathbb{C}$.

Moreover, $[f_x, f_y]_{\mathcal{B}_K} := [x, y]_{\mathcal{B}}$ defines a compatible semi-inner-product on $\mathcal{B}_K$.

The space $\mathcal{B}_K$ becomes an s.i.p. RKBS whose s.i.p. reproducing kernel is given by

$$k(t, s) = [K(t), K(s)]_{\mathcal{B}}\,, \quad t, s \in \Omega\,.$$

Indeed, for each $t \in \Omega$, the evaluation functional $E_t : \mathcal{B}_K \longrightarrow \mathbb{C}$ is continuous:

$$|E_t(f_x)| = |f_x(t)| = |[x, K(t)]_{\mathcal{B}}| \leq \|x\|_{\mathcal{B}}\|K(t)\|_{\mathcal{B}} = \|K(t)\|_{\mathcal{B}}\|f_x\|_{\mathcal{B}_K}\,.$$

Observe that, by definition, $k_t := k(t, \cdot) = f_{K(t)} \in \mathcal{B}_K$ for all $t \in \Omega$. Hence, we deduce that

$$f_x(t) = [x, K(t)]_{\mathcal{B}} = [f_x, k_t]_{\mathcal{B}_K} = [f_x, k(t, \cdot)]_{\mathcal{B}_K}\,,$$

being $k$ the s.i.p. reproducing kernel for $\mathcal{B}_K$. See Ref. [59, Theorem 10] for more details.

Note also that convergence in the norm of $\mathcal{B}_K$ implies pointwise convergence and uniform convergence in subsets of $\Omega$ where the function $t \longmapsto \|K(t)\|_{\mathcal{B}}$ is bounded.

**Lemma 4.3** *For each $t \in \Omega$, the sequence $\big\{S_n(t)\big\}_{n=1}^{\infty}$ belongs to $X_d^*$.*

***Proof*** Consider $c \in X_d$ and $t \in \Omega$. We must prove that the series $\sum_{n=1}^{\infty} c_n S_n(t)$ is convergent. Indeed, by using that for each $z \in \mathcal{B}$ the mapping $x \mapsto [x, z]_{\mathcal{B}}$ is a continuous linear functional on $\mathcal{B}$, and $\{y_n\}_{n=1}^{\infty}$ is an $X_d$-Riesz basis for $\mathcal{B}$, we get

$$\Big|\sum_{n=1}^{\infty} c_n S_n(t)\Big| = \Big|\sum_{n=1}^{\infty} c_n [y_n, K(t)]_{\mathcal{B}}\Big| = \Big|\Big[\sum_{n=1}^{\infty} c_n y_n, K(t)\Big]_{\mathcal{B}}\Big|$$

$$\leq \|K(t)\|_{\mathcal{B}} \Big\|\sum_{n=1}^{\infty} c_n y_n\Big\|_{\mathcal{B}} \leq B\|K(t)\|_{\mathcal{B}}\|c\|_{X_d}.$$

□

**Theorem 4.10 (A Kramer-Type Sampling Theorem for $\mathcal{B}_K$)** *Under the above circumstances, suppose that, for each $t \in \Omega$, we have the expansion*

$$\big[K(t)\big]^* = \sum_{n=1}^{\infty} S_n(t) x_n^*,$$

*where $\{x_n^*\}_{n=1}^{\infty} \subset \mathcal{B}^*$ is an $X_d^*$-Riesz basis for $\mathcal{B}^*$ and $S_n(t) = [y_n, K(t)]_{\mathcal{B}}$, being $\{y_n\}_{n=1}^{\infty}$ the dual $X_d$-Riesz basis for $\mathcal{B}$ of $\{x_n^*\}_{n=1}^{\infty}$.*

*Assume also the existence of sequences $\{t_m\}_{m=1}^{\infty} \subset \mathbb{C}$ and $\{a_m\}_{m=1}^{\infty} \subset \mathbb{C} \setminus \{0\}$ such that the interpolation property* (4.64) *holds. Then, the sequence $\{S_n\}_{n=1}^{\infty}$ is an $X_d$-Riesz basis for $\mathcal{B}_K$ and, for each $f \in \mathcal{B}_K$, we have the sampling expansion*

$$f(t) = \sum_{n=1}^{\infty} f(t_n) \frac{S_n(t)}{a_n}, \quad t \in \Omega.$$

*The convergence of the above series is in the norm of $\mathcal{B}_K$, and uniform on subsets of $\Omega$ where the function $t \mapsto \|K(t)\|_{\mathcal{B}}$ is bounded.*

***Proof*** First, we prove that the sequence $\{S_n\}_{n=1}^{\infty}$ is an $X_d$-Riesz basis for $\mathcal{B}_K$.

1. Consider $t \in \Omega$ and $x \in \mathcal{B}$. Then,

$$f_x(t) = [x, K(t)]_{\mathcal{B}} = \Big[\sum_{n=1}^{\infty} [x, x_n]_{\mathcal{B}}\, y_n, K(t)\Big]_{\mathcal{B}} = \sum_{n=1}^{\infty} [x, x_n]_{\mathcal{B}}\, S_n(t), \quad t \in \Omega.$$

   Hence, $\overline{\text{span}}\{S_n\}_{n=1}^{\infty} = \mathcal{B}_K$.
2. Let $c$ be a sequence in $X_d$. Since operator $\mathcal{T}_K$ is an isometry and $\{y_n\}_{n=1}^{\infty}$ is an $X_d$-Riesz basis for $\mathcal{B}$,

$$\Big\|\sum_{n=1}^{\infty} c_n S_n\Big\|_{\mathcal{B}_K} = \Big\|\sum_{n=1}^{\infty} c_n y_n\Big\|_{\mathcal{B}},$$

   and thus the series $\sum_{n=1}^{\infty} c_n S_n$ converges in $\mathcal{B}_K$ for each $c \in X_d$.

3. As operator $\mathcal{T}_K$ is an isometry, for every $c \in X_d$ we have

$$A\,\|c\|_{X_d} \le \Big\| \sum_{n=1}^{\infty} c_n\, y_n \Big\|_{\mathcal{B}} = \Big\| \sum_{n=1}^{\infty} c_n\, S_n \Big\|_{\mathcal{B}_K} \le B\,\|c\|_{X_d}\,.$$

Now, for $m \in \mathbb{N}$ we have

$$f_x(t_m) = [x, K(t_m)]_{\mathcal{B}} = \sum_{n=1}^{\infty} [x, x_n]_{\mathcal{B}}\, S_n(t_m) = a_m [x, x_m]_{\mathcal{B}}\,.$$

Thus,

$$f_x(t) = \sum_{n=1}^{\infty} [x, x_n]_{\mathcal{B}}\, S_n(t) = \sum_{n=1}^{\infty} \frac{f_x(t_n)}{a_n} S_n(t)$$

in the norm of $\mathcal{B}_K$.

The pointwise and uniform convergence comes from the fact that $\mathcal{B}_K$ is an s.i.p. RKBS. □

**Example 4.15 (Sampling in an s.i.p. RKBS Associated to the Banach Space $L^p[-\frac{1}{2}, \frac{1}{2}]$ with $p \in (1, 2]$)** Consider $p \in (1, 2]$ and its conjugate index $q \in \mathbb{R}$, i.e., $1/p + 1/q = 1$. We consider the compatible semi-inner-product for $\mathcal{B} := L^p[-\frac{1}{2}, \frac{1}{2}]$ given by

$$[f, g]_p = \|g\|_p^{2-p} \int_{-1/2}^{1/2} f(x)\,\overline{g(x)}\,|g(x)|^{p-2}\,dx\,.$$

Remember that $\mathcal{B}^* = L^q[-\frac{1}{2}, \frac{1}{2}]$. We take $X_d := \ell^q(\mathbb{Z})$ and, consequently, $X_d^* = \ell^p(\mathbb{Z})$.

Define $e_n(w) := \mathrm{e}^{2\pi i n w}$ for $n \in \mathbb{Z}$. An easy computation shows that $\|e_n\|_p = 1$. On the other hand, we have that $e_n^*(w) = \mathrm{e}^{-2\pi i n w}$ and $\|e_n^*\|_q = \|e_n\|_p = 1$ (see Ref. [59]). We also know that (see Ref. [48, p. 20]):

$$\overline{\operatorname{span}}\big\{\mathrm{e}^{-2\pi i n w} : n \in \mathbb{Z}\big\} = L^q[-\frac{1}{2}, \frac{1}{2}]\,.$$

We define the linear operator

$$\begin{aligned} U : L^p[-\tfrac{1}{2}, \tfrac{1}{2}] &\longrightarrow \mathbb{C}^{\mathbb{Z}} \\ F \quad &\longmapsto \big\{[F, e_n]_p\big\}_{n\in\mathbb{Z}} \end{aligned}.$$

The Hausdorff-Young theorem (see Ref. [60, p. 101]) ensures that $U$ is a bounded operator with values on $\ell^q(\mathbb{Z})$ (and thus a closed operator).

Hence we have that $U(L^p[-\frac{1}{2}, \frac{1}{2}])$ is a closed subspace of $\ell^q(\mathbb{Z})$ and, as well as a Banach space with the metric induced by $\ell^q(\mathbb{Z})$.

Remember that the sequence $\{\boldsymbol{\delta}_n\}_{n\in\mathbb{Z}}$, where $\boldsymbol{\delta}_n(m) = \delta_{n,m}$ is a Schauder basis for $\ell^q(\mathbb{Z})$. Thus, as $\boldsymbol{\delta}_n = U(e_n)$, we obtain that $U$ is a surjective mapping. By using Proposition 2.12 in Ref. [58], the sequence $\{e_n^*\}_{n\in\mathbb{Z}}$ is an $\ell^p(\mathbb{Z})$-Riesz basis for $\mathcal{B}^* = L^q[-1/2, 1/2]$.

Now, we define the kernel

$$\mathbb{C} \ni z \longmapsto K(z) \in L^p[-1/2, 1/2] \,:\, [K(z)](x) := \mathrm{e}^{2\pi i z x},\ x \in [-1/2, 1/2]\,.$$

Thus, we obtain the following s.i.p. RKBS

$$\mathcal{B}_K := \left\{ f(z) = \left[F, \mathrm{e}^{2\pi i z w}\right]_p,\ z \in \mathbb{C},\ \text{where}\ F \in L^p[-1/2, 1/2] \right\},$$

endowed with the norm $\|f\|_{\mathcal{B}_K} := \|F\|_{L^p[-1/2,1/2]}$.

Next, we compute $S_n(z) := [e_n, K(z)]_p$. First, observe that, if we write $z = x + iy$, then

$$\begin{aligned}\|K(z)\|_p &= \left(\int_{-1/2}^{1/2} \left|\mathrm{e}^{2\pi i z w}\right|^p dw\right)^{1/p} = \left(\int_{-1/2}^{1/2} \mathrm{e}^{-2\pi y w p} dw\right)^{1/p} \\ &= \operatorname{sinc}^{1/p}(iyp)\,.\end{aligned}$$

Thus, we have that

$$\begin{aligned}S_n(z) = [e_n, K(z)]_p &= \operatorname{sinc}^{(2-p)/p}(iyp) \int_{-1/2}^{1/2} \mathrm{e}^{2\pi i n w} \mathrm{e}^{-2\pi i z w} \mathrm{e}^{-2\pi y w (p-2)} dw \\ &= \operatorname{sinc}^{(2-p)/p}(iyp) \int_{-1/2}^{1/2} \mathrm{e}^{-2\pi i[(z-n)-iy(p-2)]w} dw \\ &= \operatorname{sinc}^{(2-p)/p}(iyp) \operatorname{sinc}\left[(z-n) - iy(p-2)\right],\quad z \in \mathbb{C},\end{aligned}$$

for $n \in \mathbb{Z}$. Moreover, $S_n(m) = \delta_{m,n}$ for every $m, n \in \mathbb{Z}$.

Finally, Theorem 4.10 gives, for any $f \in \mathcal{B}_K$, the following sampling formula:

$$f(z) = \operatorname{sinc}^{(2-p)/p}(iyp) \sum_{n=-\infty}^{\infty} f(n) \operatorname{sinc}\left[(z-n) - iy(p-2)\right], \tag{4.65}$$

where $z = x + iy \in \mathbb{C}$. The convergence of the series in (4.65) is uniform on horizontal strips of $\mathbb{C}$.

Observe that, if $p = 2$ or $z \in \mathbb{R}$, formula (4.65) coincides with the cardinal series.

For further information on the recent topic of sampling in Banach spaces, see, for instance, Refs. [3, 10, 11, 26, 27, 39, 58].

### 4.3.2 A Distributional Version of the Kramer Sampling Theorem

Following Ref. [18], a Hilbert space $\mathcal{H}$ whose elements are distributions supported on the interval $I = [0, 1]$ and for which the natural injection from $\mathcal{H}$ into $\mathcal{E}'(\mathbb{R})$ (the space of compact supported distributions) is continuous is called a *Hilbert space of distributions on I.*

In general, it is clear that a Hilbert space of distributions on $I$ having the system $\{e^{2\pi inx}\chi_I(x)\}_{n\in\mathbb{Z}}$ as a complete orthogonal system is completely determined by the sequence of positive numbers $c_n = \|e^{2\pi inx}\chi_I(x)\|_{\mathcal{H}}$, $n \in \mathbb{Z}$.

In Ref. [18], it is proved that if the sequence $\{c_n^{-1/2}e^{2\pi inx}\chi_I(x)\}_{n\in\mathbb{Z}}$ is a complete orthonormal system for a Hilbert space $\mathcal{H}$ of distributions on $I$, then we have

$$\sum_{n\in\mathbb{Z}} \frac{1}{c_n(1+n^2)} < \infty \tag{4.66}$$

and the inner product in $\mathcal{H}$ is given by

$$\langle T, S\rangle_{\mathcal{H}} = \sum_{n\in\mathbb{Z}} c_n \widehat{T}(n)\,\overline{\widehat{S}(n)}, \quad T, S \in \mathcal{H},$$

where $\widehat{T}$ and $\widehat{S}$ are, respectively, the distributional Fourier transforms of $T$ and $S$. Throughout this section, the distributional Fourier transform of $T \in \mathcal{E}'(\mathbb{R})$ is defined as

$$\widehat{T}(t) = \langle T, e^{-2\pi itx}\rangle, \quad t \in \mathbb{R}.$$

For each $m \in \mathbb{Z}$, denote $T_m(x) = c_m^{-1/2}e^{2\pi imx}\chi_I(x)$. Since

$$\begin{aligned}\widehat{T_m}(w) &= \int_0^1 c_m^{-1/2}\, e^{2\pi imx}\, e^{-2\pi iwx}\, dx \\ &= c_m^{-1/2}\, e^{-i\pi(w-m)} \frac{\sin\pi(w-m)}{\pi(w-m)}, \qquad w \in \mathbb{R},\end{aligned}$$

we get

$$\left\langle T, c_n^{-1/2}\, e^{2\pi inx}\chi_I(x)\right\rangle_{\mathcal{H}} = c_n^{1/2}\,\widehat{T}(n) \qquad \text{for all } T \in \mathcal{H} \tag{4.67}$$

and, any $T \in \mathcal{H}$ can be expanded as

$$T = \sum_{n\in\mathbb{Z}} \widehat{T}(n)\, e^{2\pi i n x} \chi_I(x)\,, \tag{4.68}$$

where the series converges both in the Hilbert space of distributions $\mathcal{H}$, and in $\mathcal{E}'(\mathbb{R})$ (see Ref. [18] for the details).

Assume that $\left\{T_n = c_n^{-1/2}\, e^{2\pi i n x} \chi_I(x)\right\}_{n\in\mathbb{Z}}$ is a complete orthonormal system for a Hilbert space $\mathcal{H}$ of distributions on $I$.

Since the embedding of $\mathcal{H}$ into $\mathcal{E}'(\mathbb{R})$ is continuous, for each test function $\phi \in \mathcal{D}(\mathbb{R})$ (the vector space of all complex-valued functions that are infinitely smooth and zero outside some finite interval endowed with its usual topology) there exists a constant $C_\phi \geq 0$ such that

$$|T(\phi)| \leq C_\phi \|T\|_{\mathcal{H}} \quad \text{for any } T \text{ in } \mathcal{H}\,.$$

In particular, if we choose $\phi_t(x) = e^{-2\pi i t x}$ on a neighborhood of $I$ then, for every $t \in \mathbb{R}$, there exists a constant $C_t$ such that

$$|\widehat{T}(t)| \leq C_t \|T\|_{\mathcal{H}} \quad \text{for any } T \text{ in } \mathcal{H}\,,$$

where $\widehat{T}$ denotes the distributional Fourier transform of $T$. This implies that, for every $t \in \mathbb{R}$, the linear functional $T \mapsto \widehat{T}(t)$ is continuous in $\mathcal{H}$.

By using the Riesz representation theorem, there exists an element $K(t)$ in $\mathcal{H}$ such that

$$\widehat{T}(t) = \langle T, K(t)\rangle_{\mathcal{H}}, \quad t \in \mathbb{R}\,.$$

Having in mind (4.67), for every $n \in \mathbb{Z}$ we have that $K(n) = c_n^{-1/2}\, T_n$. Besides,

$$\begin{aligned} S_n(t) :&= \langle T_n, K(t)\rangle_{\mathcal{H}} = \widehat{T_n}(t) \\ &= c_n^{-1/2}\, e^{-i\pi(t-n)} \frac{\sin \pi(t-n)}{\pi(t-n)}, \quad t \in \mathbb{R}\,. \end{aligned}$$

Moreover, the function $t \longmapsto \|K(t)\| = \sup_{\|T\|=1} |\widehat{T}(t)|$ is bounded on compact subsets of $\mathbb{R}$.

Indeed, by applying both sides of (4.68) on the test function $e^{-2\pi i t x}$ it follows that

$$\widehat{T}(t) = \sum_{n\in\mathbb{Z}} \widehat{T}(n) \frac{1-e^{-2\pi i t}}{2\pi i(t-n)} = \sum_{n\in\mathbb{Z}} c_n^{1/2}\, \widehat{T}(n) \frac{1-e^{-2\pi i t}}{c_n^{1/2}\, 2\pi i(t-n)}, \quad t \in \mathbb{R}\,. \tag{4.69}$$

The boundedness on compact subsets of $\mathbb{R}$ is deduced by using Cauchy-Schwarz inequality in (4.69) and taking into account (4.66) and that

$$\|T\|^2 = \sum_{n\in\mathbb{Z}} c_n\,|\widehat{T}(n)|^2\,.$$

According to Theorem 4.2, we obtain the following sampling theorem for the RKHS $\mathcal{H}_K := \mathcal{F}(\mathcal{H})$, where $\mathcal{F}$ denotes the distributional Fourier transform:

**Example 4.16 (The Space $\mathcal{H}_K$ is Associated to a Hilbert Space of Distributions)** Let $\mathcal{H}$ be a Hilbert space of distributions on $I = [0, 1]$ such that the sequence $\{c_n^{-1/2} e^{2\pi inx} \chi_I(x)\}_{n\in\mathbb{Z}}$ is a complete orthonormal system for $\mathcal{H}$. Then, under the above circumstances, any function $g \in \mathcal{H}_K$, i.e.,

$$g(t) = \langle T, K(t)\rangle_{\mathcal{H}}, \quad t \in \mathbb{R},$$

for some $T \in \mathcal{H}$, can be recovered from its samples $\{g(n)\}_{n\in\mathbb{Z}}$ by means of the sampling formula

$$\begin{aligned} g(t) &= \sum_{n\in\mathbb{Z}} c_n^{1/2}\, g(n)\, S_n(t) \\ &= \sum_{n\in\mathbb{Z}} g(n) e^{-i\pi(t-n)} \frac{\sin\pi(t-n)}{\pi(t-n)}, \quad t \in \mathbb{R}\,. \end{aligned}$$

The series converges in the $\mathcal{H}_K$-norm sense and also absolutely and uniformly on compact subsets of $\mathbb{R}$.

### Hilbert Spaces of Distributions Having a Reproducing Distribution

A distributional version of Kramer sampling theorem can be stated in terms of *Hilbert spaces of distributions having a reproducing distribution*. Let $\mathcal{D}(\mathbb{R})$ be the vector space of test functions, $\mathcal{H}$ a separable Hilbert space, and let

$$K : \mathcal{D}(\mathbb{R}) \longrightarrow \mathcal{H}$$

be a continuous conjugate linear function. For any $x \in \mathcal{H}$ the function

$$h_x(\phi) := \langle x, K(\phi)\rangle_{\mathcal{H}}, \quad \phi \in \mathcal{D}(\mathbb{R}),$$

is well-defined as a function $h_x : \mathcal{D}(\mathbb{R}) \to \mathbb{C}$. Moreover, $h_x$ is a continuous linear form because $K$ is continuous and conjugate linear, i.e., $h_x \in \mathcal{D}'(\mathbb{R})$ for each $x \in \mathcal{H}$.

Denote by $\mathcal{H}_K \subset \mathcal{D}'(\mathbb{R})$ the set of distributions obtained in this way and by $\mathcal{T}_K$ the linear transform

$$\mathcal{T}_K : \mathcal{H} \ni x \longmapsto h_x \in \mathcal{H}_K \,. \tag{4.70}$$

Note that $\mathcal{T}_K : \mathcal{H} \longrightarrow \mathcal{D}'(\mathbb{R})$ is the transpose of the function $K$. If we endow $\mathcal{H}_K$ with the norm

$$\|h\|_{\mathcal{H}_K} = \inf\{\|x\|_{\mathcal{H}} : h = \mathcal{T}_K(x)\} \,,$$

we obtain a reproducing distribution Hilbert space whose reproducing distribution is given by (see Ref. [12, p. 40]):

$$k(\phi, \psi) = \langle K(\psi), K(\phi)\rangle_{\mathcal{H}} \,, \quad \phi \,, \psi \in \mathcal{D}(\mathbb{R}) \,. \tag{4.71}$$

For each $\psi \in \mathcal{D}(\mathbb{R})$ the function $k_\psi$ defined as $k_\psi(\phi) := k(\phi, \psi)$ belongs to $\mathcal{H}_K$, and the reproducing property

$$h(\psi) = \langle h, k_\psi\rangle_{\mathcal{H}_K} = \langle h, k(\cdot, \psi)\rangle_{\mathcal{H}_K} \,, \quad \psi \in \mathcal{D}(\mathbb{R}) \,, \quad h \in \mathcal{H}_K \,, \tag{4.72}$$

holds.

The space $\mathcal{H}_K$ is nothing but the completion of the range space $\mathcal{T}_K K(\mathcal{D}(\mathbb{R}))$ which is a pre-Hilbert space with the norm defined by $\|h\| = \|K(\phi)\|_{\mathcal{H}}$, where $\phi \in \mathcal{D}(\mathbb{R})$ is such that $h = \mathcal{T}_K K(\phi)$ (see Ref. [38]).

Exactly as for a functional RKHS, the linear operator $\mathcal{T}_K$ is one-to-one if and only if the set $\{K(\phi)\}_{\phi \in \mathcal{D}(\mathbb{R})}$ is a complete set in $\mathcal{H}$.

A couple of comments are in order:

(*i*) In case $\mathcal{H} = L^2(\mathbb{R})$, the conjugate map $\overline{K}$ of $K$ which maps $\phi \mapsto \overline{K(\phi)}$ is just a *generalized stochastic process* in the sense of Refs. [12, 38].

(*ii*) Notice that, in particular, a continuous $\mathcal{H}$-valued function $K : \mathbb{R} \to \mathcal{H}$ can be considered as a continuous conjugate linear function $\mathbf{K} : \mathcal{D}(\mathbb{R}) \to \mathcal{H}$ through the formula

$$\mathbf{K}(\phi) = \int_{-\infty}^{\infty} \overline{\phi(t)}\, K(t)\, dt \,, \quad \phi \in \mathcal{D}(\mathbb{R}) \,.$$

The (Bochner) integral in this formula has always a meaning because for every fixed $\phi \in \mathcal{D}(\mathbb{R})$ the integral is defined on a set of finite measure, the $\mathcal{H}$-valued function $\overline{\phi}\, K$ is weakly Lebesgue measurable and then strongly Lebesgue measurable since $\mathcal{H}$ is separable, and $\int_{-\infty}^{\infty} \|\overline{\phi(t)} K(t)\|_{\mathcal{H}} dt < \infty$. Thus, the distributional version is indeed a generalization of the RKHS setting of Sect. 4.1.2.

The corresponding Kramer sampling theorem for the reproducing distribution Hilbert space $\mathcal{H}_K$ reads:

**Theorem 4.11 (A Distributional Kramer Sampling Theorem)** *Assume that there exist $\{\phi_n\}_{n=1}^\infty \subset \mathcal{D}(\mathbb{R})$, $\{a_n\}_{n=1}^\infty \subset \mathbb{C}\setminus\{0\}$ such that $K(\phi_n) = \overline{a}_n\, x_n$ for some Riesz basis $\{x_n\}_{n=1}^\infty$ for $\mathcal{H}$.*

*Then, the sequence of distributions $\{S_n\}_{n=1}^\infty$ given by*

$$S_n(\phi) := \langle y_n, K(\phi)\rangle_{\mathcal{H}}, \quad \phi \in \mathcal{D}(\mathbb{R}), \quad n \in \mathbb{N},$$

*where $\{y_n\}_{n=1}^\infty$ is the dual basis of $\{x_n\}_{n=1}^\infty$, forms a Riesz basis for the space $\mathcal{H}_K$.*

*Moreover, for any $h \in \mathcal{H}_K$ we obtain the sampling expansion:*

$$h(\phi) = \sum_{n=1}^\infty h(\phi_n)\, \frac{S_n(\phi)}{a_n}, \quad \phi \in \mathcal{D}(\mathbb{R}). \tag{4.73}$$

*The series in (4.73) converges in the $\mathcal{H}_K$-norm sense and, also, absolutely and uniformly on subsets of $\mathcal{D}(\mathbb{R})$ where the map $\phi \mapsto \|K(\phi)\|_{\mathcal{H}}$ is bounded.*

***Proof*** It is similar to the proof of Theorem 4.2. Indeed, from $K(\phi_n) = \overline{a}_n\, x_n$ we deduce that $\mathcal{T}_K$ is a bijective isometry and since $\mathcal{T}_K(y_m) = S_m$, $m \in \mathbb{N}$, we obtain that $\{S_m\}_{m=1}^\infty$ is a Riesz basis for $\mathcal{H}_K$ whose biorthonormal basis $\{T_m\}_{m=1}^\infty$ is given by $T_m = \mathcal{T}_K(x_m)$, $m \in \mathbb{N}$.

Expanding any $h \in \mathcal{H}_K$ in this Riesz basis, we have

$$h = \sum_{n=1}^\infty \langle h, T_n\rangle_{\mathcal{H}_K}\, S_n,$$

in the $\mathcal{H}_K$-norm sense and, consequently, pointwise in $\mathcal{D}(\mathbb{R})$ since for any $\phi \in \mathcal{D}(\mathbb{R})$ the mapping $h \mapsto h(\phi)$ is continuous on $\mathcal{H}_K$ (see Ref. [12, p. 41]).

Moreover, having in mind that $\mathcal{T}_K$ is an isometry, and that $K(\phi_n) = \overline{a}_n\, x_n$ we obtain

$$\begin{aligned}\langle h, T_n\rangle_{\mathcal{H}} &= \big\langle \mathcal{T}_K(x), \mathcal{T}_K(x_n)\big\rangle_{\mathcal{H}_K} = \langle x, x_n\rangle_{\mathcal{H}} \\ &= \Big\langle x, \frac{K(\phi_n)}{\overline{a}_n}\Big\rangle_{\mathcal{H}} = \frac{h(\phi_n)}{a_n}, \qquad n \in \mathbb{N},\end{aligned}$$

and hence the sampling expansion (4.73).

As in Theorem 4.2, the absolute convergence of the series for each $\phi \in \mathcal{D}(\mathbb{R})$ follows from the unconditional convergence of a Riesz basis expansion. The uniform convergence on subsets of $\mathcal{D}(\mathbb{R})$ where the map $\phi \mapsto \|K(\phi)\|_{\mathcal{H}}$ is bounded follows by using Cauchy-Schwarz's inequality in the reproducing property (4.72). □

**Example 4.17 (The Space $\mathcal{H}_K$ is a Hilbert Space of Distributions Having a Reproducing Distribution)** Let $\{e_n\}_{n\in\mathbb{Z}}$ be an orthonormal basis of a separable

Hilbert space $\mathcal{H}$ and let a non zero function $\phi_0 \in \mathcal{D}(\mathbb{R})$ such that $\operatorname{supp}\phi_0 \subset I$, where $I \subset \mathbb{R}$ is an arbitrary interval with length $m(I) \leq 1$. For every $n \in \mathbb{Z}$, set $\phi_n(t) := \phi_0(t-n)$, $t \in \mathbb{R}$, and consider the function $K : \mathcal{D}(\mathbb{R}) \to \mathcal{H}$ defined by

$$K(\psi) = \sum_{n\in\mathbb{Z}} \langle \phi_n, \psi \rangle_{L^2} e_n\,, \quad \psi \in \mathcal{D}(\mathbb{R})\,.$$

Note that for every fixed $\psi \in \mathcal{D}(\mathbb{R})$ there are only a finite set of non-zero terms in the above summation, so that $K(\psi) \in \mathcal{H}$ and it is easily checked that $K$ is an $\mathcal{H}$-valued continuous conjugate linear function.

Moreover,

$$K(\phi_m) = \sum_{n\in\mathbb{Z}} \langle \phi_n, \phi_m \rangle_{L^2} e_n = \|\phi_m\|^2_{L^2} e_m = \|\phi_0\|^2_{L^2} e_m\,, \quad m \in \mathbb{Z}\,.$$

In order to apply Theorem 4.11, for each $m \in \mathbb{Z}$ we compute the sampling function $S_m$:

$$S_m(\psi) = \langle e_m, K(\psi)\rangle_{\mathcal{H}} = \Big\langle e_m, \sum_{n\in\mathbb{Z}} \langle \phi_n, \psi \rangle_{L^2} e_n\Big\rangle_{\mathcal{H}} = \langle \psi, \phi_m \rangle_{L^2} \quad \psi \in \mathcal{D}(\mathbb{R})\,.$$

Therefore, for every $h \in \mathcal{H}_K$ the sampling expansion (4.73) reads as

$$h(\psi) = \frac{1}{\|\phi_0\|^2_{L^2}} \sum_{n\in\mathbb{Z}} h(\phi_n)\,\langle \psi, \phi_n \rangle_{L^2}\,, \quad \psi \in \mathcal{D}(\mathbb{R})\,,$$

where the series converges in the $\mathcal{H}_K$-norm sense and also, absolutely and uniformly on subsets of $\mathcal{D}(\mathbb{R})$ where the map $\psi \mapsto \sum_{n\in\mathbb{Z}} \big|\langle \phi_n, \psi \rangle_{L^2}\big|^2$ is bounded.

## References


1. Abramowitz, G., Stegun, I.: Handbook of Mathematical Functions. Dover, New York (1972)
2. Akhiezer, N.I.: The Classical Moment Problem. Oliver & Boyd, Edinburgh (1965)
3. Aldroubi, A., Sun, Q., Tang, W.: $p$-frames and shift-invariant subspaces of $L^p$. J. Fourier Anal. Appl. **7**(1), 1–22 (2001)
4. Annaby, M.H., E-Sayed, M.A.: Kramer-type sampling theorems associated with Fredholm integral operators. Methods Appl. Anal. **2**(1), 76–91 (1995)
5. Aronszajn, N.: Theory of reproducing kernels. Trans. Amer. Math. Soc. **68**, 337–404 (1950)
6. Berg, C.: Indeterminate moment problem and the theory of entire functions. J. Comput. Appl. Math. **65**, 27–55 (1995)
7. Berg, C., Valent, G.: Nevanlinna extremal measures for some orthogonal polynomials related to birth and death processes. J. Comput. Appl. Math. **57**, 29–43 (1995)
8. Boumenir, A., Zayed, A.: The equivalence of Kramer and Shannon sampling theorems revisited. Sampl. Theory Signal Image Process. **4**, 251–269 (2005)

9. Butzer, P.L., Nasri-Roudsari, G.: Kramer sampling theorem and its role in mathematics. In: Blackledge, J.M. (ed.) Image Processing: Mathematical Methods and Applications, pp. 49–95. Oxford University Press, Oxford, UK (1997)
10. Centeno, H., Medina, J.M.: A converse sampling theorem in reproducing kernel Banach spaces. Sampl. Theory Signal Process. Data Anal. **20**, 8 (2022)
11. Christensen, J.G.: Sampling in reproducing kernel Banach spaces on Lie groups. J. Approx. Theory **164**(1), 179–203 (2010)
12. Cristescu, R., Marinescu, G.: Applications of the Theory of Distributions. John Wiley & Sons, London (1973)
13. de Branges, L.: Hilbert Spaces of Entire Functions. Prentice-Hall, Englewood Cliffs, NJ (1968)
14. Everitt, W.N., García, A.G., Hernández-Medina, M.A.: On Lagrange-type interpolation series and analytic Kramer kernels. Results Math. **51**, 215–228 (2008)
15. Everitt, W.N., Nasri-Roudsari, G.: Interpolation and sampling theories, and linear ordinary boundary value problems. In: Higgins, J.R., Stens, R.L. (eds.) Sampling Theory in Fourier and Signal Analysis: Advanced Topics, pp. 96–129. Oxford University Press, Oxford (1999)
16. Everitt, W.N., Butzer, P., Schöttler, G.: Sturm-Liouville boundary-value problems and Lagrange interpolation series. Rend. Mat. Appl. **14**, 87–126 (1994)
17. Fernández-Moncada, P.E., García, A.G., Hernández-Medina, M.A.: The zero-removing property and Lagrange-type interpolation series. Numer. Funct. Anal. Optim. **32**, 858–876 (2011)
18. Gabardo, J.P.: Hilbert spaces of distributions having an orthogonal basis of exponentials. J. Fourier Anal. Appl. **6**(3), 277–298 (2000)
19. García, A.G., Hernández-Medina, M.A.: Sampling theorems and difference Sturm–Liouville problems. J. Differ. Equations Appl. **6**, 695–717 (2000)
20. García, A.G., Hernández-Medina, M.A.: The discrete Kramer sampling theorem and indeterminate moment problems. J. Comput. Appl. Math. **134**, 13–22 (2001)
21. García, A.G., Portal, A.: Sampling in reproducing kernel Banach spaces. Mediterr. J. Math. **10**(3), 1401–1417 (2013)
22. García, A.G., Szafraniec, F.H.: A converse of the Kramer sampling theorem. Samp. Theory Signal Image Process. **1**, 53–61 (2002)
23. García, A.G., Hernández-Medina, M.A., Muñoz-Bouzo, M.J.: The Kramer sampling theorem revisited. Acta Appl. Math. **133**(1), 87–111 (2014)
24. Giles, J.R.: Classes of semi-inner-product spaces. Trans. Amer. Math. Soc. **129**, 436–446 (1967)
25. Goffman, C., Pedrick, G.: First Course in Functional Analysis. American Mathematical Society, Providence RI (2017)
26. Han, D., Zayed, A.: Sampling expansions for functions having values in a Banach space. Proc. Amer. Math. Soc. **133**, 3597–3607 (2005)
27. Han, D., Nashed, M.Z., Sun, Q.: Sampling expansions in reproducing kernel Hilbert and Banach spaces. Numer. Funct. Anal. Optim. **30**(9–10), 971–987 (2009)
28. Hartley, R.V.: A more symmetrical Fourier analysis applied to transmission problems. Proc. Inst. Radio Eng. **30**, 144–150 (1942)
29. Higgins, J.R.: Sampling Theory in Fourier and Signal Analysis: Foundations. Oxford University Press, Oxford (1996)
30. Higgins, J.R.: A sampling principle associated with Saitoh's fundamental theory of linear transformations. In: Saitoh, S., et al. (eds.) Analytic Extension Formulas and their Applications, pp. 73–86. Kluwer Academic, Dordrecht (2001)
31. Ismail, M.E.H., Masson, D.R.: $q$-Hermite polynomials, biorthogonal rational functions, and $q$-beta integrals. Trans. Amer. Math. Soc. **346**, 63–116 (1994)
32. Kantorovich, L.V., Akilov, G.P.: Functional Analysis in Normed Spaces. Macmillan, New York (1964)
33. Karlin, S., McGregor, J.: The differential equations of birth and death processes and the Stieltjes moment problem. Trans. Amer. Math. Soc. **85**, 489–546 (1957)
34. Kramer, H.P.: A generalized sampling theorem. J. Math. Phys. **63**, 68–72 (1959)

35. Levitan, B.M., Sargsjan, I.S.: Sturm-Liouville and Dirac Operators. Kluwer Academic Publishers, Dordrecht (1991)
36. Lumer, G.: Semi-inner-product spaces. Trans. Amer. Math. Soc. **100**, 29–43 (1961)
37. Markett, C.: Kramer-type sampling theorems associated with higher-order differential equations. Sampl. Theory Signal Process. Data Anal. **21**, 11 (2023)
38. Meidan, R.: Reproducing-kernel Hilbert spaces of distributions and generalized stochastic processes. SIAM J. Math. Anal. **10**(1), 62–70 (1979)
39. Nashed, M.Z., Sun, Q.: Sampling and reconstruction of signals in a reproducing kernel subspace of $L^p(\mathbb{R}^d)$. J. Funct. Anal. **258**, 2422–2452 (2010)
40. Rudin, W.: Functional Analysis, 2nd edn. McGraw-Hill, New York (1991)
41. Saitoh, S., Sawano, Y.: Theory of Reproducing Kernels and Applications. Developments in Mathematics, vol. 44. Springer, Singapore (2016)
42. Simon, B.: The classical moment problem as a self-adjoint finite difference operator. Adv. Math. **137**, 82–203 (1998)
43. Szafraniec, F.H.: Interpolation and domination by positive definite kernels. In: Lecture Notes in Math., vol. 1014, pp. 291–295. Springer, Berlin-Heidelberg (1983)
44. Szafraniec, F.H.: The reproducing kernel Hilbert space and its multiplication operators. Operator Theory Adv. Appl. **114**, 253–263 (2000)
45. Taylor, A.E., Lay, D.C.: Introduction to Functional Analysis. John Wiley & Sons, New York (1980)
46. Weiss, P.: Sampling theorems associated with Sturm-Liouville systems. Bull. Amer. Math. Soc. **63**, 351 (1957)
47. Yao, K.: Applications of reproducing kernel Hilbert spaces–bandlimited signal models. Inform. Control **11**, 429–444 (1967)
48. Young, R.: An Introduction to Nonharmonic Fourier Series. Academic Press, San Diego (2001)
49. Zayed, A.I.: On Kramer's sampling theorem associated with general Sturm-Liouville problems and Lagrange interpolation. SIAM J. Appl. Math. **51**, 575–604 (1991)
50. Zayed, A.I.: Advances in Shannon's Sampling Theory. CRC Press, Boca Raton (1993)
51. Zayed, A.I.: A new role of Green's functions in interpolation and sampling theory. J. Math. Anal. Appl. **175**, 222–238 (1993)
52. Zayed, A.I.: Handbook of Function and Generalized Function Transformations. CRC Press, Boca Raton, FL (1996)
53. Zayed, A.I., Butzer, P.L.: Lagrange interpolation and sampling theorems. In: Marvasti, F. (ed.) Nonuniform Sampling: Theory and Practice, pp. 123–168. Kluwer Academic, Dordrecht (2001)
54. Zayed, A.I., García, A.G.: Sampling theorem associated with a Dirac operator and the Hartley transform. J. Math. Anal. Appl. **214**, 587–598 (1997)
55. Zayed, A.I., García, A.G.: New sampling formulae for the fractional Fourier transform. Signal Process. **77**, 111–114 (1999)
56. Zayed, A.I., Butzer, P.L., Hinsen, G.: On Lagrange interpolation and Kramer-type sampling theorems associated with Sturm-Liouville problems. SIAM J. Appl. Math. **50**, 893–909 (1990)
57. Zayed, A.I., E-Sayed, M.A., Annaby, M.H.: On Lagrange interpolations and Kramer's sampling theorem associated with self-adjoint boundary-value problems. J. Math. Anal. Appl. **150**(1), 269–284 (1991)
58. Zhang, H., Zhang, J.: Frames, Riesz bases, and sampling expansions in Banach spaces via semi-inner products. Appl. Comput. Harmon. Anal. **31**, 1–25 (2011)
59. Zhang, H., Xu, Y., Zhang, J.: Reproducing kernel Banach spaces for machine learning. J. Mach. Learn. Res. **10**, 2741–2775 (2009)
60. Zygmund, A.: Trigonometric Series, vol. 2. Cambridge University Press, Cambridge UK (1959)

# Chapter 5
# A Generalized Sampling Theory

This chapter can be considered as a synthesis of the sampling technique exhibited throughout this book. Firstly, we notice that most linear sampling processes can be modeled by means of a convolution system. For instance, the regular samples $\{f(n)\}_{n\in\mathbb{Z}}$ of any bandlimited function $f$ in the Paley-Wiener space $PW_\pi$ are given as

$$f(n) = (f * \mathrm{sinc})(n)\,, \quad n \in \mathbb{Z}\,,$$

where sinc denotes the cardinal sine function (1.2).

In the case of *average sampling* in the shift-invariant space $V_\varphi^2$ we have seen in Chap. 3 that

$$\langle f, \psi(\cdot - n)\rangle_{L^2(\mathbb{R})} = (f * \widetilde{\psi})(n)\,, \quad n \in \mathbb{Z}\,,$$

where $\widetilde{\psi}(t) = \overline{\psi(-t)}$ is the *average function*. Now, for each $f(t) = \sum_{n\in\mathbb{Z}} c_n\varphi(t - n)$ in $V_\varphi^2$, we have

$$\begin{aligned}\langle f, \psi(\cdot - n)\rangle &= \Big\langle \sum_{k\in\mathbb{Z}} c_n\varphi(t-k), \psi(\cdot - n)\Big\rangle = \sum_{k\in\mathbb{Z}} c_k \,\langle \varphi(\cdot - k), \psi(\cdot - n)\rangle \\ &= \sum_{k\in\mathbb{Z}} c_k\, \langle \varphi, \psi(\cdot - n + k)\rangle = \big(\mathbf{c} * \mathbf{a}\big)(n)\,, \quad n \in \mathbb{Z}\,,\end{aligned}$$

where $\mathbf{a} = \{a(k)\}_{k\in\mathbb{Z}}$ with $a(k) = \langle \varphi, \psi(\cdot - k)\rangle$, $k \in \mathbb{Z}$, is the sequence of average samples of the generator $\varphi$, and $\mathbf{c} = \{c_k\}_{k\in\mathbb{Z}}$ is the sequence of the coefficients of $f$ in $V_\varphi^2$. The sequence $\mathbf{a}$ in $\ell^2(\mathbb{Z})$ reflects the characteristics of the linear time-invariant system involved in the sampling process.

We could use $M$ average functions $\psi_j$ and to sample the functions in $V_\varphi^2$ at the group $r\mathbb{Z}$, where $r > 0$ denotes the sampling period.

 
A. García García, *The Use of Frames in Sampling Theory*,
RSME Springer Series 14, https://doi.org/10.1007/978-3-031-63242-6_5

Thus, in general, the acquisition of samples can be modeled as a filtering process in a product space $\ell^2_N(\mathbb{Z})$ or $\ell^2_N(r\mathbb{Z})$, where $N$ denotes the number of generators $\Phi = \{\varphi_1, \varphi_2, \dots, \varphi_N\}$ for the shift-invariant space $V^2_\Phi$.

Secondly, in case we deal with $M$ convolution systems $(\mathcal{L}_j f)(n) = \big(\mathbf{c} * \mathbf{a}_j\big)(n)$, $n \in \mathbb{Z}$, and $j = 1, 2, \dots, M$, the idea is to expand the sequence $\mathbf{c} = \{c_k\}_{k\in\mathbb{Z}}$ of the coefficients of $f$ in $V^2_\varphi$ following a frame expansion as

$$\mathbf{c} = \sum_{j=1}^{M} \sum_{n\in\mathbb{Z}} (\mathcal{L}_j f)(n)\, T_n \mathbf{b}_j \quad \text{in } \ell^2(\mathbb{Z}) .$$

Since we can express $\mathcal{L}_j f(n) = \langle \mathbf{c}, T_n \mathbf{a}^*_j \rangle_{\ell^2}$, where $\mathbf{a}^*_j(k) = \overline{\mathbf{a}_j(-k)}$, $k \in \mathbb{Z}$, the task is to characterize the frames for $\ell^2(\mathbb{Z})$ of the form $\{T_n \mathbf{a}^*_j\}$, and to find their dual frames $\{T_n \mathbf{b}_j\}$ having the same form. Here, $T_n$ denotes, as usual, the translation by $n \in \mathbb{Z}$ in $\ell^2(\mathbb{Z})$.

Finally, having in mind the isomorphism $\mathcal{T}_\varphi : \ell^2(\mathbb{Z}) \to V^2_\varphi$ which maps the standard orthonormal basis $\{\boldsymbol{\delta}_n\}_{n\in\mathbb{Z}}$ for $\ell^2(\mathbb{Z})$ onto the Riesz basis $\{\varphi(t-n)\}_{n\in\mathbb{Z}}$ for $V^2_\varphi$, i.e., mapping $\mathbf{c} \mapsto f$, we obtain in $V^2_\varphi$ a sampling formula as

$$f(t) = \sum_{j=1}^{M} \sum_{n\in\mathbb{Z}} (\mathcal{L}_j f)(n)\, S_j(t-n) \quad \text{in } L^2(\mathbb{R}) ,$$

where $\mathcal{T}_\varphi\big[T_n \mathbf{b}_j\big](t) = T_n\big[\mathcal{T}_\varphi \mathbf{b}_j\big](t) = T_n[S_j](t) = S_j(t-n)$.

It is necessary to justify the above commutation, where $T_n$ denotes both the translation by $n \in \mathbb{Z}$ in $\ell^2(\mathbb{Z})$ and in $V^2_\varphi$. The above series also converges absolutely and uniformly on $\mathbb{R}$ since the shift-invariant subspace $V^2_\varphi$ is a RKHS.

In order to avoid an excess of oversampling we can also consider the sequence of samples $\big\{\mathcal{L}_j f(rn)\big\}_{n\in\mathbb{Z};\, j=1,2,\dots,M}$ taken at the group $r\mathbb{Z}$, where $r \in \mathbb{N}$ denotes a positive sampling period. In this case a sampling expansion as the one above could be obtained in a similar way, but necessarily with $M \geq r$.

Gathering all the pieces we have explained until now, in a general setting:

1. We may consider a unitary representation $\{U(g)\}_{g\in G}$ of a countable discrete abelian group $(G, +)$ on a separable Hilbert space $\mathcal{H}$ which is nothing but a homomorphism between the group $G$ and the group of unitary operators on $\mathcal{H}$.
2. We may sample the elements in a $U$-invariant subspace $\mathcal{V}_\Phi$, with a set of $N$ generators $\Phi = \{\varphi_1, \varphi_2, \dots, \varphi_N\}$ (see Eq. (5.1)), by means of $M$ convolution systems $\mathcal{L}_m$, $m = 1, 2, \dots, M$, defined in $\ell^2_N(G)$ by

$$\mathcal{L}_m f(g) = \sum_{n=1}^{N} (a_{m,n} * x_n)(g) , \quad g \in G ,$$

for $f = \sum_{n=1}^{N} \sum_{g\in G} x_n(g)\, U(g)\varphi_n$ in $\mathcal{V}_\Phi$.

3. Finally, we should try to obtain, under appropriate hypotheses, sampling formulas in $\mathcal{V}_\Phi$ like

$$f = \sum_{m=1}^{M} \sum_{g \in G} \mathcal{L}_m f(g)\, U(g) S_m \quad \text{in } \mathcal{H}$$

by using suitable dual frames in the product space $\ell^2_N(G)$.

All these steps will be carried out throughout this chapter. Besides, as we will see, this general setting covers most of the known sampling examples. This chapter is broadly in line with Ref. [12].

In general, we may consider a countable discrete abelian group $G$ other than $\mathbb{Z}$ or $r\mathbb{Z}$. We will also consider other settings involving non-abelian groups, expressed as semi-direct or direct products of groups. This is the case, for instance, of crystallographic groups.

Working in locally compact abelian groups is not just a unified way of dealing with the classical groups $\mathbb{R}^d$, $\mathbb{Z}^d$, $\mathbb{T}^d$, $\mathbb{Z}_s^d$: signal processing often involves products of these groups which are also locally compact abelian groups. For example, multichannel video signal involves the group $\mathbb{Z}^d \times \mathbb{Z}_s$, where $d$ is the number of channels and $s$ the number of pixels of each image. For the interested reader in locally compact abelian groups we recommend Refs. [4, 5, 7, 8].

## 5.1 Data Samples as a Filtering Process

Let $\mathcal{H}$ be a separable Hilbert space and let $\mathcal{U}(\mathcal{H})$ denote the group of unitary operators on $\mathcal{H}$. Given a countable discrete abelian group $(G, +)$, let

$$G \ni g \longmapsto U(g) \in \mathcal{U}(\mathcal{H})$$

be a homomorphism of groups. That is, it satisfies:

$$U(g + g') = U(g)\, U(g') \quad \text{for all } g, g' \in G\,.$$

As a consequence, $U(0) = I_{\mathcal{H}}$ and $U(-g) = U^{-1}(g) = U^*(g)$ for any $g \in G$.

We will say that $\{U(g)\}_{g \in G}$ is a *unitary representation* of the countable discrete abelian group $(G, +)$ on $\mathcal{H}$ (see Definition 5.1 below).

Given a set $\Phi = \{\varphi_1, \varphi_2, \ldots, \varphi_N\}$ of generators in $\mathcal{H}$ we consider the $U$-invariant subspace of $\mathcal{H}$ defined as

$$\mathcal{V}_\Phi := \overline{\operatorname{span}}_{\mathcal{H}}\{U(g)\varphi_n\}_{g \in G;\, n=1,2,\ldots,N}\,.$$

Assuming that $\{U(g)\varphi_n\}_{g\in G;\, n=1,2,\dots,N}$ is a Riesz sequence in $\mathcal{H}$, i.e., a Riesz basis for $\mathcal{V}_\Phi$, this subspace can be expressed as

$$\mathcal{V}_\Phi = \Big\{\sum_{n=1}^{N}\sum_{g\in G} x_n(g)U(g)\varphi_n \;:\; x_n \in \ell^2(G);\; n = 1, 2, \dots, N\Big\}. \tag{5.1}$$

Let us motivate the sampling approach followed in this chapter by means of a couple of examples:

(1) Given $M$ elements $\psi_m \in \mathcal{H}$, $m = 1, 2, \dots, M$, which do not belong necessarily to $\mathcal{V}_\Phi$, for any $f \in \mathcal{V}_\Phi$ we define, for $m = 1, 2, \dots, M$, its generalized average samples as

$$\mathcal{L}_m f(g) := \big\langle f, U(g)\psi_m\big\rangle_{\mathcal{H}}, \quad g \in G. \tag{5.2}$$

These samples can be expressed as the output of a convolution system. Indeed, for any $f = \sum_{n=1}^{N}\sum_{h\in G} x_n(h)U(h)\varphi_n$ in $\mathcal{V}_\Phi$ and $m = 1, 2, \dots, M$, one immediately gets

$$\begin{aligned}\mathcal{L}_m f(g) &= \Big\langle \sum_{n=1}^{N}\sum_{h\in G} x_n(h)U(h)\varphi_n, U(g)\psi_m\Big\rangle \\ &= \sum_{n=1}^{N}\sum_{h\in G} x_n(h)\big\langle\varphi_n, U(g-h)\psi_m\big\rangle \\ &= \sum_{n=1}^{N}(a_{m,n} * x_n)(g), \qquad g \in G,\end{aligned} \tag{5.3}$$

with $a_{m,n}(g) = \langle\varphi_n, U(g)\psi_m\rangle_{\mathcal{H}}$, $g \in G$. Notice that each $a_{m,n}$ belongs to $\ell^2(G)$ since the sequence $\{U(g)\varphi_n\}_{g\in G;\, n=1,2,\dots,N}$ is, in particular, a Bessel sequence in $\mathcal{H}$.

(2) Suppose now that $\mathcal{H} = L^2(\mathbb{R}^d)$ and consider $M$ fixed points $t_m \in \mathbb{R}^d$, $m = 1, 2, \dots, M$. For each $f \in \mathcal{V}_\Phi$ we formally define its samples, for $m = 1, 2, \dots, M$, as

$$\mathcal{L}_m f(g) := \big[U(-g)f\big](t_m), \quad g \in G. \tag{5.4}$$

For $f = \sum_{n=1}^{N} \sum_{h\in G} x_n(h)U(h)\varphi_n$ in $\mathcal{V}_\Phi$ (as we will see later on, see Proposition 5.6, we may assume that $f$ is a continuous function) we have

$$\begin{aligned}\mathcal{L}_m f(g) &= \Big[U(-g)\sum_{n=1}^{N}\sum_{h\in G} x_n(h)U(h)\varphi_n\Big](t_m)\\ &= \sum_{n=1}^{N}\sum_{h\in G} x_n(h)\big[U(h-g)\varphi_n\big](t_m)\\ &= \sum_{n=1}^{N}(a_{m,n} * x_n)(g), \qquad g \in G,\end{aligned}$$

with $a_{m,n}(g) = \big[U(-g)\varphi_n\big](t_m),\ g \in G$.

Under mild hypotheses (see Sect. 5.3.3), one can obtain that $\mathcal{V}_\Phi$ is a reproducing kernel Hilbert space of continuous functions in $L^2(\mathbb{R}^d)$ where the samples (5.4) with corresponding $a_{m,n} \in \ell^2(G)$ are well-defined, and yielding pointwise sampling in $\mathcal{V}_\Phi$.

The above two situations comprise most of the regular sampling (average or pointwise) appearing in the mathematical or engineering literature, as we will see in Sects. 5.3.2–5.3.4.

Consequently, one can think of a sampling process in a subspace $\mathcal{V}_\Phi$ as $M$ expressions like (5.3), i.e., a convolution system $\mathcal{A}$ defined in the product Hilbert space $\ell^2_N(G) := \ell^2(G) \times \cdots \times \ell^2(G)$ ($N$ times) by means of a matrix

$$A = [a_{m,n}] \in \mathcal{M}_{M\times N}(\ell^2(G)),$$

that is, a $M \times N$ matrix with entries in $\ell^2(G)$, as

$$\mathcal{A}(\mathbf{x}) = \begin{pmatrix} a_{1,1} * x_1 + \cdots + a_{1,N} * x_N \\ a_{2,1} * x_1 + \cdots + a_{2,N} * x_N \\ \vdots \\ a_{M,1} * x_1 + \cdots + a_{M,N} * x_N \end{pmatrix}, \quad \mathbf{x} = (x_1, x_2, \ldots, x_N)^\top \in \ell^2_N(G).$$

Using a matrix convolution notation we write

$$\big[\mathcal{A}(\mathbf{x})\big](g) = (A * \mathbf{x})(g) = \sum_{g'\in G} A(g-g')\,\mathbf{x}(g'), \quad g \in G,$$

where the $m$-th entry of $(A * \mathbf{x})(g)$, $g \in G$, is

$$\big[\mathcal{A}(\mathbf{x})\big]_m(g) = \sum_{n=1}^{N} (a_{m,n} * x_n)(g)\,, \quad g \in G\,,$$

with $x_n$ denoting the $n$-th entry of $\mathbf{x} \in \ell^2_N(G)$.

The main aim in this chapter is to recover, in a stable way, any $\mathbf{x} \in \ell^2_N(G)$, or equivalently, the corresponding $f = \sum_{n=1}^{N}\sum_{g\in G} x_n(g)U(g)\varphi_n \in \mathcal{V}_\Phi$, from the sequence of vector data

$$\mathcal{L}f(g) := \big(\mathcal{L}_1 f(g), \mathcal{L}_2 f(g), \ldots, \mathcal{L}_M f(g)\big)^\top = \big[\mathcal{A}(\mathbf{x})\big](g)\,, \quad g \in G\,,$$

that is, from the output $\mathcal{A}(\mathbf{x})$ of the convolution system $\mathcal{A}$ with associated matrix $A$, in case the vector sampling $\mathcal{L}f \in \ell^2_M(G)$.

## 5.2 Convolution Systems on Discrete Abelian Groups

This section is devoted to collect some known results on discrete convolution systems, and to prove some new ones needed in the sequel. First, we recall some needed results on the Fourier transform on $\ell^2(G)$, where $G$ denotes a countable discrete abelian group.

### 5.2.1 *A Brief on Harmonic Analysis on Discrete Abelian Groups*

Let $(G,+)$ be a *countable discrete abelian group* and let $\mathbb{T} = \{z \in \mathbb{C} : |z| = 1\}$ be the unidimensional torus. We say that $\xi : G \mapsto \mathbb{T}$ is a *character* of $G$ if

$$\xi(g+g') = \xi(g)\,\xi(g') \quad \text{for all } g, g' \in G\,.$$

We denote $\xi(g) := \langle g, \xi\rangle$. By defining $(\xi + \xi')(g) = \xi(g)\,\xi'(g)$, $g \in G$, the set of characters $\widehat{G}$ is a group, called the *dual group* of $G$. Since the group $G$ is discrete, its dual group $\widehat{G}$ is compact (see Ref. [7, Prop. 4.4]).

In particular, it is known that $\widehat{\mathbb{Z}} \cong \mathbb{T}$, with $\langle n, z\rangle = z^n$, and $\widehat{\mathbb{Z}}_s \cong \mathbb{Z}_s := \mathbb{Z}/s\mathbb{Z}$, with $\langle n, m\rangle = W_s^{nm}$, where $W_s = e^{2\pi i/s}$.

There exists a unique measure the *Haar measure* $\mu$ on $\widehat{G}$, satisfying

$$\mu(\xi + E) = \mu(E) \quad \text{for every Borel set } E \subset \widehat{G} \text{ and } \mu(\widehat{G}) = 1.$$

We use the notation

$$\int_{\widehat{G}} X(\xi)d\xi = \int_{\widehat{G}} X(\xi)d\mu(\xi).$$

If $G = \mathbb{Z}$,

$$\int_{\widehat{G}} X(\xi)d\xi = \int_{\mathbb{T}} X(z)dz = \frac{1}{2\pi}\int_0^{2\pi} X(e^{iw})dw,$$

and if $G = \mathbb{Z}_s$,

$$\int_{\widehat{G}} X(\xi)d\xi = \int_{\mathbb{Z}_s} X(n)dn = \frac{1}{s}\sum_{n\in\mathbb{Z}_s} X(n).$$

If $G_1, G_2, \ldots, G_d$ are abelian discrete groups, then the dual group of the product group is

$$(G_1 \times G_2 \times \ldots \times G_d)^{\wedge} \cong \widehat{G}_1 \times \widehat{G}_2 \times \ldots \times \widehat{G}_d$$

with

$$\big\langle (g_1, g_2, \ldots, g_d),\, (\xi_1, \xi_2 \ldots, \xi_d)\big\rangle = \langle g_1, \xi_1\rangle \langle g_2, \xi_2\rangle \cdots \langle g_d, \xi_d\rangle.$$

If $G = \mathbb{Z}^d \times \mathbb{Z}_s$, then $\widehat{G} = \mathbb{T}^d \times \mathbb{Z}_s$ and

$$\int_{\widehat{G}} X(\xi)d\xi = \int_{\mathbb{T}^d\times\mathbb{Z}_s} X(z,n)dzdn = \frac{1}{2\pi s}\sum_{n\in\mathbb{Z}_s}\int_0^{2\pi} X(e^{i\omega}, n)d\omega.$$

For $x \in \ell^1(G)$ its *Fourier transform* is defined as

$$X(\xi) = \widehat{x}(\xi) := \sum_{g\in G} x(g)\overline{\langle g, \xi\rangle} = \sum_{g\in G} x(g)\langle -g, \xi\rangle, \quad \xi \in \widehat{G}.$$

The Plancherel theorem extends uniquely the Fourier transform on $\ell^1(G)\cap\ell^2(G)$ to a unitary operator from $\ell^2(G)$ onto $L^2(\widehat{G})$. For the details see, for instance, Ref. [7].

In the more frequent cases we have:

- For $G = \mathbb{Z}$ and $x \in \ell^2(\mathbb{Z})$, $\quad \widehat{x}(z) = \sum_{n\in\mathbb{Z}} x(n)\, z^{-n}, \quad z \in \mathbb{T}.$

- For $G = \mathbb{Z}^d$ and $x \in \ell^2(\mathbb{Z}^d)$,

$$\widehat{x}(z) = \sum_{\mathbf{n}\in\mathbb{Z}^d} x(n)\, z^{-n}, \quad z = (z_1, z_2, \ldots, z_d) \in \mathbb{T}^d .$$

Here, $n = (n_1, n_2, \ldots, n_d) \in \mathbb{Z}^d$ and $z^n = (z_1^{n_1}, z_1^{n_1}, \ldots, z_1^{n_1})$.

- For $G = \mathbb{Z}_s$ and $x \in \ell^2(\mathbb{Z}_s)$, $\widehat{x}(m) = \sum_{n\in\mathbb{Z}_s} x(n)\, W_s^{-nm}, \quad m \in \mathbb{Z}_s$.
- For $\mathbb{Z}^d \times \mathbb{Z}_s$ and $x \in \ell^2(\mathbb{Z}^d \times \mathbb{Z}_s)$,

$$\widehat{x}(z, m) = \sum_{(n,k)\in\mathbb{Z}^d\times\mathbb{Z}_s} x(n, k)\, z^{-n}\, W_s^{-km}, \quad (z, m) \in \mathbb{T}^d \times \mathbb{Z}_s .$$

The following lemma will be used in the sequel:

**Lemma 5.1** *Let $a, b \in \ell^2(G)$ such that the product $\widehat{a}(\xi)\,\widehat{b}(\xi) \in L^2(\widehat{G})$. Then the convolution $a * b \in \ell^2(G)$ and*

$$\big(\widehat{a * b}\big)(\xi) = \widehat{a}(\xi)\,\widehat{b}(\xi), \quad \text{a.e. } \xi \in \widehat{G} .$$

***Proof*** By using the Plancherel theorem (see Ref. [7]) and denoting $\widetilde{b}(g) = \overline{b(-g)}$, $g \in G$, we obtain

$$\begin{aligned}\big(a * b\big)(g) &= \sum_{h\in G} a(h)\, b(g-h) = \big\langle a, \widetilde{b}(\cdot - g)\big\rangle_{\ell^2(N)} = \big\langle \widehat{a}, \widehat{\widetilde{b}(\cdot - g)}\big\rangle_{L^2(\widehat{G})} \\ &= \int_{\widehat{G}} \widehat{a}(\xi)\, \overline{\widehat{\widetilde{b}}(\xi)}\, \overline{\langle -g, \xi\rangle}\, d\mu_{\widehat{G}}(\xi) = \int_{\widehat{G}} \widehat{a}(\xi)\, \widehat{b}(\xi)\, \overline{\langle -g, \xi\rangle}\, d\mu_{\widehat{G}}(\xi) .\end{aligned}$$

Since $\big\{\langle -g, \xi\rangle\big\}_{g\in G}$ is an orthonormal basis for $L^2(\widehat{G})$ (see Ref. [7, Theorem 4.26]) and we are assuming that $\widehat{a}(\xi)\,\widehat{b}(\xi) \in L^2(\widehat{G})$, we obtain that $a * b \in \ell^2(G)$ and

$$\widehat{a}(\xi)\,\widehat{b}(\xi) = \sum_{g\in G} \big(a * b\big)(g)\langle -g, \xi\rangle, \quad \text{a.e. } \xi \in \widehat{G} ,$$

that is, $\widehat{a} \cdot \widehat{b} = \widehat{a * b}$ in $L^2(\widehat{G})$. □

## 5.2.2 Lattices on $\mathbb{R}^d$

Firstly, we consider a lattice $\Lambda = P\mathbb{Z}^d \subset \mathbb{Z}^d$ of $\mathbb{R}^d$, where $P$ denotes a $p \times p$ matrix with integer entries and $\det P > 0$. Note that without loss of generality we

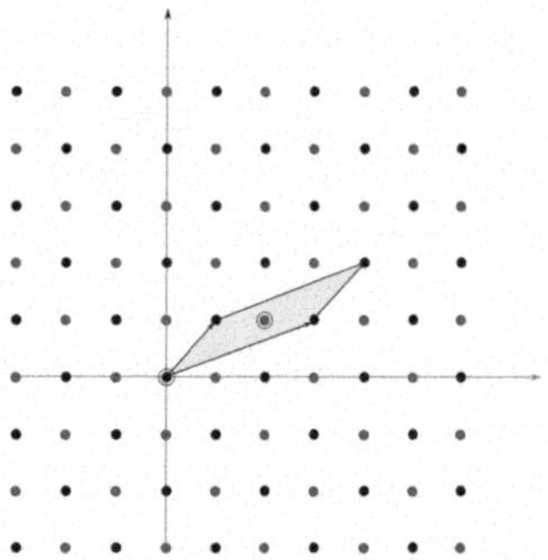

**Fig. 5.1** Decomposition of $\mathbb{Z}^2$ following the lattice $\Lambda = P\mathbb{Z}^2$

can assume that $\det P > 0$; otherwise we can consider $P' = PE$ where $E$ is some $d \times d$ matrix with $\det E = -1$.

Thus, we get an additive subgroup of $\mathbb{Z}^d$ with index $\det P$, i.e., the quotient group $\mathbb{Z}^d/\Lambda$ has order $\det P$.

The set $P\,[0, 1)^d$ is the *fundamental parallelepided* associated with $\Lambda$ and the set $\mathbb{Z}^d \cap P\,[0, 1)^d$ has exactly $\det P$ points $\{t_1 = 0, t_2, \cdots, t_{\det P}\}$.

To prove this, it is known that the volume of $P\,[0, 1)^d$ is $\det P$. Besides, we can fit $\det P$ hypercubes with unit-length sides into this parallelepiped. Since there is only one integer per hypercube, we will have $\det P$ integers in $P\,[0, 1)^d$.

Thus, the cosets in $\mathbb{Z}^d/\Lambda$ are precisely the sets $\{t_i + \Lambda\}$, $i = 1, 2, \cdots, \det P$. See the details in Ref. [18, Section 12.2].

Figure 5.1 shows an example in $\mathbb{Z}^2$ corresponding to the matrix $P = \begin{pmatrix} 3 & 1 \\ 1 & 1 \end{pmatrix}$ with $\det P = 2$. In this case we have $\mathbb{Z}^2 = \Lambda \cup \big[(2, 1)^\top + \Lambda\big]$.

In general, we could consider a full rank lattice $\Lambda = P\mathbb{Z}^d$ where $P$ a $d \times d$ matrix with $\det P \neq 0$. A function $f$ is periodic with respect to the lattice $\Lambda$ if

$$f(t + \lambda) = f(t) \quad \text{for all } t \in \mathbb{R}^d \text{ and for all } \lambda \in \Lambda\,.$$

Given a lattice $\Lambda$ in $\mathbb{R}^d$, some sampling examples in a $\Lambda$-shift-invariant subspace $V_\Phi^\Lambda$ with a set of $N$ stable generators $\Phi = \{\varphi_1, \varphi_2, \cdots, \varphi_N\} \subset L^2(\mathbb{R}^d)$, which can be expressed as

$$V_\Phi^\Lambda = \Big\{\sum_{n=1}^{N}\sum_{\lambda\in\Lambda} c_n(\lambda)\,\varphi_n(t-\lambda) \;:\; \{c_n(\lambda)\} \in \ell^2(\Lambda)\,;\; n = 1, 2, \cdots, N\Big\},$$

are presented in this chapter. See also Chap. 7.

Associated with a lattice $\Lambda$ in $\mathbb{R}^d$ we have its *dual lattice* $(P^{-1})^\top\mathbb{Z}^d$ which appears when we define the Fourier transform in $\ell^2(\Lambda)$. Indeed, following the theory of LCA groups (see Refs. [4, 7]), the dual $\widehat{\Lambda}$ is identified with $\mathbb{R}^d/\Lambda^\perp$, where $\Lambda^\perp$ is the *annihilator group*

$$\Lambda^\perp = \big\{x \in \mathbb{R}^d \;:\; e^{2\pi i\, x\cdot\lambda} = 1 \;\text{ for all } \lambda \in \Lambda\big\}\,.$$

Observe that $\Lambda^\perp$ coincides with the dual lattice $(P^{-1})^\top\mathbb{Z}^d$ above defined, that is, $\Lambda^\perp = (P^{-1})^\top\mathbb{Z}^d = P^{-T}\mathbb{Z}^d$, where $P^{-\top}$ denotes the matrix $(P^{-1})^\top$. Since $\Lambda$ is discrete its dual group $\widehat{\Lambda}$ is compact.

The *Fourier transform of the sequence* $c \in \ell^1(\Lambda)$ is defined as

$$\mathcal{F}_\Lambda(c)(\dot{x}) := \sum_{\lambda\in\Lambda} c(\lambda)\, e^{2\pi i\,\lambda\cdot x}, \quad \dot{x} \in \mathbb{R}^d/\Lambda^\perp,$$

where $\dot{x}$ denotes the image of $x$ under the natural quotient map $\mathbb{R}^d \longrightarrow \mathbb{R}^d/\Lambda^\perp$.

The Fourier transform in $\ell^1(\Lambda)\cap\ell^2(\Lambda)$ extends uniquely to a unitary mapping

$$\mathcal{F}_\Lambda : \ell^2(\Lambda) \longrightarrow L^2(\widehat{\Lambda})$$

which is the *Fourier transform in* $\ell^2(\Lambda)$.

Notice that the sequence $\big\{\sqrt{\det P}\, e^{2\pi i\alpha^\top P^\top x}\big\}_{\alpha\in\mathbb{Z}^d}$ is an orthonormal basis for $L^2\big(P^{-\top}[0,1)^d\big)$, where we are using the matrix notation $\alpha^\top P^\top x$ for $P\alpha\cdot x$. Observe that $P^{-\top}[0,1)^d$ is the fundamental parallelepided associated with the dual lattice $\Lambda^\perp$.

In the context of time-frequency analysis, when we define the Fourier transform in $\ell^2(\Lambda)$, where $\Lambda$ is a lattice in $\mathbb{R}^{2d}$, it appears its *adjoint lattice* $\Lambda^\circ$ (see Sect. 7.3 in Chap. 7).

### 5.2.3 The Convolution Systems Theory

In this section we consider bounded convolution operators as

$$\begin{aligned} \mathcal{A} : \ell^2_N(G) &\longrightarrow \ell^2_M(G) \\ \mathbf{x} &\longmapsto \mathcal{A}(\mathbf{x}) = A * \mathbf{x}, \end{aligned} \tag{5.5}$$

associated with a matrix $A \in \mathcal{M}_{M\times N}\big(\ell^2(G)\big)$. As usual, the translation by $g\in G$ of any $\mathbf{x}\in\ell^2_N(G)$ is denoted by $T_g\mathbf{x}(h) = \mathbf{x}(h-g)$, $h\in G$.

**Proposition 5.1** *The above operator $\mathcal{A}$ is a well-defined bounded operator from $\ell^2_N(G)$ into $\ell^2_M(G)$ if and only if $\widehat{A}\in\mathcal{M}_{M\times N}\big(L^\infty(\widehat{G})\big)$, where*

$$\widehat{A}(\xi) := \big[\widehat{a}_{m,n}(\xi)\big], \quad a.e.\ \xi\in\widehat{G}.$$

***Proof*** Assume first that $\mathcal{A}$ is a well-defined bounded operator from $\ell^2_N(G)$ into $\ell^2_M(G)$. Denoting by $a_{m,n}$ the $(m,n)$-entry of $A$ and $\mathbf{e}_n$ the $n$-th column of the identity matrix $\mathbb{I}_N$, we have that for any $x\in\ell^2(G)$

$$\|a_{m,n} * x\|_{\ell^2(G)} = \big\|[A * x\,\mathbf{e}_n]_m\big\|_{\ell^2(G)} \le \|\mathcal{A}\|\,\|x\,\mathbf{e}_n\|_{\ell^2_N(G)} = \|\mathcal{A}\|\,\|x\|_{\ell^2(G)}. \tag{5.6}$$

On the other hand, let us assume that $C$ is a number such that $|\widehat{a}_{m,n}(\xi)| > C$ for all $\xi$ in a set $\Omega$ of positive measure. Since $\widehat{G}$ is compact, the characteristic function $\chi_\Omega$ of $\Omega$, belongs to $L^2(\widehat{G})$ and then $x = \mathcal{F}^{-1}\chi_\Omega \in \ell^2(G)$.

By using Lemma 5.1, we obtain that for such $C$ and $x$ it is satisfied

$$\|a_{m,n} * x\|_{\ell^2(G)} = \|\widehat{a_{m,n} * x}\|_{L^2(\widehat{G})} = \|\widehat{a}_{m,n} \cdot \chi_\Omega\|_{L^2(\widehat{G})}$$
$$\geq C\|\chi_\Omega\|_{L^2(\widehat{G})} = C\|x\|_{\ell^2(G)},$$

and then, from (5.6), we get $C \leq \|\mathcal{A}\|$. Thus, we have proved that

$$|\widehat{a}_{m,n}(\xi)| \leq \|\mathcal{A}\| \quad \text{a.e. } \xi \in \widehat{G}$$

and hence, that $\widehat{a}_{m,n} \in L^\infty(\widehat{G})$.

Conversely, we assume that $\widehat{A} \in \mathcal{M}_{M\times N}\big(L^\infty(\widehat{G})\big)$. By applying Lemma 5.1 we obtain that $\widehat{a_{m,n} * x} = \widehat{a}_{m,n} \cdot \widehat{x}$ for all $x \in \ell^2(G)$. Then, for any $x \in \ell^2(G)$

$$\|a_{m,n} * x\|_{\ell^2(G)} = \|\widehat{a_{m,n} * x}\|_{L^2(\widehat{G})} = \|\widehat{a}_{m,n} \cdot \widehat{x}\|_{L^2(\widehat{G})}$$
$$\leq \|\widehat{a}_{m,n}\|_{L^\infty(\widehat{G})}\|x\|_{\ell^2(G)},$$

from which easily follows that operator $\mathcal{A}$ is bounded. □

Under the above circumstances we say that $\mathcal{A}$ is a *bounded convolution operator*, and the matrix $\widehat{A}$ is the unique matrix-valued function $\widehat{A}(\xi)$ which satisfies

$$\widehat{\mathcal{A}(\mathbf{x})} = \widehat{A} \cdot \widehat{\mathbf{x}} \quad \text{for each } \mathbf{x} \in \ell^2_N(G).$$

We refer to $\widehat{A}$ as the *transfer matrix* of the convolution operator $\mathcal{A}$.

Observe that a bounded convolution system $\mathcal{A} : \ell^2_N(G) \to \ell^2_M(G)$ with associated matrix $A = [a_{m,n}] \in \mathcal{M}_{M\times N}(\ell^2(G))$ commutes with translations, i.e.,

$$\mathcal{A}\, T_g = T_g\, \mathcal{A} \quad \text{for all } g \in G.$$

In engineering jargon, the system $\mathcal{A}$ is a bounded *linear time-invariant* (LTI) *multi-input multi-output* (MIMO) system.

From now on, $\lambda_{\max}[B]$ (respectively $\lambda_{\min}[B]$) will denote the largest eigenvalue (respectively the smallest eigenvalue) of a matrix $B$.

**Proposition 5.2** *Let $\mathcal{A} : \ell^2_N(G) \longrightarrow \ell^2_M(G)$ be a bounded convolution operator with transfer matrix $\widehat{A}$. Then:*

*(a) The adjoint operator $\mathcal{A}^*$ is a bounded convolution operator with transfer matrix $\widehat{A}^*$, the adjoint matrix (transpose conjugate) of $\widehat{A}$, i.e., $\widehat{A}^*(\xi) = [\widehat{A}(\xi)]^*$, a.e. $\xi \in \widehat{G}$ (in the sequel $\widehat{A}(\xi)^*$).*

*(b) If* $\mathcal{B} : \ell^2_M(G) \to \ell^2_K(G)$ *is another bounded convolution operator with transfer matrix* $\widehat{B}$*, then the composition* $\mathcal{BA} : \ell^2_N(G) \to \ell^2_K(G)$ *is a bounded convolution operator with transfer matrix the matrix product* $\widehat{B} \cdot \widehat{A}$*.*

*(c) The operator norm of* $\mathcal{A}$ *is*

$$\|\mathcal{A}\| = \operatorname*{ess\,sup}_{\xi \in \widehat{G}} \|\widehat{A}(\xi)\|_2 \,,$$

*where* $\|\cdot\|_2$ *denotes the spectral norm of the matrix.*

*(d)* $\mathcal{A}$ *is injective with a closed range if and only if* $\operatorname{ess\,inf}_{\xi \in \widehat{G}} \det[\widehat{A}(\xi)^* \widehat{A}(\xi)] > 0$.

*(e)* $\mathcal{A}$ *is onto if and only if* $\operatorname{ess\,inf}_{\xi \in \widehat{G}} \det[\widehat{A}(\xi) \widehat{A}(\xi)^*] > 0$.

*(f)* $\mathcal{A}$ *is an isomorphism if and only if* $M = N$ *and* $\operatorname{ess\,inf}_{\xi \in \widehat{G}} |\det \widehat{A}(\xi)| > 0$*. In this case,* $\mathcal{A}^{-1}$ *is a bounded convolution operator with transfer matrix* $(\widehat{A})^{-1}$ *and operator norm*

$$\|\mathcal{A}^{-1}\| = \Big( \operatorname*{ess\,inf}_{\xi \in \widehat{G}} \lambda_{\min}[\widehat{A}(\xi)^* \widehat{A}(\xi)] \Big)^{-1/2}.$$

***Proof***

(a) Using Proposition 5.1, for each $\mathbf{x} \in \ell^2_N(G)$ and $\mathbf{y} \in \ell^2_M(G)$ we have

$$\begin{aligned} \langle \widehat{\mathbf{x}}, \widehat{\mathcal{A}^*\mathbf{y}} \rangle_{L^2_N(\widehat{G})} &= \langle \mathbf{x}, \mathcal{A}^*\mathbf{y} \rangle_{\ell^2_N(G)} = \langle \mathcal{A}\mathbf{x}, \mathbf{y} \rangle_{\ell^2_M(G)} \\ &= \langle \widehat{A} \cdot \widehat{\mathbf{x}}, \widehat{\mathbf{y}} \rangle_{L^2_M(\widehat{G})} = \langle \widehat{\mathbf{x}}, \widehat{A}^* \cdot \widehat{\mathbf{y}} \rangle_{L^2_N(\widehat{G})} \,. \end{aligned}$$

Hence $\widehat{\mathcal{A}^*\mathbf{y}} = \widehat{A}^* \cdot \widehat{\mathbf{y}}$ for all $\mathbf{y} \in \ell^2_N(G)$, and the result follows from Proposition 5.1.

(b) For each $\mathbf{x} \in \ell^2_N(G)$ we have $\widehat{\mathcal{BA}(\mathbf{x})} = \widehat{B} \cdot \widehat{\mathcal{A}(\mathbf{x})} = \widehat{B} \cdot \widehat{A} \cdot \widehat{\mathbf{x}}$. Since the entries of $\widehat{A}$ and $\widehat{B}$ belong to $L^\infty(\widehat{G})$, we get that $\widehat{B} \cdot \widehat{A} \in \mathcal{M}_{K \times N}(L^\infty(\widehat{G}))$, and the result follows from Proposition 5.1.

(c) The result is proved in Ref. [17, Corollary 3.8] for the case $N = M$. In short, let $B_{LTI}(\ell^2_N(G))$ denote the space of bounded $LTI$ endomorphisms of $\ell^2_N(G)$. There, the mapping

$$B_{LTI}(\ell^2_N(G)) \ni \mathcal{B} \longmapsto \widehat{B} \in \mathcal{M}_N(L^\infty(\widehat{G}))\,,$$

proves that $B_{LTI}(\ell^2_N(G))$ is a $C^*$-subalgebra of $B(\ell^2_N(G))$ which is $*$-isometric to the $C^*$-algebra $\mathcal{M}_N(L^\infty(\widehat{G}))$.
Thus,

$$\|\mathcal{B}\| = \|\widehat{B}\|_{\mathcal{M}_N(L^\infty(\widehat{G}))} = \operatorname*{ess\,sup}_{\xi \in \widehat{G}} \|\widehat{B}(\xi)\|_2 \,.$$

Hence, the above result for $\mathcal{A} : \ell^2_N(G) \longrightarrow \ell^2_M(G)$ gives

$$\|\mathcal{A}\|^2 = \|\mathcal{A}^*\mathcal{A}\| = \operatorname*{ess\,sup}_{\xi\in\widehat{G}} \|\widehat{A}(\xi)^*\widehat{A}(\xi)\|_2 = \operatorname*{ess\,sup}_{\xi\in\widehat{G}} \|\widehat{A}(\xi)\|_2^2 .$$

Recall that the *spectral norm* of an $N \times N$ matrix is the operator norm induced by the Euclidean norm in $\mathbb{C}^N$, and coincides with the largest singular value of the matrix (see Ref. [15]).

(d) A bounded operator $\mathcal{A}$ between Hilbert spaces is injective with a closed range if and only if the operator $\mathcal{A}^*\mathcal{A}$ is invertible. By using (a) and (b), we have that $\mathcal{A}^*\mathcal{A}$ is a bounded convolution operator with transfer matrix $\widehat{A}(\xi)^*\widehat{A}(\xi)$, and the result follows from Ref. [17, Theorem 3.11].
In short, the referred result says that an operator $\mathcal{B} \in B_{LTI}(\ell^2_N(G))$ with (square) transfer matrix $\widehat{B}$ is invertible if and only if $\widehat{B}$ is invertible in $\mathcal{M}_N(L^\infty(\widehat{G}))$. The last assertion is equivalent to $\operatorname{ess\,inf}_{\xi\in\widehat{G}} |\det \widehat{B}(\xi)| > 0$.
Thus, for $\mathcal{A} : \ell^2_N(G) \to \ell^2_M(G)$, the operator $\mathcal{A}^*\mathcal{A}$ is invertible if and only if

$$\operatorname*{ess\,inf}_{\xi\in\widehat{G}} \det[\widehat{A}(\xi)^*\widehat{A}(\xi)] > 0 .$$

(e) A bounded operator $\mathcal{A}$ is onto if and only if its adjoint operator $\mathcal{A}^*$ is injective with a closed range. From (a), the transfer matrix of the adjoint operator $\mathcal{A}^*$ is $\widehat{A}^*$. Thus, the result follows from (d).

(f) This characterization is a consequence of (d) and (e). Indeed, having in mind that $B_{LTI}(\ell^2_N(G))$ is a $C^*$-subalgebra of $B(\ell^2_N(G))$ which is $*$-isometric to the $C^*$-algebra $\mathcal{M}_N(L^\infty(\widehat{G}))$, the inverse operator $\mathcal{A}^{-1}$ is a bounded convolution operator with transfer matrix $(\widehat{A}(\xi))^{-1}$ and operator norm

$$\begin{aligned}\|\mathcal{A}^{-1}\|^{-2} &= \|(\widehat{A}(\xi))^{-1}\|_{\mathcal{M}_N(L^\infty(\widehat{G}))} = \Big[\operatorname*{ess\,sup}_{\xi\in\widehat{G}} \|(\widehat{A}(\xi))^{-1}\|_2\Big]^{-2} \\ &= \Big[\operatorname*{ess\,sup}_{\xi\in\widehat{G}} \lambda_{\min}^{-1}[\widehat{A}(\xi)^*\widehat{A}(\xi)]\Big]^{-1} \\ &= \operatorname*{ess\,inf}_{\xi\in\widehat{G}} \lambda_{\min}[\widehat{A}(\xi)^*\widehat{A}(\xi)] .\end{aligned}$$

See the details in Ref. [17, Theorem 3.11].

□

Observe that from (a) the matrix associated with the adjoint operator $\mathcal{A}^*$ is not the adjoint matrix of $A$, but the one defined by means of the *involution*

$$A^* = \big[a^*_{m,n}\big]^\top \in \mathcal{M}_{N\times M}\big(\ell^2(G)\big), \text{ where } \quad a^*_{m,n}(g) := \overline{a_{m,n}(-g)}, \quad g \in G. \tag{5.7}$$

Indeed, since $\widehat{a^*_{m,n}}(\xi) = \overline{\widehat{a}_{m,n}(\xi)}$, we have $\widehat{A^*}(\xi) = \widehat{A}(\xi)^* = \widehat{A}^*(\xi)$, a.e. $\xi \in \widehat{G}$.

## Convolution Systems, Frames of Translates for $\ell^2_N(G)$, and Their Structure-Preserving Dual Frames

A convolution operator $\mathcal{B} : \ell^2_M(G) \longrightarrow \ell^2_N(G)$ with matrix $B$ in $\mathcal{M}_{N\times M}(\ell^2(G))$ can be written in terms of the $M$ columns $\mathbf{b}_1, \mathbf{b}_2 \dots, \mathbf{b}_M$ of $B$ as

$$\mathcal{B}(\mathbf{x}) = B * \mathbf{x} = \sum_{m=1}^{M} \sum_{g\in G} x_m(g) T_g \mathbf{b}_m\,, \quad \mathbf{x} \in \ell^2_M(G)\,, \tag{5.8}$$

where $T_g\mathbf{b}_m = \mathbf{b}_m(\cdot - g)$ denotes the translation operator for each $m = 1, 2, \dots, M$.

In other words, the convolution operator $\mathcal{B}$ is the *synthesis operator* of the sequence $\{T_g\mathbf{b}_m\}_{g\in G;\, m=1,2,\dots,M}$ in $\ell^2_N(G)$. Thus, the results in Propositions 5.1 and 5.2 can be transferred to the sequence $\{T_g\mathbf{b}_m\}_{g\in G;\, m=1,2,\dots,M}$ (interchanging $M$ by $N$).

For instance, since a sequence in a Hilbert space is a Bessel sequence if and only if its synthesis operator is bounded and, in this case, its optimal Bessel bound is the square of the synthesis operator norm, from Propositions 5.1 and 5.2 we get

**Proposition 5.3** *The sequence $\{T_g\mathbf{b}_m\}_{g\in G;\, m=1,2,\dots,M}$ is a Bessel sequence for $\ell^2_N(G)$ if and only if the transfer matrix $\widehat{B}$ belongs to $\mathcal{M}_{N\times M}\big(L^\infty(\widehat{G})\big)$. In this case the optimal Bessel bound is $\beta_B = \operatorname{ess\,sup}_{\xi\in\widehat{G}} \|\widehat{B}(\xi)\|_2^2$.*

Let $\mathbf{a}^*_m$ denote the $m$-th column of the matrix $A^*$, the associated matrix of $\mathcal{A}^*$, given in (5.7). The convolution operator $\mathcal{A} : \ell^2_N(G) \to \ell^2_M(G)$ can also be written as

$$[\mathcal{A}(\mathbf{x})]_m(g) = [A * \mathbf{x}]_m(g) = \langle \mathbf{x}, T_g\mathbf{a}^*_m\rangle_{\ell^2_N(G)}\,. \tag{5.9}$$

In other words, the convolution operator $\mathcal{A}$ is the *analysis operator* for the sequence $\{T_g\mathbf{a}^*_m\}_{g\in G;\, m=1,2,\dots,M}$ in $\ell^2_N(G)$.

**Proposition 5.4** *Assume that the transfer matrix $\widehat{A} \in \mathcal{M}_{M\times N}(L^\infty(\widehat{G}))$. Then, the following results hold:*

*(a) The sequence $\{T_g\mathbf{a}^*_m\}_{g\in G;\, m=1,2,\dots,M}$ is a frame for $\ell^2_N(G)$ if and only if*

$$\operatorname*{ess\,inf}_{\xi\in\widehat{G}} \det\big[\widehat{A}(\xi)^*\widehat{A}(\xi)\big] > 0.$$

*In this case, the optimal frame bounds are*

$$\alpha_A = \operatorname*{ess\,inf}_{\xi\in\widehat{G}} \lambda_{\min}[\widehat{A}(\xi)^*\widehat{A}(\xi)] \quad \textit{and} \quad \beta_A = \operatorname*{ess\,sup}_{\xi\in\widehat{G}} \lambda_{\max}[\widehat{A}(\xi)^*\widehat{A}(\xi)]\,.$$

*(b) The sequence* $\{T_g \mathbf{a}_m^*\}_{g\in G;\, m=1,2,\dots,M}$ *is a Riesz basis for* $\ell^2_N(G)$ *if and only if*

$$N = M \quad \textit{and} \quad \operatorname*{ess\,inf}_{\xi\in\widehat{G}} \big|\det[\widehat{A}(\xi)]\big| > 0\,.$$

***Proof***

*(a)* From Proposition 5.3, the sequence $\{T_g \mathbf{a}_m^*\}_{g\in G;\, m=1,2,\dots,M}$ is a Bessel sequence of $\ell^2_N(G)$. Notice that its corresponding transfer matrix is $\widehat{A}^*$ and $\widehat{A}$ belongs to $\mathcal{M}_{M\times N}(L^\infty(\widehat{G}))$.

Since a Bessel sequence is a frame if and only if its analysis operator $\mathcal{A}$ is injective with a closed range (see Corollary 1.1), the result is a consequence of Proposition 5.2(d).

Moreover, since the optimal upper frame bound $\beta_A$ is the squared norm of the analysis operator $\mathcal{A}$, and the optimal lower frame bound $\alpha_A$ is the reciprocal of the norm of the inverse of the frame operator $\mathcal{A}^*\mathcal{A}$, from Proposition 5.2 we get

$$\beta_A = \|\mathcal{A}\|^2 = \operatorname*{ess\,sup}_{\xi\in\widehat{G}} \|\widehat{A}(\xi)\|_2^2 = \operatorname*{ess\,sup}_{\xi\in\widehat{G}} \lambda_{\max}[\widehat{A}(\xi)^*\widehat{A}(\xi)],$$

$$\alpha_A = \|(\mathcal{A}^*\mathcal{A})^{-1}\|^{-1} = \big(\operatorname*{ess\,inf}_{\xi\in\widehat{G}} \lambda_{\min}[\widehat{A}(\xi)^*\widehat{A}(\xi)\widehat{A}(\xi)^*\widehat{A}(\xi)]\big)^{1/2}$$

$$= \operatorname*{ess\,inf}_{\xi\in\widehat{G}} \lambda_{\min}[\widehat{A}(\xi)^*\widehat{A}(\xi)]\,.$$

*(b)* The Bessel sequence $\{T_g \mathbf{a}_m^*\}_{g\in G;\, m=1,2,\dots,M}$ is a Riesz basis for $\ell^2_N(G)$ if and only if its synthesis operator $\mathcal{A}^*$ is an isomorphism. Hence, the result is a consequence of Proposition 5.2(f).

□

**Proposition 5.5** *Assume that* $\widehat{A} \in \mathcal{M}_{M\times N}(L^\infty(\widehat{G}))$ *and* $\widehat{B} \in \mathcal{M}_{N\times M}(L^\infty(\widehat{G}))$. *Then the sequences* $\{T_g \mathbf{a}_n^*\}_{g\in G;\, n=1,2,\dots,M}$ *and* $\{T_g \mathbf{b}_n\}_{g\in G;\, n=1,2,\dots,M}$ *form a pair of dual frames for* $\ell^2_N(G)$ *if and only if*

$$\widehat{B}(\xi)\,\widehat{A}(\xi) = I_N\,, \quad \textit{a.e. } \xi \in \widehat{G}\,.$$

***Proof*** Having in mind that the analysis operator of $\{T_g \mathbf{a}_n^*\}_{g\in G;\, n=1,2,\dots,M}$ is $\mathcal{A}$ and that the synthesis operator of $\{T_g \mathbf{b}_n\}_{g\in G;\, n=1,2,\dots,M}$ is $\mathcal{B}$, we obtain that these two Bessel sequences form a pair of dual frames if and only if $\mathcal{B}\mathcal{A} = I_{\ell^2_N(G)}$ (see Proposition 1.11) or, equivalently, $\widehat{B}(\xi)\,\widehat{A}(\xi) = I_N$, a.e. $\xi \in \widehat{G}$ (see Proposition 5.2(b)). □

## 5.3 Regular Sampling in $U$-Invariant Subspaces

Suppose that $g \longmapsto U(g)$ is a unitary representation of the countable discrete abelian group $G$ on a separable Hilbert space $\mathcal{H}$, and assume that for a fixed set of generators $\Phi = \{\varphi_1, \varphi_2, \dots, \varphi_N\}$ in $\mathcal{H}$ the sequence $\big\{U(g)\varphi_n\big\}_{g\in G;\, n=1,2,\dots,N}$ is a Riesz sequence for $\mathcal{H}$.

In this regard, we state a result borrowed from Ref. [17, Theorem 4.2]. See also Refs. [2, 3, 13, 16].

Namely, for $m, n = 1, 2, \dots, N$, we consider the sequences

$$a_{m,n}(g) := \langle \varphi_n, U(g)\varphi_m\rangle_{\mathcal{H}}, \quad g \in G,$$

and assume that the matrix $A = [a_{m,n}]$ belongs to $\mathcal{M}_N\big(\ell^2(G)\big)$. Next, we consider the *Gram matrix-valued function* $\mathcal{G}(\xi) := \widehat{A}^\top(\xi)$, a.e. $\xi \in \widehat{G}$, which belongs to $\mathcal{M}_N\big(L^2(\widehat{G})\big)$. Thus, we have the following result:

**Theorem 5.1** *The sequence $\big\{U(g)\varphi_n\big\}_{g\in G;\, n=1,2,\dots,N}$ is a Riesz sequence for $\mathcal{H}$ if and only if*

$$\mathcal{G} \in \mathcal{M}_N\big(L^\infty(\widehat{G})\big) \quad \textit{and} \quad \operatorname*{ess\,inf}_{\xi\in\widehat{G}} \det[\mathcal{G}(\xi)] > 0\,.$$

*In this case, the optimal Riesz bounds are given by*

$$\operatorname*{ess\,inf}_{\xi\in\widehat{G}} \lambda_{min}[\mathcal{G}(\xi)] \quad \textit{and} \quad \operatorname*{ess\,sup}_{\xi\in\widehat{G}} \lambda_{max}[\mathcal{G}(\xi)]\,.$$

In the particular but important case of shift-invariant subspaces $V_\Phi^2$ in $L^2(\mathbb{R}^d)$, in order to be the sequence $\big\{\varphi_n(t-\alpha)\big\}_{\alpha\in\mathbb{Z}^d;\, n=1,2,\dots,N}$ a Riesz basis for $V_\Phi^2$, a necessary and sufficient condition was stated in Sect. 3.3.2 of Chap. 3.

### 5.3.1 A General Sampling Result

In case $\big\{U(g)\varphi_n\big\}_{g\in G;\, n=1,2,\dots,N}$ is a Riesz sequence for $\mathcal{H}$ the $U$-invariant subspace $\mathcal{V}_\Phi := \overline{\operatorname{span}}_{\mathcal{H}}\{U(g)\varphi_n\}_{g\in G;\, n=1,2,\dots,N}$ in $\mathcal{H}$ can be expressed as

$$\mathcal{V}_\Phi = \Big\{\sum_{n=1}^{N}\sum_{g\in G} x_n(g)U(g)\varphi_n \;:\; x_n \in \ell^2(G);\; n = 1, 2, \dots, N\Big\}. \tag{5.10}$$

For a given matrix $A = [a_{m,n}] \in \mathcal{M}_{M\times N}\big(\ell^2(G)\big)$, we consider the sequence of the vector samples $\big\{\mathcal{L}f(g)\big\}_{g\in G}$ of any $f = \sum_{n=1}^{N}\sum_{g\in G} x_n(g)U(g)\varphi_n \in \mathcal{V}_\Phi$ defined by

$$
\begin{aligned}
\mathcal{L}f(g) :=& \big(\mathcal{L}_1 f(g), \mathcal{L}_2 f(g), \dots, \mathcal{L}_M f(g)\big)^\top \\
=& (A * \mathbf{x})(g) = \big[\mathcal{A}(\mathbf{x})\big](g)\,, \qquad g \in G\,.
\end{aligned}
\tag{5.11}
$$

Assume that the matrix $\widehat{A} \in \mathcal{M}_{M\times N}(L^\infty(\widehat{G}))$ and $\operatorname{ess\,inf}_{\xi\in\widehat{G}} \det\big[\widehat{A}(\xi)^*\widehat{A}(\xi)\big] > 0$.

Since $[\mathcal{A}(\mathbf{x})]_m(g) = \langle \mathbf{x}, T_g\mathbf{a}_m^*\rangle_{\ell^2_N(G)}$, the optimal frame bounds given in Proposition 5.4 provide relevant information about the stability of the recovering. Namely,

$$
\alpha_A\,\|\mathbf{x}\|^2 \le \sum_{m=1}^{M}\sum_{g\in G} |\mathcal{L}_m f(g)|^2 \le \beta_A\,\|\mathbf{x}\|^2 \quad \text{for all } \mathbf{x}\in \ell^2_N(G)\,,
$$

where

$$
\alpha_A = \operatorname*{ess\,inf}_{\xi\in\widehat{G}} \lambda_{\min}[\widehat{A}(\xi)^*\widehat{A}(\xi)] \quad\text{and}\quad \beta_A = \operatorname*{ess\,sup}_{\xi\in\widehat{G}} \lambda_{\max}[\widehat{A}(\xi)^*\widehat{A}(\xi)]\,.
$$

Moreover, denoting by $\alpha_\Phi$ and $\beta_\Phi$ the Riesz bounds for $\big\{U(g)\varphi_n\big\}_{g\in G;\, n=1,2,\dots,N}$ we get

$$
\alpha_\Phi\alpha_A\,\|f\|^2 \le \sum_{m=1}^{M}\sum_{g\in G} |\mathcal{L}_m f(g)|^2 \le \beta_\Phi\beta_A\,\|f\|^2 \quad \text{for all } f\in\mathcal{V}_\Phi.
$$

Now, for the recovering any $f\in\mathcal{V}_\Phi$ from its generalized samples (5.11), the idea is to find a $N\times M$ matrix $\widehat{B}\in\mathcal{M}_{N\times M}(L^\infty(\widehat{G}))$ such that

$$
\widehat{B}(\xi)\,\widehat{A}(\xi) = I_N\,, \quad a.e.\ \xi\in\widehat{G}\,.
$$

In other words, the corresponding convolution operator $\mathcal{B}(\mathbf{x}) = B * \mathbf{x}$ should satisfy $\mathbf{x} = \mathcal{B}\mathcal{A}(\mathbf{x}) = \mathcal{B}(\mathcal{L}f)$, that is

$$
\mathbf{x} = B * \mathcal{L}f \quad\text{and}\quad f = \sum_{n=1}^{N}\sum_{g\in G} x_n(g)U(g)\varphi_n\,, \quad f\in\mathcal{V}_\Phi\,.
$$

Moreover, an explicit structured sampling formula can be obtained. Namely, the recovering of $\mathbf{x} = B * \mathcal{L}f$ from the samples $\mathcal{L}f$ can be written as an expansion in terms of a pair of dual frames (see Eqs. (5.8) and (5.9) and Proposition 5.5)

$$\begin{aligned}\mathbf{x} &= \sum_{m=1}^{M}\sum_{g\in G}\langle \mathbf{x}, T_g\mathbf{a}_m^*\rangle_{\ell^2_N(G)}\, T_g\mathbf{b}_m \\ &= \sum_{m=1}^{M}\sum_{g\in G}\mathcal{L}_m f(g)\, T_g\mathbf{b}_m \quad \text{in } \ell^2_N(G).\end{aligned} \tag{5.12}$$

Besides, we consider the natural isomorphism

$$\mathcal{T}_{U,\Phi} : \ell^2_N(G) \longrightarrow \mathcal{V}_\Phi \tag{5.13}$$

which maps the standard orthonormal basis $\{\boldsymbol{\delta}_{g,n}\}_{g\in G;\, n=1,2,\ldots,N}$ for $\ell^2_N(G)$ onto the Riesz basis $\{U(g)\varphi_n\}_{g\in G;\, n=1,2,\ldots,N}$ for $\mathcal{V}_\Phi$. Thus, for any $\mathbf{b} = (b_1, b_2, \ldots, b_N)$ in $\ell^2_N(G)$ we have $\mathcal{T}_{U,\Phi}\mathbf{b} = \sum_{n=1}^{N}\sum_{g\in G} b_n(g)\, U(g)\varphi_n$.

The above isomorphism satisfies the *shifting property*:

$$\mathcal{T}_{U,\Phi}\big(T_g\mathbf{b}\big) = U(g)\big(\mathcal{T}_{U,\Phi}\mathbf{b}\big) \quad \text{for each } g \in G \text{ and } \mathbf{b} \in \ell^2_N(G). \tag{5.14}$$

Finally, for each $f = \mathcal{T}_{U,\Phi}\mathbf{x} \in \mathcal{V}_\Phi$, applying the isomorphism $\mathcal{T}_{U,\Phi}$ on (5.12) and the shifting property (5.14) we obtain the sampling expansion in $\mathcal{V}_\Phi$

$$\begin{aligned}f &= \sum_{m=1}^{M}\sum_{g\in G}\mathcal{L}_m f(g)\, \mathcal{T}_{U,\Phi}\big(T_g\mathbf{b}_m\big) \\ &= \sum_{m=1}^{M}\sum_{g\in G}\mathcal{L}_m f(g)\, U(g)\big(\mathcal{T}_{U,\Phi}\mathbf{b}_m\big) \\ &= \sum_{m=1}^{M}\sum_{g\in G}\mathcal{L}_m f(g)\, U(g)[S_m] \quad \text{in } \mathcal{H}\end{aligned} \tag{5.15}$$

where the reconstruction elements $S_m$ are given by

$$S_m = \mathcal{T}_{U,\Phi}\mathbf{b}_m \in \mathcal{V}_\Phi\,, \quad m = 1, 2, \ldots, M\,,$$

and the sequence $\{U(g)S_m\}_{g\in G;\, m=1,2,\ldots,M}$ is a frame for $\mathcal{V}_\Phi$.

In fact, in the $U$-invariant subspace $\mathcal{V}_\Phi$ the following sampling theorem holds:

**Theorem 5.2** *Let $A = [a_{m,n}] \in \mathcal{M}_{M\times N}\big(\ell^2(G)\big)$ be the matrix defining the samples $\mathcal{L}f(g)$, $g \in G$, for each $f \in \mathcal{V}_\Phi$ as in (5.11), and assume that its transfer matrix $\widehat{A}$ has all its entries in $L^\infty(\widehat{G})$. Then, the following statements are equivalent:*

*(a) The constant* $\delta_A := \operatorname*{ess\,inf}_{\xi\in\widehat{G}} \det[\widehat{A}(\xi)^*\widehat{A}(\xi)] > 0$.

*(b) There exist constants $0 < \alpha \le \beta$ such that*

$$\alpha\|f\|^2 \le \sum_{m=1}^{M}\sum_{g\in G} |\mathcal{L}_m f(g)|^2 \le \beta\|f\|^2\,, \quad f \in \mathcal{V}_\Phi\,.$$

*(c) There exists a matrix $\widehat{B} \in \mathcal{M}_{N\times M}\big(L^\infty(\widehat{G})\big)$ such that*

$$\widehat{B}(\xi)\,\widehat{A}(\xi) = I_N\,, \quad \text{a.e. } \xi \in \widehat{G}.$$

*(d) There exists a matrix $B \in \mathcal{M}_{N\times M}(\ell^2(G))$ with $\widehat{B} \in \mathcal{M}_{N\times M}(L^\infty(\widehat{G}))$, such that*

$$\mathbf{x} = B * \mathcal{L}f \quad \text{and} \quad f = \sum_{n=1}^{N}\sum_{g\in G} x_n(g)U(g)\varphi_n\,, \quad f \in \mathcal{V}_\Phi\,.$$

*In other words, there exists a bounded convolution system $\mathcal{B} : \ell^2_M(G) \to \ell^2_N(G)$ such that $\mathcal{B}\mathcal{A} = \mathcal{I}_{\ell^2_N(G)}$.*

*(e) There exist $M$ elements $S_m \in \mathcal{V}_\Phi$ such that the sequence $\big\{U(g)S_m\big\}_{g\in G;\, m=1,2,\dots,M}$ is a frame for $\mathcal{V}_\Phi$ and for each $f \in \mathcal{V}_\Phi$ the reconstruction formula*

$$f = \sum_{m=1}^{M}\sum_{g\in G} \mathcal{L}_m f(g)\,U(g)S_m \quad \text{in } \mathcal{H} \tag{5.16}$$

*holds.*

*(f) There exists a frame $\big\{S_{g,m}\big\}_{g\in G;\, m=1,2,\dots,M}$ for $\mathcal{V}_\Phi$ such that for each $f \in \mathcal{V}_\Phi$ the expansion*

$$f = \sum_{m=1}^{M}\sum_{g\in G} \mathcal{L}_m f(g)\, S_{g,m} \quad \text{in } \mathcal{H}$$

*holds.*

*In this case, the reconstruction elements* $\{S_m\}_{m=1,2,\dots,M}$ *in* $\mathcal{V}_\Phi$ *in formula* (5.16) *are necessarily obtained from the columns* $\mathbf{b}_1, \mathbf{b}_2, \dots, \mathbf{b}_M$ *of a matrix* $B$ *satisfying condition* $(c)$, *i.e.,*

$$S_m = \mathcal{T}_{U,\Phi}\mathbf{b}_m = \sum_{n=1}^{N}\sum_{g\in G} b_{n,m}\, U(g)\varphi_n\,, \quad m = 1, 2, \dots, M\,.$$

***Proof*** First we note that, since $\beta_{_A} < \infty$, condition $\delta_{_A} > 0$ is equivalent to condition $\alpha_{_A} > 0$.

Next we prove that conditions $(a)$ and $(b)$ are equivalent. Indeed, since $\mathcal{T}_{U,\Phi}$ is an isomorphism, condition (b) is equivalent to the existence of $0 < \alpha_1 \le \beta_1$ such that

$$\alpha_1 \|\mathbf{x}\|^2 \le \sum_{m=1}^{M}\sum_{g\in G} |\mathcal{L}_m f(g)|^2 \le \beta_1 \|\mathbf{x}\|^2\,, \quad \mathbf{x} \in \ell^2_N(G)\,.$$

Due to Eq. (5.9), this is equivalent to be the sequence $\big\{T_g\mathbf{a}^*_m\big\}_{g\in G;\, m=1,2,\dots,M}$ a frame for $\ell^2_N(G)$. Therefore, the result follows from Proposition 5.4.

Assume now that condition $(a)$ holds. Then, from Proposition 5.2, the operator $\mathcal{A}^*\mathcal{A}$ is invertible, and $\mathcal{B} := (\mathcal{A}^*\mathcal{A})^{-1}\mathcal{A}^*$ is a bounded convolution operator satisfying $\mathcal{B}\mathcal{A} = I_{\ell^2_N(G)}$. From Proposition 5.2, its transfer matrix satisfies the requirement in condition $(c)$.

If the transfer matrix $\widehat{B}$ satisfies condition $(c)$, then the bounded convolution operator $\mathcal{B}$ whose transfer matrix is $\widehat{B}$ satisfies $\mathcal{B}\mathcal{A} = I_{\ell^2_N(G)}$ and, from Proposition 5.2, condition $(d)$.

We have proved that condition $(d)$ implies a sampling expansion like (5.16), where $S_m = \mathcal{T}_{U,\Phi}\mathbf{b}_m$, $m = 1, 2, \dots, M$, and $\mathbf{b}_1, \dots, \mathbf{b}_M$ are the columns of a matrix $B$ satisfying condition $(d)$.

Besides, the sequence

$$\big\{U(g)S_m\big\}_{g\in G;\, m=1,2,\dots,M} = \mathcal{T}_{U,\Phi}\big\{T_g\mathbf{b}_m\big\}_{g\in G;\, m=1,2,\dots,M}$$

is a frame since (5.12) is a frame expansion in $\ell^2_N(G)$ and $\mathcal{T}_{U,\Phi}$ an isomorphism. This proves condition $(e)$ which trivially implies condition $(f)$.

Finally, condition $(f)$ implies $(a)$. Applying $\mathcal{T}^{-1}_{U,\Phi}$ to the formula in $(f)$ we obtain that $\big\{T_g\mathbf{a}^*_m\big\}_{g\in G;\, m=1,2,\dots,M}$ and $\{\mathcal{T}^{-1}_{U,\Phi}S_{g,m}\}_{g\in G;\, m=1,2,\dots,M}$ form a pair of dual frames for $\ell^2_N(G)$ (see Proposition 1.11); in particular, by using Proposition 5.4(a) we obtain that $\delta_{_A} > 0$. □

Some comments on the above theorem are in order:

(*i*) All the possible solutions of $\widehat{B}(\xi)\widehat{A}(\xi) = I_N$ a.e. $\xi \in \widehat{G}$ with entries in $L^\infty(\widehat{G})$ are given in terms of the Moore-Penrose pseudo-inverse

$$\widehat{A}(\xi)^\dagger = \big[\widehat{A}(\xi)^*\widehat{A}(\xi)\big]^{-1}\widehat{A}(\xi)^*\,, \quad \text{a.e. } \xi \in \widehat{G}\,,$$

by means of the $N \times M$ matrices

$$\widehat{B}(\xi) := \widehat{A}(\xi)^\dagger + C(\xi)\big[I_M - \widehat{A}(\xi)\widehat{A}(\xi)^\dagger\big]\,, \quad \text{a.e. } \xi \in \widehat{G}\,,$$

where $C(\xi)$ denotes any $N \times M$ matrix with entries in $L^\infty(\widehat{G})$ (see (3.16)).

(*ii*) Since $\mathbf{x} = B * \mathcal{L}f$, from Proposition 5.1 we have that

$$\|\mathbf{x}\|^2_{\ell^2_N(G)} \le C \sum_{m=1}^{M}\sum_{g\in G} |\mathcal{L}_m f(g)|^2\,,$$

where $C = \operatorname{ess\,sup}_{\xi\in\widehat{G}} \|\widehat{B}(\xi)\|_2^2$. The best possible bound we can get is

$$C = \alpha_A^{-1} = \operatorname*{ess\,inf}_{\xi\in\widehat{G}} \lambda_{\min}[\widehat{A}(\xi)^*\widehat{A}(\xi)]^{-1}$$

which corresponds with the choice $\widehat{B} = \widehat{A}^\dagger$ or, equivalently, $B = (A^*A)^{-1}A^*$, the pseudo-inverse of $A$ (see Ref. [4]).

(*iii*) Observe that condition (*b*) in Theorem 5.2 shows that the $\ell^2$-norm of the sequence of samples $\big\{\mathcal{L}_n f(g)\big\}_{g\in G;\, n=1,2,\ldots,N}$ defines an equivalent norm in the subspace $\mathcal{V}_\Phi$.

Notice that in Theorem 5.2 necessarily $M \ge N$, where $N$ is the number of generators in $\mathcal{V}_\Phi$. In case $M = N$, we have:

**Corollary 5.1** *In case $M = N$, assume that the transfer matrix $\widehat{A}(\xi)$ has all entries in $L^\infty(\widehat{G})$. The following statements are equivalent:*

1. *The constant* $\operatorname*{ess\,inf}_{\xi\in\widehat{G}} \big|\det[\widehat{A}(\xi)]\big| > 0$.
2. *There exist $N$ unique elements $S_n$, $n = 1, 2, \ldots, N$, in $\mathcal{V}_\Phi$ such that the associated sequence $\big\{U(g)S_n\big\}_{g\in G;\, n=1,2,\ldots,N}$ is a Riesz basis for $\mathcal{V}_\Phi$ and the sampling formula*

$$f = \sum_{n=1}^{N}\sum_{g\in G} \mathcal{L}_n f(g)\, U(g)S_n \qquad \text{in } \mathcal{H}$$

*holds for each $f \in \mathcal{V}_\Phi$.*

*Moreover, the interpolation property* $\mathcal{L}_n S_{n'}(g) = \delta_{n,n'}\delta_{g,0_G}$ *holds, where* $g \in G$ *and* $n, n' = 1, 2, \ldots, N$.

***Proof*** In this case, the square matrix $\widehat{A}(\xi)$ is invertible and the result comes out from Proposition 5.4(b). The uniqueness of the coefficients in a Riesz basis expansion gives the interpolation property. □

**Example 5.1 (A More General Framework)** A slightly more general setting is motivated by condition $(f)$ in Theorem 5.2. Namely, let $F := \{f_{g,n}\}_{g\in G;\, n=1,2,\ldots,N}$ be a Riesz sequence in a separable Hilbert space $\mathcal{H}$, and let $\mathcal{V}_F := \overline{\text{span}}_{\mathcal{H}}\{f_{g,n}\}_{g\in G;\, n=1,2,\ldots,N}$ be its associated subspace, that is,

$$\mathcal{V}_F = \Big\{\sum_{n=1}^{N}\sum_{g\in G} x_n(g)\, f_{g,n} \;:\; x_n \in \ell^2(G);\; n = 1, 2, \ldots, N\Big\}.$$

Given a matrix $A = [a_{m,n}] \in \mathcal{M}_{M\times N}\big(\ell^2(G)\big)$, for each $f = \sum_{n=1}^{N}\sum_{g\in G} x_n(g)\, f_{g,n}$ in $\mathcal{V}_F$ we define its data samples $\mathcal{L}f$ by means of the matrix $A$ and the element $\mathbf{x} = (x_1, x_2, \ldots, x_N)^\top$ in $\ell^2_N(G)$ associated to $f$ as

$$\mathcal{L}f(g) := \big(\mathcal{L}_1 f(g), \mathcal{L}_2 f(g), \ldots, \mathcal{L}_M f(g)\big)^\top = (A * \mathbf{x})(g)\,, \quad g \in G\,.$$

As before, the aim is the stable recovery of any $f \in \mathcal{V}_F$ from data $\mathcal{L}f \in \ell^2_M(G)$.

Under the hypotheses on the matrix $A$ assumed in Theorem 5.2 there exists a frame $\{S_{g,m}\}_{g\in G;\, m=1,2,\ldots,M}$ for $\mathcal{V}_F$ such that for each $f \in \mathcal{V}_F$ the reconstruction formula

$$f = \sum_{m=1}^{M}\sum_{g\in G} \mathcal{L}_m f(g)\, S_{g,m} \quad \text{in } \mathcal{H}$$

holds. Moreover, there exist $\mathbf{b}_m$ in $\ell^2_N(G)$, $m = 1, 2, \ldots, M$, such that

$$S_{g,m} = \mathcal{T}_F\big(T_g \mathbf{b}_m\big)\,, \quad g \in G \text{ and } m = 1, 2, \ldots, M$$

where $\mathcal{T}_F : \ell^2_N(G) \longrightarrow \mathcal{V}_F$ stands for the natural isomorphism which maps the standard orthonormal basis $\{\boldsymbol{\delta}_{g,n}\}_{g\in G;\, n=1,2,\ldots,N}$ for $\ell^2_N(G)$ onto the given Riesz basis $\{f_{g,n}\}_{g\in G;\, n=1,2,\ldots,N}$ for $\mathcal{V}_F$. Besides, the sequence $\{S_{g,m}\}_{g\in G;\, m=1,2,\ldots,M}$ is a frame for $\mathcal{V}_F$.

Since the subspace $\mathcal{V}_F$ has not any *a priori* structure, the same occurs for the reconstruction functions $S_{g,m}$, for $g \in G$ and $m = 1, 2, \ldots, M$.

### 5.3.2 Some Examples from Usual Sampling Settings

Next, we illustrate the sampling result in Theorem 5.2 with some examples concerning average sampling:

**Example 5.2 (Oversampled Average Sampling in $V_\Phi^2$)** Choose $\mathcal{H} := L^2(\mathbb{R}^d)$, $G := \mathbb{Z}^d$ and the unitary representation of $G$ on $L^2(\mathbb{R}^d)$ given by

$$\big[U(p)f\big](t) := f(t-p)\,, \quad t \in \mathbb{R}^d \text{ and } p \in \mathbb{Z}^d\,.$$

Under the hypotheses in Theorem 5.2 for the average samples given by (5.2), i.e., for the associated $M \times N$ matrix $A = [a_{m,n}]$ whose $(m,n)$-entry is

$$a_{m,n}(p) = \big\langle \varphi_n, \psi_m(\cdot - p)\big\rangle_{L^2(\mathbb{R}^d)}\,, \quad p \in \mathbb{Z}^d\,,$$

we obtain an oversampled average sampling in the classical shift-invariant subspace $V_\Phi^2$ of $L^2(\mathbb{R}^d)$ described as

$$V_\Phi^2 = \Big\{\sum_{n=1}^{N}\sum_{p\in\mathbb{Z}^d} x_n(p)\,\varphi_n(t-p) \,:\, x_n \in \ell^2(\mathbb{Z}^d);\ n = 1, 2, \ldots, N\Big\}\,.$$

For each $f \in V_\Phi^2$ a sampling expansion having the form

$$f(t) = \sum_{m=1}^{M}\sum_{p\in\mathbb{Z}^d} \big\langle f, \psi_m(\cdot - p)\big\rangle_{L^2(\mathbb{R}^d)}\, S_m(t-p) \quad \text{in } L^2(\mathbb{R}^d)\,,$$

holds, for some sampling functions $S_m \in V_\Phi^2$, $m = 1, 2, \ldots, M$. Moreover, the sequence $\{S_m(t-p)\}_{p\in\mathbb{Z}^d;\, m=1,2,\ldots,M}$ is a frame for $V_\Phi^2$.

Under mild hypotheses (see Sect. 5.3.3), the space $V_\Phi^2$ is a reproducing kernel Hilbert space. As a consequence of the RKHS setting the convergence of the series in the $L^2(\mathbb{R}^d)$-norm sense implies pointwise convergence which is uniform on $\mathbb{R}^d$. The pointwise convergence is also absolute.

As we will see later (see Sect. 5.3.3), this oversampling can be reduced by sampling on a sub-lattice $P\mathbb{Z}^d$ of $\mathbb{Z}^d$, where $P$ denotes a $d \times d$ matrix with integer entries and positive determinant.

**Example 5.3 (The Case where the Group G is the Semi-direct Product of Two Groups)** The case where the group $G$ is the semi-direct product of two groups can be easily reduced to the situation described in this chapter. Suppose that $(k,h) \mapsto U(k,h)$ is a unitary representation of the semi-direct product group $G = K \rtimes_\sigma H$ (or, in particular, the direct product $G = K \times H$) on a separable Hilbert space $\mathcal{H}$, where $K$ is a countable discrete group and $H$ a finite, not necessarily abelian group.

The subscript $\sigma$ denotes the action of the group $H$ on the group $K$, i.e., a homomorphism

$$\begin{aligned} \sigma : H &\longrightarrow Aut(K) \\ h &\longmapsto \sigma_h \end{aligned}$$

The composition law in $G$ is

$$(k_1, h_1)\,(k_2, h_2) := (k_1 + \sigma_{h_1}(k_2), h_1 h_2) \quad \text{for } (k_1, h_1),\ (k_2, h_2) \in G$$

where we are using an additive notation in $K$. The unity in $G$ is $1_G = (0_k, 1_H)$ and the inverse of $(k, h) \in G$ is $(k, h)^{-1} = (-\sigma_{h^{-1}}(k), h^{-1})$.

In general, the group $G = K \rtimes_\sigma H$ is not abelian. In case $\sigma_h \equiv Id_K$ for each $h \in H$ we recover the direct product group $G = K \times H$.

Assume that for a fixed $\varphi \in \mathcal{H}$ the sequence $\big\{U(k, h)\varphi\big\}_{(k,h)\in G}$ is a Riesz sequence for $\mathcal{H}$. Thus, the $U$-invariant subspace in $\mathcal{H}$ spanned by $\big\{U(k, h)\varphi\big\}_{(k,h)\in G}$ can be described as

$$\mathcal{V}_\varphi = \Big\{ \sum_{(k,h)\in G} x(k, h)\, U(k, h)\varphi \ :\ \{x(k, h)\}_{(k,h)\in G} \in \ell^2(G) \Big\}.$$

Assume that the order of the group $H$ is $N$. Since

$$U(k, h)\varphi = U[(k, 1_H)(0_K, h)]\varphi = U(k, 1_H)\varphi_h\,,$$

where $\varphi_h := U(0_K, h)\varphi$ for $h \in H$, the subspace $\mathcal{V}_\varphi$ coincides with the subspace $\mathcal{V}_\Phi$ generated by the set $\Phi = \{\varphi_1, \varphi_2, \dots, \varphi_N\}$, i.e.,

$$\mathcal{V}_\Phi = \Big\{ \sum_{n=1}^{N} \sum_{k\in K} x_n(k) U(k, 1_H)\varphi_n \ :\ x_n \in \ell^2(N);\ n = 1, 2, \dots, N \Big\},$$

where $x_n(k) := x(k, h_n)$ and $\varphi_n := \varphi_{h_n}$, $n = 1, 2, \dots, N$.

For $M$ fixed elements $\psi_m \in \mathcal{H}$, $m = 1, 2, \dots, M$, not necessarily in $\mathcal{V}_\Phi$, we consider, for each $f \in \mathcal{V}_\Phi$, its generalized average samples defined as

$$\mathcal{L}_m f(k) := \big\langle f, U(k, 1_H)\, \psi_m \big\rangle_{\mathcal{H}}, \quad k \in K\,, \quad m = 1, 2, \dots, M\,. \tag{5.17}$$

Notice that these samples are a particular case of samples (5.2). Then, under the hypotheses in Theorem 5.2 on the matrix $A = [a_{m,n}]$, where

$$a_{m,n}(k) = \big\langle \varphi_n, U(k, 1_H)\psi_m \big\rangle_{\mathcal{H}} = \big\langle \varphi, U[(-k, h_n)^{-1}]\psi_m \big\rangle_{\mathcal{H}}, \quad k \in K\,,$$

there exist $M$ elements $S_m \in \mathcal{V}_\Phi$ such that the sequence $\{U(k, 1_H)S_m\}_{k\in K;\, m=1,2,\dots,M}$ is a frame for $\mathcal{V}_\Phi$, and for each $f \in \mathcal{V}_\Phi$ we have the reconstruction formula

$$f = \sum_{m=1}^{M}\sum_{k\in K} \mathcal{L}_m f(k)\, U(k, 1_H)S_m \quad \text{in } \mathcal{H}. \tag{5.18}$$

**Example 5.4 (The Particular Case where G is a Crystallographic Group)** An important particular case of the above example is given by the crystallographic groups. The Euclidean motion group $E(d)$ is the semi-direct product $\mathbb{R}^d \rtimes_\sigma O(d)$ corresponding to the homomorphism $\sigma : O(d) \to Aut(\mathbb{R}^d)$ given by $\sigma_\gamma(x) = \gamma x$, where $\gamma \in O(d)$ and $x \in \mathbb{R}^d$; $O(d)$ denotes the orthogonal group of order $d$.

The composition law on $E(d) = \mathbb{R}^d \rtimes_\sigma O(d)$ reads

$$(x, \gamma)\cdot(x', \gamma') = (x + \gamma x', \gamma\gamma')\,.$$

Let $P$ be a non-singular $d \times d$ matrix and $\Gamma$ a finite subgroup of $O(d)$ of order $N$ such that $\gamma(P\mathbb{Z}^d) = P\mathbb{Z}^d$ for each $\gamma \in \Gamma$, i.e., the lattice $P\mathbb{Z}^d$ is invariant under the matrices in $\Gamma$.

We consider the crystallographic group

$$C_{P,\Gamma} := P\mathbb{Z}^d \rtimes_\sigma \Gamma\,,$$

and its quasi regular representation (see Ref. [2]) on $L^2(\mathbb{R}^d)$

$$U(p, \gamma)f(t) = f[\gamma^\top(t-p)]\,, \quad p \in P\mathbb{Z}^d,\ \gamma \in \Gamma \text{ and } f \in L^2(\mathbb{R}^d).$$

For a fixed $\varphi \in L^2(\mathbb{R}^d)$ such that the sequence $\{U(p, \gamma)\varphi\}_{(p,\gamma)\in C_{P,\Gamma}}$ is a Riesz sequence for $L^2(\mathbb{R}^d)$ we consider the $U$-invariant subspace in $L^2(\mathbb{R}^d)$

$$\mathcal{V}_\varphi = \Big\{ \sum_{(p,\gamma)\in C_{P,\Gamma}} x(p,\gamma)\,\varphi[\gamma^\top(t-p)] \ :\ \{x(p,\gamma)\} \in \ell^2(C_{P,\Gamma})\Big\}\,. \tag{5.19}$$

Choosing $M$ functions $\psi_m \in L^2(\mathbb{R}^d)$, $m = 1, 2, \dots, M$, we consider the average samples of $f \in \mathcal{V}_\varphi$

$$\mathcal{L}_m f(p) = \langle f, U(p, I)\psi_m\rangle = \langle f, \psi_m(\cdot - p)\rangle\,, \quad p \in P\mathbb{Z}^d\,.$$

Denoting $\{\gamma_1 = I, \gamma_2, \dots, \gamma_N\}$ the elements of the group $\Gamma$, under the hypotheses of Theorem 5.2 on the matrix $A = [a_{m,n}]$ where

$$a_{m,n}(p) = \big\langle \varphi(t), \psi_m(\gamma_n t - p)\big\rangle_{L^2(\mathbb{R}^d)}\,, \quad p \in P\mathbb{Z}^d\,,$$

there exist $M \geq N$ sampling functions $S_m \in \mathcal{V}_\varphi$, $m = 1, 2, \ldots, M$, such that the sequence $\{S_m(\cdot - p)\}_{p \in P\mathbb{Z}^d;\, m=1,2,\ldots,M}$ is a frame for $\mathcal{V}_\varphi$, and the sampling expansion

$$f(t) = \sum_{m=1}^{M} \sum_{p \in P\mathbb{Z}^d} \big\langle f, \psi_m(\cdot - p) \big\rangle_{L^2(\mathbb{R}^d)} S_m(t - p) \quad \text{in } L^2(\mathbb{R}^d) \tag{5.20}$$

for any $f \in \mathcal{V}_\varphi$ holds.

If the generator $\varphi$ is continuous in $\mathbb{R}^d$ and the function $t \mapsto \sum_{p \in \mathbb{Z}^d} |\varphi(t - p)|^2$ is bounded on $\mathbb{R}^d$, a standard argument shows that $\mathcal{V}_\varphi$ is a RKHS of bounded continuous functions in $L^2(\mathbb{R}^d)$ (see, for instance, Proposition 5.6 infra). As a consequence, convergence in $L^2(\mathbb{R}^d)$-norm implies pointwise convergence which is uniform on $\mathbb{R}^d$. The pointwise convergence is also absolute.

**Example 5.5 (Average Sampling Formulas in $\ell^2_N$-Spaces)** Taking $G = \mathbb{Z}$, notice that formula (5.12) can be considered as an average sampling formula in $\ell^2_N(\mathbb{Z})$. Indeed, in the product space $\ell^2_N(\mathbb{Z})$ we consider the sampling process:

$$\ell^2_N(\mathbb{Z}) \ni \mathbf{x} = (x_1, x_2, \ldots, x_N)^\top \longmapsto A * \mathbf{x} \in \ell^2_M(\mathbb{Z}),$$

given by a $M \times N$ matrix $A = [a_{m,n}]$ with entries in $\ell^2(\mathbb{Z})$.

In case the hypotheses in Proposition 5.5 are satisfied, then there exist $M$ elements $\mathbf{b}_m$ in $\ell^2_N(\mathbb{Z})$, where necessarily $M \geq N$, such that

$$\mathbf{x}(n) = \sum_{m=1}^{M} \sum_{k \in \mathbb{Z}} \big(\mathbf{a}_m * \mathbf{x}\big)(k)\, \mathbf{b}_m(n - k), \quad n \in \mathbb{Z},$$

where $\mathbf{a}_m \in \ell^2_N(\mathbb{Z})$ denotes the $m$-th row of the matrix $A$.

### 5.3.3 *The Case of Pointwise Samples whenever $\mathcal{H} = L^2(\mathbb{R}^d)$*

Assume here that $G$ is a countable discrete subgroup of a locally compact abelian (LCA) group $\widetilde{G}$ and let $\widetilde{G} \ni t \longmapsto U(t) \in \mathcal{U}(L^2(\widetilde{G}))$ be a unitary representation of $\widetilde{G}$ on $L^2(\widetilde{G})$.

**Definition 5.1** A unitary representation of a locally compact abelian group $\widetilde{G}$ on a Hilbert space $\mathcal{H}$ is a map $\widetilde{G} \ni t \longmapsto U(t) \in \mathcal{U}(\mathcal{H})$ satisfying:

1. $U(t + t') = U(t)\, U(t')$ for all $t, t' \in \widetilde{G}$, and
2. the map $t \mapsto U(t)x$ is continuous for each $x \in \mathcal{H}$.

Notice that in case of a discrete countable group, the second property always holds and, as a consequence, this property was omitted in Sect. 5.1. As said before, a unitary representation of a countable discrete group $G$ is just a homomorphism between the groups $G$ and $\mathcal{U}(\mathcal{H})$.

Let $\mathcal{V}_\Phi$ be the corresponding $U$-invariant subspace of $\mathcal{H} = L^2(\widetilde{G})$ given in (5.10), i.e.,

$$\mathcal{V}_\Phi = \Big\{ \sum_{n=1}^{N} \sum_{g\in G} x_n(g) U(g)\varphi_n \ : \ x_n \in \ell^2(G);\ n = 1, 2, \dots, N \Big\}.$$

For any $f \in \mathcal{V}_\Phi$ we consider the samples defined in Eq. (5.4) from $M$ fixed points $t_m \in \widetilde{G}$, $m = 1, 2, \dots, M$, i.e.,

$$\mathcal{L}_m f(g) := \big[U(-g)f\big](t_m), \quad g \in G,\ m = 1, 2, \dots, M. \tag{5.21}$$

Let $A = [a_{m,n}]$ be the $M \times N$ matrix with entries

$$a_{m,n}(g) = \big[U(-g)\varphi_n\big](t_m), \quad g \in G.$$

Assume that the sequence $\{[U(g)\varphi_n](t)\}_{g\in G}$ belongs to $\ell^2(G)$ for each $t \in \widetilde{G}$ and $n = 1, 2, \dots, N$. Then the matrix $A$ has entries in $\ell^2(G)$.

Moreover, if the function $[U(g)\varphi_n](t)$, $t \in \widetilde{G}$, is continuous on $\widetilde{G}$ for any $g \in G$ and $n = 1, 2, \dots, N$, and the condition

$$\sup_{t\in\widetilde{G}} \sum_{g\in G} \big|[U(g)\varphi_n](t)\big|^2 < +\infty, \quad n = 1, 2, \dots, N, \tag{5.22}$$

holds, then the subspace $\mathcal{V}_\Phi$ is a reproducing kernel Hilbert space of continuous bounded functions in $L^2(\widetilde{G})$. In fact, it is a necessary and sufficient condition as the following result shows.

**Proposition 5.6** *For any sequence $\{x_n(g)\}_{g\in G;\, n=1,2,\dots,N} \in \ell^2_N(G)$ the series*

$$\sum_{n=1}^{N} \sum_{g\in G} x_n(g)\, [U(g)\varphi_n](t)$$

*converges pointwise to a continuous bounded function on $\widetilde{G}$ if and only if for each $g \in G$ and $n = 1, 2, \dots, N$, the function $U(g)\varphi_n$ is continuous on $\widetilde{G}$, and condition (5.22) holds.*

***Proof*** The sufficient condition is a straightforward consequence of the Cauchy-Schwarz inequality and the Weierstrass M-test.

To prove the necessary condition we follow the arguments in Ref. [19]. Indeed, notice first that choosing the delta sequences in $\ell^2_N(G)$ we deduce that each function $U(g)\varphi_n$ is continuous on $\mathbb{R}^d$.

For each fixed $t \in \widetilde{G}$, since the series $\sum_{n=1}^N \sum_{g\in G} x_n(g)\,[U(g)\varphi_n](t)$ converges for all $\{x_n(g)\}_{g\in G;\, n=1,2,\dots,N}$ in $\ell^2_N(G)$, we obtain that the series

$$\sum_{n=1}^{N}\sum_{g\in G}\big|[U(g)\varphi_n](t)\big|^2$$

is convergent for each $t \in \widetilde{G}$ (see Ref. [14]).

Moreover, for each fixed $t \in \widetilde{G}$, the functional $\Omega_t : \ell^2_N(G) \longrightarrow \mathbb{C}$, defined as

$$\Omega_t(\mathbf{x}) := \sum_{n=1}^{N}\sum_{g\in G} x_n(g)\,[U(g)\varphi_n](t)$$

for each $\mathbf{x} = \{x_n(g)\}_{g\in G;\, n=1,2,\dots,N}$ in $\ell^2_N(G)$, is bounded with norm

$$\|\Omega_t\|^2 = \sum_{n=1}^{N}\sum_{g\in G}\big|[U(g)\varphi_n](t)\big|^2 .$$

(See, for instance, Ref. [14, p. 145]).

Next, for fixed $\mathbf{x} = \{x_n(g)\}_{g\in G;\, n=1,2,\dots,N} \in \ell^2_N(G)$ we consider its associated function $f_{\mathbf{x}}(t) := \sum_{n=1}^N \sum_{g\in G} x_n(g)\,[U(g)\varphi_n](t)$, $t \in \mathbb{R}^d$. Since $f_{\mathbf{x}}$ is bounded on $\widetilde{G}$, we get

$$\sup_{t\in\widetilde{G}} |\Omega_t(\mathbf{x})| = \sup_{t\in\widetilde{G}} |f_{\mathbf{x}}(t)| < +\infty .$$

Hence, from the Banach-Steinhaus theorem we conclude that

$$\sup_{t\in\widetilde{G}} \|\Omega_t\| = \sup_{t\in\widetilde{G}} \Big(\sum_{n=1}^{N}\sum_{g\in G}\big|[U(g)\varphi_n](t)\big|^2\Big)^{1/2} < +\infty .$$

□

Observe that, whenever $\mathcal{H} = L^2(\mathbb{R}^d)$ and $[U(p)f](t) := f(t-p)$, $t \in \mathbb{R}^d$, $p \in \mathbb{Z}^d$, the samples defined in (5.4) read

$$\mathcal{L}_m f(p) = \big[U(-p)f\big](t_m) = f(p+t_m), \quad p \in \mathbb{Z}^d \text{ and } m = 1, 2, \dots, M.$$

**Example 5.6 (Oversampled Pointwise Sampling in a Shift-Invariant Subspace $V_\Phi^2$)** Choose $\mathcal{H} := L^2(\mathbb{R}^d)$, $G := \mathbb{Z}^d$ and $\big[U(p)f\big](t) := f(t-p)$, $t \in \mathbb{R}^d$ and $p \in \mathbb{Z}^d$. Thus, under the hypotheses in Theorem 5.2 on the matrix $A = [a_{m,n}]$ with entries

$$a_{m,n}(p) = \varphi_n(t_m + p)\,, \quad p \in \mathbb{Z}^d\,,$$

we obtain an oversampled pointwise sampling formula in the shift-invariant subspace $V_\Phi^2$ of $L^2(\mathbb{R}^d)$. Namely:

For each $f \in V_\Phi^2$ a sampling expansion having the form

$$f(t) = \sum_{m=1}^{M} \sum_{p \in \mathbb{Z}^d} f(p + t_m)\, S_m(t-p)\,, \quad t \in \mathbb{R}^d\,,$$

holds, for some functions $S_m \in V_\Phi^2$, $m = 1, 2, \ldots, M$.

The convergence of the series is in $L^2(\mathbb{R}^d)$-norm. In case the hypotheses in Proposition 5.6 are satisfied, the above series converges also absolutely and uniformly on $\mathbb{R}^d$.

**Example 5.7 (Pointwise Sampling Associated with a Crystallographic Group)** In case of considering the quasi regular representation of the crystallographic group $C_{P,\Gamma} = P\mathbb{Z}^d \rtimes_\sigma \Gamma$ defined, for each $f \in \mathcal{V}_\varphi$, in (5.19), the samples given by (5.4) read

$$\mathcal{L}_m f(p) = \big[U(-p, I)f\big](t_m) = f(p + t_m)\,, \quad p \in P\mathbb{Z}^d \text{ and } m = 1, 2, \ldots, M\,,$$

where $t_m$, $m = 1, 2, \ldots, M$, are $M$ fixed points in $\mathbb{R}^d$.

Under the hypotheses in Theorem 5.2 on the matrix $A = [a_{m,n}]$ with entries

$$a_{m,n}(p) = \varphi[\gamma_n^\top (t_m - p)]\,, \quad p \in P\mathbb{Z}^d\,,$$

there exist $M$ functions $S_m \in \mathcal{V}_\varphi$, $m = 1, 2, \ldots, M$, such that for each $f \in \mathcal{V}_\varphi$ the sampling formula

$$f(t) = \sum_{m=1}^{M} \sum_{p \in P\mathbb{Z}^d} f(p + t_m)\, S_m(t-p)\,, \quad t \in \mathbb{R}^d\,,$$

holds.

If the generator $\varphi$ is continuous in $\mathbb{R}^d$ and the function $t \mapsto \sum_{p \in \mathbb{Z}^d} |\varphi(t-p)|^2$ is bounded on $\mathbb{R}^d$, then the convergence of the series in the $L^2(\mathbb{R}^d)$-norm sense implies pointwise convergence which is absolute and uniform on $\mathbb{R}^d$.

**Example 5.8 (Sampling in $\Lambda$-Shift-Invariant Subspaces of $L^2(\mathbb{R}^d)$)** Let $\Lambda$ be a full rank lattice in $\mathbb{R}^d$. As in Example 5.2, for any function $f$ in the $\Lambda$-shift-invariant subspace

$$V_\Phi^\Lambda := \Big\{ \sum_{n=1}^{N} \sum_{\lambda \in \Lambda} c_n(\lambda)\, \varphi_n(t-\lambda) \ :\ \{c_n(\lambda)\} \in \ell^2(\Lambda)\,;\ n = 1, 2, \cdots, N \Big\}\,,$$

we can consider $M$ sequences of the average samples defined by

$$\mathcal{L}_m f(\lambda) := \big\langle f, \psi_m(\cdot - \lambda) \big\rangle_{L^2(\mathbb{R}^d)}, \quad \lambda \in \Lambda\,, \quad m = 1, 2, \cdots, M\,.$$

Under the hypotheses in Theorem 5.2 on the associated $M \times N$ matrix $A = [a_{m,n}]$ whose $(m, n)$-entry is

$$a_{m,n}(\lambda) = \big\langle \varphi_n, \psi_m(\cdot - \lambda) \big\rangle_{L^2(\mathbb{R}^d)}, \quad \lambda \in \Lambda\,,$$

we obtain an average sampling formula in $\Lambda$-shift-invariant subspace $V_\Phi^\Lambda$ of $L^2(\mathbb{R}^d)$. Namely, for each $f \in V_\Phi^\Lambda$ a sampling expansion having the form

$$f(t) = \sum_{m=1}^{M} \sum_{\lambda \in \Lambda} \big\langle f, \psi_m(\cdot - \lambda) \big\rangle_{L^2(\mathbb{R}^d)}\, S_m(t-\lambda) \quad \text{in } L^2(\mathbb{R}^d)$$

holds, for some sampling functions $S_m \in V_\Phi^\Lambda$, $m = 1, 2, \ldots, M$. Moreover, the sequence $\{S_m(t-\lambda)\}_{\lambda \in \Lambda;\, m=1,2,\ldots,M}$ is a frame for $V_\Phi^\Lambda$.

Next, as in Example 5.6, we could consider, in $V_\Phi^\Lambda$, the samples

$$\mathcal{L}_m f(\lambda) = \big[U(-\lambda) f\big](t_m) = f(\lambda + t_m)\,, \quad \lambda \in \Lambda \text{ and } m = 1, 2, \ldots, M\,,$$

where $t_m$, $m = 1, 2, \cdots, M$, are $M$ fixed points in $\mathbb{R}^d$. Thus, under hypotheses in Theorem 5.2 on the matrix $A = [a_{m,n}]$ where

$$a_{m,n}(\lambda) = \varphi_n(t_m + \lambda)\,, \quad \lambda \in \Lambda\,,$$

we obtain a pointwise sampling formula in the $\Lambda$-shift-invariant subspace $V_\Phi^\Lambda$ of $L^2(\mathbb{R}^d)$. Namely, for each $f \in V_\Phi^\Lambda$ a sampling expansion having the form

$$f(t) = \sum_{m=1}^{M} \sum_{\lambda \in \Lambda} f(\lambda + t_m)\, S_m(t-\lambda)\,, \quad t \in \mathbb{R}^d\,,$$

holds, for some functions $S_m \in V_\Phi^\Lambda$, $m = 1, 2, \ldots, M$.

In case the hypotheses in Proposition 5.6 are satisfied, the convergence of the series in $L^2(\mathbb{R}^d)$-norm implies pointwise convergence which is uniform on $\mathbb{R}^d$. The pointwise convergence of the series is also absolute.

A necessary and sufficient condition to be $\{\varphi_n(t-\lambda)\}_{\lambda\in\Lambda;\, n=1,2,\dots,N}$ a Riesz basis for $V_\Phi^\Lambda$ can be stated in terms of the matrix-valued function:

$$G_\Phi^\Lambda(x) := \sum_{\lambda\in\Lambda^\perp} \widehat{\Phi}(\lambda+x)\,\overline{\widehat{\Phi}(\lambda+x)}^\top, \qquad \text{a.e. } x\in\mathbb{R}^d,$$

where $\widehat{\Phi} = \big(\widehat{\varphi}_1, \widehat{\varphi}_2, \dots, \widehat{\varphi}_N\big)^\top$, and $\Lambda^\perp$ denotes the dual lattice of $\Lambda$. Namely, there exist two constants $0 < m \le M$ such that

$$m\,\mathbb{I}_N \le G_\Phi^\Lambda(x) \le M\,\mathbb{I}_N, \qquad \text{a.e. } x\in\mathbb{R}^d.$$

### 5.3.4 *Sampling in a Subgroup*

Let $(G,+)$ be a countable discrete LCA group, and let $H$ be a subgroup of $G$ with finite index $L$. We fix a set $\{g_1, g_2, \dots, g_L\}$ of representatives of the cosets of $H$, i.e., the group $G$ can be decomposed as

$$G = (g_1+H)\cup(g_2+H)\cup\cdots\cup(g_L+H) \text{ with } (g_l+H)\cap(g_{l'}+H) = \varnothing \text{ for } l\neq l'.$$

Given a unitary representation $g\longmapsto U(g)$ of the group $G$ on a separable Hilbert space $\mathcal{H}$ and a set of generators $\Phi = \{\varphi_1, \varphi_2, \dots, \varphi_N\}$ in $\mathcal{H}$, we consider the subspace $\mathcal{V}_\Phi = \overline{\operatorname{span}}_{\mathcal{H}}\{U(g)\varphi_n\}_{g\in G;\, n=1,2,\dots,N}$ in $\mathcal{H}$.

In case $\{U(g)\varphi_n\}_{g\in G;\, n=1,2,\dots,N}$ is a Riesz sequence in $\mathcal{H}$, it can be expressed as

$$\begin{aligned}\mathcal{V}_\Phi &= \Big\{\sum_{n=1}^N\sum_{g\in G} x_n(g)\,U(g)\varphi_n \;:\; x_n\in\ell^2(G)\Big\}\\ &= \Big\{\sum_{n=1}^N\sum_{l=1}^L\sum_{h\in H} x_n(g_l+h)\,U(g_l+h)\varphi_n\Big\},\end{aligned}$$

where the sequence

$$\mathbf{x}(h) := \big(x_{11}(h), \dots, x_{1L}(h), x_{21}(h), \dots, x_{2L}(h), \dots, x_{N1}(h), \dots, x_{NL}(h)\big)^\top,$$

belongs to $\ell^2_{NL}(H)$, with $x_{nl}(h) := x_n(g_l+h)$. From now on we consider a new index $nl$, going from 11 to $NL$ following the above indicated order, i.e., the lexicographic order in the double index $nl$.

Next, for $M$ fixed elements $\psi_m \in \mathcal{H}$, $m = 1, 2, \ldots, M$, not necessarily in $\mathcal{V}_\Phi$, for each $f \in \mathcal{V}_\Phi$ we define its generalized average samples

$$\mathcal{L}_m f(h) = \langle f, U(h)\psi_m \rangle_{\mathcal{H}}, \quad h \in H \text{ and } m = 1, 2, \ldots, M. \tag{5.23}$$

For each $f = \sum_{n=1}^{N} \sum_{l=1}^{L} \sum_{k \in H} x_n(g_l + k)\, U(g_l + k)\varphi_n$ in $\mathcal{V}_\Phi$, the samples in (5.23) can be expressed as

$$\begin{aligned}\mathcal{L}_m f(h) &= \sum_{n=1}^{N} \sum_{l=1}^{L} \sum_{k \in H} x_n(g_l + k)\langle \varphi_n, U(h - g_l - k)\psi_m \rangle \\ &= \sum_{n=1}^{N} \sum_{l=1}^{L} \big(a_{m,nl} *_H x_{nl}\big)(h),\end{aligned}$$

where, for $m = 1, 2, \ldots, M$, $n = 1, 2, \ldots, N$, and $l = 1, 2, \ldots, L$, we have

$$a_{m,nl}(h) := \langle \varphi_n, U(h - g_l)\psi_m \rangle_{\mathcal{H}}, \quad h \in H.$$

Notice that each $a_{m,nl} \in \ell^2(H)$. The subscript $*_H$ means convolution over the subgroup $H$.

If we consider the $M \times NL$ matrix $A = [a_{m,nl}]$, the hypotheses in Theorem 5.2 on the matrix $A$ prove, with slight differences, that necessarily $M \geq NL$, and there exists a frame sequence $\{T_h \mathbf{b}_m\}_{h \in H;\, m=1,2,\ldots,M}$ for $\ell^2_{NL}(H)$ which is a dual frame of $\{T_h \mathbf{a}^*_m\}_{h \in H;\, m=1,2,\ldots,M}$. Recall that $\mathbf{a}^*_m$ denotes the $m$-th column of the matrix $A^*$ given in (5.7).

To be precise, for $m = 1, 2, \ldots, M$, we have that

$$\mathcal{L}_m f(h) = \langle \mathbf{x}, T_h \mathbf{a}^*_m \rangle_{\ell^2_{NL}(H)}, \quad h \in H,$$

where $\mathbf{x}$ is the sequence in $\ell^2_{NL}(H)$ representing $f \in \mathcal{V}_\Phi$.

Thus, for any $\mathbf{x} \in \ell^2_{NL}(H)$ we have

$$\begin{aligned}\mathbf{x} &= \sum_{m=1}^{M} \sum_{h \in H} \langle \mathbf{x}, T_h \mathbf{a}^*_m \rangle_{\ell^2_{NL}(H)} T_h \mathbf{b}_m \\ &= \sum_{m=1}^{M} \sum_{h \in H} \mathcal{L}_m f(h)\, T_h \mathbf{b}_m \quad \text{in } \ell^2_{NL}(H).\end{aligned} \tag{5.24}$$

Now, the natural isomorphism

$$\mathcal{T}_{U,\Phi} : \ell^2_{NL}(H) \longrightarrow \mathcal{V}_\Phi\,, \tag{5.25}$$

which maps the standard orthonormal basis $\{\delta_{h,nl}\}$ for $\ell^2_{NL}(H)$ on the Riesz basis $\{U(g_l+h)\varphi_n\}$ for $\mathcal{V}_\Phi$, satisfies the *shifting property* $\mathcal{T}_{U,\Phi}(T_h\mathbf{b}) = U(h)(\mathcal{T}_{U,\Phi}\mathbf{b})$ for each $h \in H$ and $\mathbf{b} \in \ell^2_{NL}(H)$.

Hence, applying the isomorphism $\mathcal{T}_{U,\Phi}$ in (5.24) we obtain that any $f = \mathcal{T}_{U,\Phi}\mathbf{x}$ in $\mathcal{V}_\Phi$ can be recovered from data $\{\mathcal{L}_m f(h)\}_{h\in H;\, m=1,2,\ldots,M}$ by means of the sampling formula

$$f = \sum_{m=1}^{M}\sum_{h\in H} \mathcal{L}_m f(h)\, U(h) S_m \quad \text{in } \mathcal{H} \tag{5.26}$$

for some sampling functions $S_m = \mathcal{T}_{U,\Phi}\mathbf{b}_m \in \mathcal{V}_\Phi$, $m = 1, 2, \ldots, M$. Moreover, the sequence $\{U(h)S_m\}_{h\in H;\, m=1,2,\ldots,M}$ is a frame for $\mathcal{V}_\Phi$.

**Example 5.9 (Average Sampling in the Shift-Invariant Subspace $V^2_\Phi$)** Consider $\mathcal{H} := L^2(\mathbb{R}^d)$, $G := \mathbb{Z}^d$ and the unitary representation of $\mathbb{Z}^d$ on $L^2(\mathbb{R}^d)$

$$[U(p)f](t) := f(t-p)\,,\, t \in \mathbb{R}^d \text{ and } p \in \mathbb{Z}^d\,.$$

Let $P\mathbb{Z}^d$ be a sub-lattice in $\mathbb{Z}^d$, where $P$ denotes a $d \times d$ matrix of integer entries with positive determinant $L := \det P$.

Under the hypotheses in Theorem 5.2 on the matrix $A = [a_{m,nl}]$ with entries

$$a_{m,nl}(p) = \langle \varphi_n(t), \psi_m(t-p+g_l)\rangle_{L^2(\mathbb{R}^d)}\,, \quad p \in P\mathbb{Z}^d\,,$$

the sampling formula (5.26) gives an average sampling formula in the classical shift-invariant subspace $V^2_\Phi$ of $L^2(\mathbb{R}^d)$. Namely, for each $f \in V^2_\Phi$ the sampling formula (5.26) reads

$$f(t) = \sum_{m=1}^{M}\sum_{p\in P\mathbb{Z}^d} \langle f, \psi_m(\cdot - p)\rangle_{L^2(\mathbb{R}^d)}\, S_m(t-p)\,, \quad t \in \mathbb{R}^d\,,$$

for some sampling functions $S_m \in V^2_\Phi$, $m = 1, 2, \ldots, M$, with $M \geq LN$. Moreover, the sampling sequence $\{S_m(t-p)\}_{p\in P\mathbb{Z}^d;\, m=1,2,\ldots,M}$ is a frame for $V^2_\Phi$.

The convergence of the series in the $L^2(\mathbb{R}^d)$-norm sense implies pointwise convergence which is absolute and uniform on $\mathbb{R}^d$.

In particular, in the motivating example at the beginning of the chapter, we consider the sequence of samples $\{\mathcal{L}_j f(rn)\}_{n\in\mathbb{Z};\, j=1,2,\ldots,s}$ of $f \in V^2_\varphi$ at the sub-lattice $r\mathbb{Z}$ of $\mathbb{Z}$.

Under the hypotheses in Theorem 5.2 on the $s \times r$ matrix $A = [a_{j,l}]$ with entries

$$a_{j,l}(rk) = \langle \varphi, \psi_j(\cdot - rk + l) \rangle\,, \quad k \in \mathbb{Z}\,,$$

for $j = 1, 2, \ldots, s$ and $l = 0, 1, \ldots, r-1$, there exist a frame for $V_\varphi^2$ having the form $\{S_j(t - rn)\}_{n \in \mathbb{Z};\, j=1,2,\ldots,s}$ such that for each $f \in V_\varphi^2$, the sampling formula

$$f(t) = \sum_{j=1}^{s} \sum_{n \in \mathbb{Z}} (\mathcal{L}_j f)(rn)\, S_j(t - rn) \quad \text{in } L^2(\mathbb{R})\,,$$

holds. The convergence of the series is also absolute and uniform on $\mathbb{R}$.

Recall that the Fourier transform of any sequence $\mathbf{c}$ in $\ell^2(r\mathbb{Z})$ is given by the Fourier series

$$\mathcal{F}\mathbf{c}(x) = \sum_{k \in \mathbb{Z}} c(rk)\, e^{2\pi i rkx}\,, \quad x \in [0, 1/r]\,.$$

**Example 5.10 (Pointwise Sampling in a Shift-Invariant Subspace $V_\Phi^2$)** In case $\mathcal{H} = L^2(\mathbb{R}^d)$, in the shift-invariant subspace $V_\Phi^2$ we consider the samples

$$\mathcal{L}_m f(h) := \big[U(-h) f\big](t_m)\,, \quad h \in H\,, \quad m = 1, 2, \ldots, M\,,$$

for $M$ fixed points $t_m \in \mathbb{R}^d$.

For each $f = \sum_{n=1}^{N} \sum_{l=1}^{L} \sum_{k \in H} x_n(g_l + k)\, U(g_l + k)\varphi_n$ in $\mathcal{V}_\Phi$ the above samples can be expressed as

$$\begin{aligned}
\mathcal{L}_m f(h) &= \sum_{n=1}^{N} \sum_{l=1}^{L} \sum_{k \in H} x_n(g_l + k)\big[U(k - h) U(g_l)\varphi_n\big](t_m) \\
&= \sum_{n=1}^{N} \sum_{l=1}^{L} \big(a_{m,nl} *_H x_{nl}\big)(h)\,,
\end{aligned}$$

where

$$a_{m,nl}(h) := \big[U(-h + g_l)\varphi_n\big](t_m)\,, \quad h \in H\,,$$

for $m = 1, 2, \ldots, M$, $n = 1, 2, \ldots, N$ and $l = 1, 2, \ldots, L$.

In particular, we can sample any function in a shift-invariant subspace $V_\Phi^2$ on a sub-lattice $P\mathbb{Z}^d$ of $\mathbb{Z}^d$, where $P$ denotes a $d \times d$ matrix of integer entries with positive determinant $L := \det P$, by using the shift operator.

Under the hypotheses in Theorem 5.2 on the $M \times NL$ matrix $A = [a_{m,nl}]$ with entries

$$a_{m,nl}(p) = \varphi_n(t_m + p - g_l)\,, \quad p \in P\mathbb{Z}^d\,,$$

there exist $M \geq NL$ sampling functions $S_m \in V_\Phi^2$, $m = 1, 2, \ldots, M$, such that we recover any $f \in V_\Phi^2$ from its samples

$$\left\{\mathcal{L}_m f(p) := f(p + t_m)\right\}_{p \in P\mathbb{Z}^d;\, m=1,2,\ldots,M}$$

by means of the sampling formula

$$f(t) = \sum_{m=1}^{M} \sum_{p \in P\mathbb{Z}^d} f(p + t_m)\, S_m(t - p)\,, \quad t \in \mathbb{R}^d\,.$$

Moreover, the sampling sequence $\{S_m(t - p)\}_{p \in P\mathbb{Z}^d;\, m=1,2,\ldots,M}$ is a frame for $V_\Phi^2$.

The convergence of the series in the $L^2(\mathbb{R}^d)$-norm sense implies pointwise convergence which is absolute and uniform on $\mathbb{R}^d$.

**Example 5.11 (The Case of U-invariant Spaces)** The results announced in Sect. 3.3.3 for $U$-invariant subspaces of a Hilbert space $\mathcal{H}$ (and, in particular, for shift-invariant subspaces in $L^2(\mathbb{R})$) hold as a consequence of Theorem 5.2 once the required hypotheses on the corresponding matrix $A$ are satisfied.

Indeed, we take a sampling period $r \in \mathbb{N}$ and consider the sequence of samples $\{\mathcal{L}_j x(rk)\}_{k \in \mathbb{Z};\, j=1,2,\ldots,s}$, where

$$\mathcal{L}_j x(rk) = \langle x, U^{rk} b_j \rangle_{\mathcal{H}}\,, \quad k \in \mathbb{Z}\,,$$

for some fixed $b_j \in \mathcal{H}$, $j = 1, 2, \ldots, s$.

Under the appropriate hypotheses in Theorem 5.2 on the Fourier transform in $\ell^2(r\mathbb{Z})$ of the entries of the $s \times Lr$ matrix $A = [a_{j,li}]$, with entries

$$a_{j,li}(rk) = \langle a_l, U^{rk-i} h_j \rangle\,, \quad k \in \mathbb{Z}\,,$$

where $j = 1, 2, \ldots, s$, $l = 1, 2, \ldots, L$, and $i = 0, 1, \cdots, r - 1$, there exist $s$ elements $c_j$ in $\mathcal{A}_\mathbf{a}$, $j = 1, 2, \ldots, s$, where necessarily $s \geq Lr$, such that the sequence $\left\{U^{rk} c_j\right\}_{k \in \mathbb{Z};\, j=1,2,\ldots,s}$ is a frame for $\mathcal{A}_\mathbf{a}$, and the sampling formula

$$x = \sum_{j=1}^{s} \sum_{k \in \mathbb{Z}} \mathcal{L}_j x(rk)\, U^{rk} c_j \quad \text{in } \mathcal{H}\,,$$

holds for any $x \in \mathcal{A}_\mathbf{a}$. Remind that $L$ denotes the number of generators in $\mathcal{A}_\mathbf{a}$.

In particular if $s = L$ whenever $r = 1$, or $s = Lr$ whenever $r > 0$ in the above case, then the corresponding sequences $\{U^k c_j\}_{k\in\mathbb{Z};\, j=1,2,\dots,s}$ or $\{U^{rk} c_j\}_{k\in\mathbb{Z};\, j=1,2,\dots,s}$ are Riesz bases for $\mathcal{A}_{\mathbf{a}}$.

Besides, since the expansion following a Riesz basis is unique, the interpolation properties

$$\mathcal{L}_{j'} c_j(k) = \delta_{j,j'}\,\delta_{k,0} \quad \text{or} \quad \mathcal{L}_{j'} c_j(rk) = \delta_{j,j'}\,\delta_{k,0}\,,$$

where $k \in \mathbb{Z}$ and $j, j' = 1, 2, \dots, s$, hold.

**Example 5.12 (An Average Sampling Formula in $\ell^2(\mathbb{Z})$)** Here we consider $\mathcal{H} = \ell^2(\mathbb{Z})$ and a sampling period $r \in \mathbb{N}$. Any sequence $x \in \ell^2(\mathbb{Z})$ can be splitted as

$$\begin{aligned} x(n) &= \sum_{m\in\mathbb{Z}} x(m)\,\delta(n-m) \\ &= \sum_{l=0}^{r-1}\sum_{k\in\mathbb{Z}} x(rk+l)\,\delta(n-l-rk) \\ &= \sum_{l=0}^{r-1}\sum_{k\in\mathbb{Z}} x(rk+l)\,T_{rk}\delta(n-l)\,, \quad n \in \mathbb{Z}\,. \end{aligned}$$

For $l = 0, 1, \dots, r-1$, we denote $x_l(rk) := x(rk+l)\,,\ k \in \mathbb{Z}$.

Thus we have written the shift-invariant space $\ell^2(\mathbb{Z})$ as a shift-invariant space in $\ell^2(r\mathbb{Z})$ with $r$ generators $\delta(\cdot - l)$, $l = 0, 1, \dots, r-1$. Recall that $\delta(n-l) = \delta_{n,l}$.

In the present case the isomorphism $\mathcal{T}_\delta : \ell^2_r(r\mathbb{Z}) \longmapsto \ell^2(\mathbb{Z})$ in (5.25) reads:

$$\mathbf{x} = \big(x(rk), x(rk+1), \dots, x(rk+r-1)\big)^\top \overset{\mathcal{T}_\delta}{\longmapsto} x(n) = \sum_{l=0}^{r-1}\sum_{k\in\mathbb{Z}} x(rk+l)\,T_{rk}\delta(n-l)\,.$$

Given a $M \times r$ matrix $A = [a_{m,l}]$ with entries in $\ell^2(r\mathbb{Z})$, for each sequence $x \in \ell^2(\mathbb{Z})$ we consider its samples

$$\begin{aligned} \mathcal{L}_m x(rk) &= \sum_{l=0}^{r-1} \big(a_{m,l} * x_l\big)(rk) \\ &= \big(\mathbf{a}_m * \mathbf{x}\big)(rk)\,, \quad k \in \mathbb{Z} \text{ and } m = 1, 2, \dots, M\,, \end{aligned}$$

where $\mathbf{a}_m = \big(a_{m,0}(rk), a_{m,1}(rk+1), \dots, a_{m,l-1}(rk+r-1)\big)^\top$ denotes the $m$-th row of the matrix $A$.

Under the hypotheses in Theorem 5.2 on the matrix $A$, there exist $M$ vectors

$$\mathbf{b}_m = \big(b_m(rk), b_m(rk+1), \ldots, b_m(rk+r-1)\big)^\top \in \ell^2_r(r\mathbb{Z}), \quad m = 1, 2, \ldots, M,$$

such that

$$\mathbf{x} = \sum_{m=1}^{M} \sum_{k\in\mathbb{Z}} \mathcal{L}_m x(rk)\, T_{rk}\mathbf{b}_m \quad \text{in } \ell^2_r(r\mathbb{Z}).$$

Applying the isomorphism $\mathcal{T}_\delta$ in the above formula, we obtain the following sampling formula in $\ell^2(\mathbb{Z})$:

$$x(n) = \sum_{m=1}^{M} \sum_{k\in\mathbb{Z}} \mathcal{L}_m x(rk)\, S_m(n-rk), \quad n \in \mathbb{Z},$$

where $S_m = \mathcal{T}_\delta \mathbf{b}_m$, $m = 1, 2, \ldots, M$, that is, the sequence

$$S_m(n) = \sum_{l=0}^{r-1} \sum_{k\in\mathbb{Z}} b_m(rk+l)\, T_{rk}\delta(n-l), \quad n \in \mathbb{Z}.$$

Observe that the sequence of samples $\big\{\mathcal{L}_m x(rk)\big\}_{k\in\mathbb{Z};\, m=1,2,\ldots,M}$ considered in this example is, in engineering jargon, nothing but the output of a filter-bank.

Finally, it is worth noticing that the sampling results exhibited in these sections are related with the sampling results obtained in Refs. [1, 6, 9–11].

## References


1. Aldroubi, A., Sun, Q., Tang, W.S.: Convolution, average sampling, and a Calderon resolution of the identity for shift-invariant spaces. J. Fourier Anal. Appl. **11**(2), 215–244 (2005).
2. Barbieri, D., Hernández, E., Parcet, J.: Riesz and frame systems generated by unitary actions of discrete groups. Appl. Comput. Harmon. Anal. **39**(3), 369–399 (2015)
3. Cabrelli, C., Paternostro, V.: Shift-invariant spaces on LCA groups. J. Funct. Anal. **258**, 2034–2059 (2010)
4. Christensen, O.: An Introduction to Frames and Riesz Bases, 2nd edn. Birkhäuser, Boston (2016)
5. Deitmar, A., Echterhoff, S.: Principles of Harmonic Analysis, 2nd edn. Universitext. Springer, Berlin (2014)
6. Fernández-Morales, H.R., García, A.G., Hernández-Medina, M.A., Muñoz-Bouzo, M.J.: Generalized sampling: from shift-invariant to $U$-invariant spaces. Anal. Appl. **13**(3), 303–329 (2015)
7. Folland, G.B.: A Course in Abstract Harmonic Analysis, CRC Press, Boca Raton, FL (1995)
8. Führ, H.: Abstract Harmonic Analysis of Continuous Wavelet Transform. Springer, Berlin-Heidelberg (2005)

9. García, A.G., Pérez-Villalón, G.: Dual frames in $L^2(0, 1)$ connected with generalized sampling in shift-invariant spaces. Appl. Comput. Harmon. Anal. **20**(3), 422–433 (2006)
10. García, A.G., Pérez-Villalón, G.: Multivariate generalized sampling in shift-invariant spaces and its approximation properties. J. Math. Anal. Appl. **355**, 397–413 (2009)
11. García, A.G., Hernández-Medina, M.A., Pérez-Villalón, G.: Generalized sampling in shift-invariant spaces with multiple stable generators. J. Math. Anal. Appl. **337**, 69–84 (2008)
12. García, A.G., Hernández-Medina, M.A., Pérez-Villalón, G.: Convolution systems on discrete abelian groups as a unifying strategy in sampling theory. Results Math. **75**, 40 (2020)
13. Goodman, T.N., Lee, S.L., Tang, W.S.: Wavelet bases for a set of commuting unitary operators. Adv. Comput. Math. **1**(1), 109–126 (1993)
14. Heuser, H.: Functional Analysis, John Wiley, New York (1982)
15. Horn, R.A., Johnson, C.R.: Matrix Analysis. Cambridge University Press, Cambridge UK (1999)
16. Jia, R.Q., Micchelli, C.A.: Using the refinement equations for the construction of pre-waveles II: Powers of two. In: Laurent, P.J., Le Méhauté, A., Schumaker, L.L. (eds.) Curves and Surfaces, pp. 209–246. Academic Press, Boston (1991)
17. Pérez-Villalón, G.: Discrete convolution operators and Riesz systems generated by actions of abelian groups. Ann. Funct. Anal. **11**, 285–297 (2020)
18. Vaidyanathan, P.P.: Multirate Systems and Filter Banks. Prentice-Hall, Englewood Hills, NJ (1993)
19. Zhou, X., Sun, W.: On the Sampling Theorem for Wavelet Subspaces. J. Fourier Anal. Appl. **5**(4), 347–354 (1999)

# Chapter 6
# Finite Frames Related to Sampling in Finite-Dimensional $U$-Invariant Subspaces

A generalization of shift-invariant subspaces in $L^2(\mathbb{R})$ was proposed at the end of Chap. 3 in the following sense: Let $U : \mathcal{H} \to \mathcal{H}$ be a unitary operator in a separable Hilbert space $\mathcal{H}$; for a fixed $a \in \mathcal{H}$, consider the closed subspace given by $\mathcal{V}_a := \overline{\text{span}}\{U^n a,\ n \in \mathbb{Z}\}$. In case that the (infinite) sequence $\{U^n a\}_{n\in\mathbb{Z}}$ is a Riesz sequence in $\mathcal{H}$ we obtain the $U$-invariant subspace in $\mathcal{H}$

$$\mathcal{V}_a = \Big\{\sum_{n\in\mathbb{Z}} \alpha_n\, U^n a : \{\alpha_n\}_{n\in\mathbb{Z}} \in \ell^2(\mathbb{Z})\Big\}.$$

There, the sequence of generalized samples $\{(\mathcal{L}_j x)(rm)\}_{m\in\mathbb{Z};\, j=1,2,\dots,s}$ of $x \in \mathcal{V}_a$, where $r \in \mathbb{N}$ denotes the sampling period, is obtained from $s$ elements $b_j \in \mathcal{H}$, not necessarily in $\mathcal{V}_a$, as

$$(\mathcal{L}_j x)(rm) := \langle x, U^{rm} b_j\rangle_{\mathcal{H}}, \quad m \in \mathbb{Z};\ j = 1, 2, \dots, s\,.$$

Under appropriate hypotheses we have seen in Chap. 5 that there exist frames in $\mathcal{V}_a$, having the form $\{U^{rm} c_j\}_{m\in\mathbb{Z};\, j=1,2,\dots,s}$, where $c_j \in \mathcal{V}_a$ for $j = 1, 2, \dots, s$, such that for each $x \in \mathcal{V}_a$ we get a sampling expansion as

$$x = \sum_{j=1}^{s}\sum_{m\in\mathbb{Z}} \mathcal{L}_j x(rm)\, U^{rm} c_j \quad \text{in } \mathcal{H}.$$

In Refs. [3, 9] it was implicitly assumed that the stationary sequence $\{U^n a\}_{n\in\mathbb{Z}}$ in $\mathcal{H}$ has infinitely many different elements. It could happen that for some $a \in \mathcal{H}$ there exists $N \in \mathbb{N}$ such that $U^N a = a$, i.e., 1 is an eigenvalue of the unitary operator $U^N$ with eigenvector $a$.

In this case, $\mathcal{V}_a$ is just the finite-dimensional subspace of $\mathcal{H}$ spanned by the set $\{a, Ua, U^2 a, \dots, U^{N-1} a\}$. The main aim in this chapter is to derive a sampling

 
A. García García, *The Use of Frames in Sampling Theory*,
RSME Springer Series 14, https://doi.org/10.1007/978-3-031-63242-6_6

theory in $\mathcal{V}_a$ involving dual finite frames. In general, we will consider several generators with different orders for the above subspace.

In applications, frames in finite-dimensional spaces are required. Since in finite dimension, a frame is nothing but a spanning set of vectors, finite frames require control of certain condition numbers, and over the spectrum of certain matrices; see finite frames in Sect. 1.2. Thus, frame theory, firstly considered in applied harmonic analysis, meets matrix analysis and numerical linear algebra.

In this chapter a sampling theory for multiple generated finite $U$-invariant subspaces of a Hilbert space $\mathcal{H}$ is developed; as usual, $U$ denotes a unitary operator on $\mathcal{H}$.

Thus, we denote the set of generators as $\mathbf{a} := \{a_1, a_2, \dots, a_L\}$ and their respective orders as $N_1, N_2, \dots, N_L$, that is, $U^{N_i} a_i = a_i$, $i = 1, 2, \dots, L$. The $U$-invariant subspace $\mathcal{V}_\mathbf{a}$ is defined by

$$\mathcal{V}_\mathbf{a} := \operatorname{span}\left\{a_i, Ua_i, U^2 a_i, \dots, U^{N_i-1} a_i\right\}_{i=1}^{L}.$$

Linear independence of the above vectors gives an $N_1 + N_2 + \cdots + N_L$ dimensional subspace in $\mathcal{H}$ described as

$$\mathcal{V}_\mathbf{a} = \Big\{ \sum_{i=1}^{L} \sum_{k=0}^{N_i-1} \alpha_k^i \, U^k a_i \ : \ \alpha_k^i \in \mathbb{C};\ 0 \le k \le N_i - 1,\ 1 \le i \le L \Big\}.$$

These spaces are very natural in signal theory. As an academic example, consider the three dimensional subspace $\mathcal{A}$ of all 4-periodic signals depicted in Fig. 6.1.

Signals in $\mathcal{A}$ are obtained by a symmetrization process. If we consider the shift operator

$$U : \{x(m)\} \longmapsto \{x(m-2)\},$$

the $U$-invariant subspace generated by the two 4-periodic signals $a_1 = (1, 0, 0, 0)$ and $a_2 = (0, 1, 0, 1)$ coincides with $\mathcal{A}$. Indeed, they have orders 2 and 1 respectively, i.e., $U^2 a_1 = a_1$ and $U a_2 = a_2$; besides, the set $\{a_1, Ua_1, a_2\}$ forms a basis for $\mathcal{A}$.

For the signal depicted in Fig. 6.1, the $\Delta$-red is its component in $a_1$, the $\circ$-black is its component in $Ua_1$, and the $\diamond$-blue is its component in $a_2$.

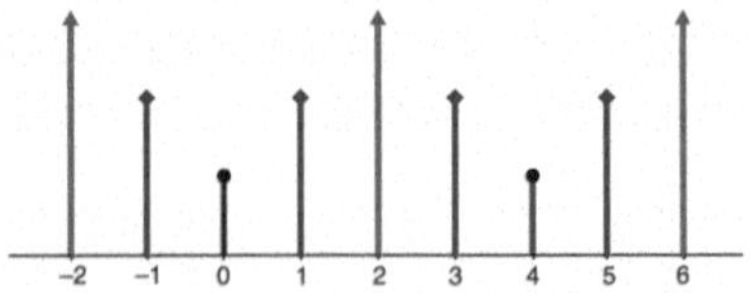


**Fig. 6.1** Symmetric extension of a finite signal

Formally, a sampling formula in $\mathcal{V}_{\mathbf{a}}$ is an expression such as

$$x = \sum_{j=1}^{s} \sum_{n=0}^{\ell-1} \mathcal{L}_j x(rn)\, U^{rn} c_j\,, \quad x \in \mathcal{V}_{\mathbf{a}}\,, \tag{6.1}$$

where $c_j \in \mathcal{V}_{\mathbf{a}}$, $j = 1, 2, \ldots, s$, and $\{\mathcal{L}_j x(rn)\}_{\substack{j=1,2,\ldots,s \\ n=0,1,\ldots,\ell-1}}$ denotes the set of samples of $x \in \mathcal{V}_{\mathbf{a}}$.

The meaning of all terms appearing in formula (6.1) is the following: the sequence of $s\ell$ generalized samples $\{(\mathcal{L}_j x)(rn)\}_{\substack{j=1,2,\ldots,s \\ n=0,1,\ldots,\ell-1}}$ of $x \in \mathcal{V}_a$, where $r \in \mathbb{N}$ denotes the sampling period, is obtained from $s$ elements $b_j \in \mathcal{H}$ as

$$(\mathcal{L}_j x)(rn) := \langle x, U^{rn} b_j \rangle_{\mathcal{H}}\,, \quad n = 0, 1, \ldots, \ell - 1\,. \tag{6.2}$$

For each $j = 1, 2, \ldots, s$, the number of samples is $\ell = N/r$, where $N$ is the least common multiple of the orders $N_1, N_2, \ldots, N_L$, and the sampling period $r$ divides $N$.

The sequence $\{U^{rn} c_j\}_{\substack{j=1,2,\ldots,s \\ n=0,1,\ldots,\ell-1}}$ in formula (6.1) should be a frame, i.e., a spanning set for $\mathcal{V}_{\mathbf{a}}$. In next section, which is broadly in line with Refs. [4, 5], we go into the details.

## 6.1 Sampling in Finite-Dimensional $U$-Invariant Subspaces

Let $\mathbf{a} := \{a_1, a_2, \ldots, a_L\}$ be a fixed subset of $\mathcal{H}$, and assume that there exist positive integers $N_1, N_2, \ldots, N_L$ such that $U^{N_i} a_i = a_i$, $i = 1, 2, \ldots, L$; suppose each $N_i$ is the smallest index with this property. We refer $N_i$ as the order of the element $a_i$.

Next, we consider the finite dimensional subspace of $\mathcal{H}$

$$\mathcal{V}_{\mathbf{a}} := \operatorname{span}\{a_i, U a_i, U^2 a_i, \ldots, U^{N_i - 1} a_i\}_{i=1}^{L}\,.$$

Whenever the involved vectors are linearly independent, the $(N_1 + N_2 + \cdots + N_L)$-dimensional subspace $\mathcal{V}_{\mathbf{a}}$ of $\mathcal{H}$ can be described as

$$\mathcal{V}_{\mathbf{a}} = \Big\{ \sum_{i=1}^{L} \sum_{k=0}^{N_i - 1} \alpha_k^i\, U^k a_i \;:\; \alpha_k^i \in \mathbb{C};\; 0 \le k \le N_i - 1,\; 1 \le i \le L \Big\}. \tag{6.3}$$

Thus, we refer to $\mathcal{V}_{\mathbf{a}}$ as a $U$-invariant subspace of $\mathcal{H}$ with $L$ generators $a_1, a_2, \ldots, a_L$ with respective orders $N_1, N_2, \ldots, N_L$.

The choice $N :=$ l.c.m. $(N_1, N_2, \ldots, N_L)$ yields the smallest positive integer $N$ such that $U^N x = x$ for all $x \in \mathcal{V}_{\mathbf{a}}$.

In order to characterize the linear independence in $\mathcal{H}$ of the set

$$\big\{a_i, Ua_i, U^2a_i, \ldots, U^{N_i-1}a_i\big\}_{i=1}^{L}$$

we consider the square matrix of order $N_1 + N_2 + \cdots + N_L$

$$\mathbf{C_a} := \begin{pmatrix} \mathbf{C}_{a_1,a_1} & \mathbf{C}_{a_2,a_1} & \cdots & \mathbf{C}_{a_L,a_1} \\ \mathbf{C}_{a_1,a_2} & \mathbf{C}_{a_2,a_2} & \cdots & \mathbf{C}_{a_L,a_2} \\ \vdots & \vdots & \ddots & \vdots \\ \mathbf{C}_{a_1,a_L} & \mathbf{C}_{a_2,a_L} & \cdots & \mathbf{C}_{a_L,a_L} \end{pmatrix},$$

where for $1 \le i, j \le L$ each block $\mathbf{C}_{a_i,a_j}$ denotes the $N_j \times N_i$ matrix

$$\mathbf{C}_{a_i,a_j} := \Big(\langle U^l a_i, U^k a_j\rangle\Big)_{0\le k\le N_j-1;\ 0\le l\le N_i-1}.$$

**Proposition 6.1** *The set of vectors* $\big\{a_i, Ua_i, U^2a_i, \ldots, U^{N_i-1}a_i\big\}_{i=1}^{L}$ *is linearly independent in* $\mathcal{H}$ *if and only if* $\det \mathbf{C_a} \neq 0$.

***Proof*** Assume that the set is linearly independent. If $\det \mathbf{C_a} = 0$ then there exists a nonzero vector $\boldsymbol{\lambda}$ in $\mathbb{C}^{N_1+N_2+\cdots+N_L}$

$$\boldsymbol{\lambda} = \big(\lambda_{1,0}, \lambda_{1,1}, \ldots, \lambda_{1,N_1-1}, \lambda_{2,0}, \lambda_{2,1}, \ldots, \lambda_{2,N_2-1}, \\ \ldots, \lambda_{L,0}, \lambda_{L,1}, \ldots, \lambda_{L,N_L-1}\big)^{\top}$$

such that $\mathbf{C_a}\boldsymbol{\lambda} = 0$.

Consider the vector $b = \sum_{i=1}^{L}\sum_{k=0}^{N_i-1} \lambda_{i,k} U^k a_i$; then, condition $\mathbf{C_a}\boldsymbol{\lambda} = 0$ implies that $b$ is orthogonal to every vector $U^r a_s$ with $1 \le s \le L$ and $0 \le r \le N_s - 1$. Thus $b = 0$, which implies $\boldsymbol{\lambda} = 0$, a contradiction.

Conversely, if $\sum_{i=1}^{L}\sum_{k=0}^{N_i-1} \lambda_{i,k} U^k a_i = 0$ for some $\boldsymbol{\lambda} \neq 0$ then the inner products

$$\Big\langle \sum_{i=1}^{L}\sum_{k=0}^{N_i-1} \lambda_{i,k} U^k a_i, U^r a_s \Big\rangle$$

vanish for every $1 \le s \le L$ and $0 \le r \le N_s - 1$. This implies that $\mathbf{C_a}\boldsymbol{\lambda} = 0$ and, consequently $\det \mathbf{C_a} = 0$. □

In the one-generator case, the linear independence condition can be given in terms of the *circulant matrix* of auto-covariances $\mathbf{R}_a$ defined by

$$\mathbf{R}_a := \begin{pmatrix} R_a(0) & R_a(1) \dots & R_a(N-1) \\ R_a(1) & R_a(2) \dots & R_a(0) \\ \vdots & \vdots \quad \ddots & \vdots \\ R_a(N-1) & R_a(0) \dots & R_a(N-2) \end{pmatrix},$$

where $R_a(k) := \langle U^k a, a\rangle_{\mathcal{H}}$, $k \in \mathbb{Z}$, denote the $N$-periodic sequence of auto-covariances of the stationary sequence $\{U^n a\}_{n\in\mathbb{Z}}$.

Namely, the set of vectors $\{a, Ua, U^2a, \dots, U^{N-1}a\}$ is linearly independent in $\mathcal{H}$ if and only if $\det \mathbf{R}_a \neq 0$.

**The Isomorphism $\mathcal{T}_{\mathbf{N},\mathbf{a}}$ connecting $\mathbb{C}^{N_1+N_2+\cdots+N_L}$ with $\mathcal{V}_\mathbf{a}$**

As a recurrent fact throughout this book, our technique consists of mapping suitable frame expansions in $\mathbb{C}^{N_1+N_2+\cdots+N_L}$ into the desired sampling formulas (6.1) in $\mathcal{V}_\mathbf{a}$. This will be done by means of the following isomorphism:

$$\begin{aligned} \mathcal{T}_{\mathbf{N},\mathbf{a}} : \mathbb{C}^{N_1+N_2+\cdots+N_L} &\longrightarrow \mathcal{V}_\mathbf{a} \\ \boldsymbol{\alpha} = \Big((\alpha_k^1)_{k=0}^{N_1-1}, (\alpha_k^2)_{k=0}^{N_2-1}, \dots, (\alpha_k^L)_{k=0}^{N_L-1}\Big)^\top &\longmapsto x = \sum_{i=1}^{L}\sum_{k=0}^{N_i-1} \alpha_k^i\, U^k a_i\,, \end{aligned} \tag{6.4}$$

where $\mathbf{N} := (N_1, N_2, \dots, N_L)$ denotes the involved orders.

Besides, this isomorphism $\mathcal{T}_{\mathbf{N},\mathbf{a}}$ has, as usual, a crucial *shifting property*. To state it, consider any vector $\mathbf{T}_0$ in $\mathbb{C}^{N_1+N_2+\cdots+N_L}$ denoted by

$$\mathbf{T}_0 = \big(\mathbf{T}_0^1, \mathbf{T}_0^2, \dots, \mathbf{T}_0^L\big)^\top, \tag{6.5}$$

where each block, denoted by $\mathbf{T}_0^i = \big(T^i(0), T^i(1), \dots, T^i(N_i-1)\big)$, $1 \le i \le L$, is a row vector of dimension $N_i$. Set $N = \text{l.c.m.}\,(N_1, N_2, \dots N_L)$ and $1 \le m \le N-1$.

Starting from $\mathbf{T}_0$ we define a new vector $\mathbf{T}_{N-m}$ in $\mathbb{C}^{N_1+N_2+\cdots+N_L}$ given by

$$\mathbf{T}_{N-m} := \big(\mathbf{T}_{N-m}^1, \mathbf{T}_{N-m}^2, \dots, \mathbf{T}_{N-m}^L\big)^\top, \tag{6.6}$$

where each block

$$\mathbf{T}_{N-m}^i = \big(T^i(N-m), T^i(N-m+1), \dots, T^i(N-m+N_i-1)\big)$$

is obtained by assuming an $N_i$-periodic character in each $T^i(\cdot)$, $i = 1, 2, \dots, L$. Thus we obtain

**Proposition 6.2** *Consider the vectors* $\mathbf{T}_0$ *and* $\mathbf{T}_{N-m}$, $1 \le m \le N-1$, *in* $\mathbb{C}^{N_1+N_2+\cdots+N_L}$ *given by* (6.5) *and* (6.6) *respectively. Then the following shifting property holds*

$$\mathcal{T}_{\mathbf{N},\mathbf{a}}(\mathbf{T}_{N-m}) = U^m\big(\mathcal{T}_{\mathbf{N},\mathbf{a}}(\mathbf{T}_0)\big)\,.$$

***Proof*** We have that

$$\mathcal{T}_{\mathbf{N},\mathbf{a}}\big(\mathbf{T}_{N-m}\big) = \sum_{i=1}^{L}\sum_{k=0}^{N_i-1} T^i(N-m+k)\,U^k a_i\,.$$

For each $i = 1, 2, \dots, L$, the change of index $p = N - m + k$ and the $N_i$-periodic character of the corresponding $i$ part of $\mathbf{T}_0$ give

$$\begin{aligned}
\sum_{k=0}^{N_i-1} T^i(N-m+k)\,U^k a_i &= \sum_{p=N-m}^{N-m+N_i-1} T^i(p)\,U^{p-N+m} a_i \\
&= \sum_{p=N-m}^{N-m+N_i-1} T^i(p)\,U^{p+m} a_i = \sum_{q=0}^{N_i-1} T^i(q)\,U^{q+m} a_i \\
&= U^m\Big(\sum_{q=0}^{N_i-1} T^i(q)\,U^q a_i\Big)\,.
\end{aligned}$$

Collecting all the pieces we get the desired result. □

**An Expression for the Generalized Samples**

From now on we consider a sampling period $r$ which divides $N$, where

$$N = \text{l.c.m.}\,(N_1, N_2, \dots N_L) \quad \text{and} \quad \ell := N/r\,.$$

Given a choice of $s$ elements $b_j \in \mathcal{H}$, $j = 1, 2, \dots, s$, for each $x \in \mathcal{V}_\mathbf{a}$ we consider its $s\ell$ generalized samples $\big\{\mathcal{L}_j x(rn)\big\}_{\substack{j=1,2,\dots,s\\ n=0,1,\dots,\ell-1}}$ with sampling period $r$ defined by

$$\mathcal{L}_j x(rn) := \langle x, U^{rn} b_j\rangle_{\mathcal{H}}, \quad n = 0, 1, \dots, \ell-1 \text{ and } j = 1, 2, \dots, s\,. \tag{6.7}$$

Note that the samples are $N$-periodic. We refer $\mathcal{L}_j$ as the $U$-system associated with $b_j \in \mathcal{H}$.

The goal in this chapter is to recover any $x = \sum_{i=1}^{L}\sum_{k=0}^{N_i-1} \alpha_k^i\, U^k a_i \in \mathcal{V}_{\mathbf{a}}$ from its finite sequence of samples $\big\{\mathcal{L}_j x(rn)\big\}_{\substack{j=1,2,\dots,s\\ n=0,1,\dots,\ell-1}}$ by means of suitable frame expansions in the subspace $\mathcal{V}_{\mathbf{a}}$.

First, we obtain a convenient expression for the sample $\mathcal{L}_j x(rn)$. Namely:

$$
\begin{aligned}
\mathcal{L}_j x(rn) &= \langle x, U^{rn} b_j\rangle_{\mathcal{H}} = \sum_{i=1}^{L}\sum_{k=0}^{N_i-1} \alpha_k^i\, \langle U^k a_i, U^{rn} b_j\rangle_{\mathcal{H}} \\
&= \sum_{i=1}^{L} \Big\langle \sum_{k=0}^{N_i-1} \alpha_k^i\, \mathbf{e}_k^i, \sum_{k=0}^{N_i-1} \overline{\langle U^k a_i, U^{rn} b_j\rangle_{\mathcal{H}}}\, \mathbf{e}_k^i \Big\rangle_{\mathbb{C}^{N_i}} \\
&= \sum_{i=1}^{L} \big\langle \boldsymbol{\alpha}^i, \mathbf{G}_{j,n}^i \big\rangle_{\mathbb{C}^{N_i}} ,
\end{aligned}
$$

where $\{\mathbf{e}_k^i\}_{k=0}^{N_i-1}$ denotes the canonical basis for $\mathbb{C}^{N_i}$, $\boldsymbol{\alpha}^i := (\alpha_0^i, \alpha_1^i, \dots, \alpha_{N_i-1}^i)^\top$ belongs to $\mathbb{C}^{N_i}$, and $\mathbf{G}_{j,n}^i := \sum_{k=0}^{N_i-1} \overline{\langle U^k a_i, U^{rn} b_j\rangle_{\mathcal{H}}}\, \mathbf{e}_k^i \in \mathbb{C}^{N_i}$.

Finally, if we consider the vectors in $\mathbb{C}^{N_1+N_2+\cdots+N_L}$

$$
\boldsymbol{\alpha} := (\boldsymbol{\alpha}^1, \boldsymbol{\alpha}^2, \dots, \boldsymbol{\alpha}^L)^\top \quad \text{and} \quad \mathbf{G}_{j,n} := (\mathbf{G}_{j,n}^1, \mathbf{G}_{j,n}^2, \dots, \mathbf{G}_{j,n}^L)^\top
$$

(identifying the vector $\mathbf{G}_{j,n}^i$ with its coordinates in $\mathbb{C}^{N_i}$), then we get the expression for the samples

$$
\mathcal{L}_j x(rn) = \big\langle \boldsymbol{\alpha}, \mathbf{G}_{j,n} \big\rangle_{\mathbb{C}^{N_1+N_2+\cdots+N_L}} , \tag{6.8}
$$

where $j = 1, 2, \dots, s$ and $n = 0, 1, \dots, \ell - 1$.

As a consequence of the above expression, having in mind the equivalence between spanning sets and frames (see Proposition 1.12) we obtain:

**Proposition 6.3** *Any $\boldsymbol{\alpha} \in \mathbb{C}^{N_1+N_2+\cdots+N_L}$ can be recovered from the sequence of samples $\big\{\mathcal{L}_j x(rn)\big\}_{\substack{j=1,2,\dots,s\\ n=0,1,\dots,\ell-1}}$ if and only if the set of vectors $\big\{\mathbf{G}_{j,n}\big\}_{\substack{j=1,2,\dots,s\\ n=0,1,\dots,\ell-1}}$ forms a spanning set for $\mathbb{C}^{N_1+N_2+\cdots+N_L}$, that is, a frame for $\mathbb{C}^{N_1+N_2+\cdots+N_L}$.*

In particular, we derive the necessary condition on the number $s$ of $U$-systems:

$$
s\ell \ge N_1 + N_2 + \cdots + N_L .
$$

Via the isomorphism $\mathcal{T}_{\mathbf{N},\mathbf{a}}$ we can restate the above result as:

**Corollary 6.1** *Any element $x \in \mathcal{V}_{\mathbf{a}}$ can be recovered from the sequence of its generalized samples $\{\mathcal{L}_j x(rn)\}_{\substack{j=1,2,\dots,s\\ n=0,1,\dots,\ell-1}}$ if and only if the set of vectors $\{\mathbf{G}_{j,n}\}_{\substack{j=1,2,\dots,s\\ n=0,1,\dots,\ell-1}}$ is a frame for $\mathbb{C}^{N_1+N_2+\cdots+N_L}$.*

The *cross-covariance* between the stationary sequences $\{U^n a_i\}_{n\in\mathbb{Z}}$ and $\{U^n b_j\}_{n\in\mathbb{Z}}$ is the $N_i$-periodic sequence defined in Ref. [8] as

$$R_{a_i,b_j}(m) := \langle U^m a_i, b_j\rangle_{\mathcal{H}}, \quad m \in \mathbb{Z}. \tag{6.9}$$

The choice of $N = \text{l.c.m.}\,(N_1, N_2, \dots, N_L)$ allows to write, for each $i = 1, 2, \dots, L$,

$$\begin{aligned}\mathbf{G}^i_{j,n} &= \sum_{k=0}^{N_i-1} \overline{\langle U^{k-rn} a_i, b_j\rangle_{\mathcal{H}}}\,\mathbf{e}^i_k = \sum_{k=0}^{N_i-1} \overline{\langle U^{N+k-rn} a_i, b_j\rangle_{\mathcal{H}}}\,\mathbf{e}^i_k \\ &= \sum_{k=0}^{N_i-1} \overline{R_{a_i,b_j}}(N+k-rn)\,\mathbf{e}^i_k \in \mathbb{C}^{N_i}.\end{aligned}$$

Therefore, in terms of coordinates with respect to the canonical basis for $\mathbb{C}^{N_1+N_2+\cdots+N_L}$ we have

$$\mathbf{G}_{j,n} = \left(\mathbf{G}^1_{j,n}, \mathbf{G}^2_{j,n}, \dots, \mathbf{G}^L_{j,n}\right)^{\top} \in \mathbb{C}^{N_1+N_2+\cdots+N_L},$$

where, for each $i = 1, 2, \dots, L$, $\mathbf{G}^i_{j,n}$ denotes the row vector (of dimension $N_i$):

$$\mathbf{G}^i_{j,n} = \left(\overline{R_{a_i,b_j}}(N-rn), \overline{R_{a_i,b_j}}(N-rn+1), \dots, \overline{R_{a_i,b_j}}(N-rn+N_i-1)\right).$$

Next, for each $x \in \mathcal{V}_{\mathbf{a}}$ we denote the vector of its samples by

$$\mathcal{L}_{\text{sam}}x := \left(\mathcal{L}_1 x, \mathcal{L}_2 x, \dots, \mathcal{L}_s x\right)^{\top} \in \mathbb{C}^{s\ell},$$

where, for each $j = 1, 2, \dots, s$, the row vector

$$\mathcal{L}_j x := \left(\mathcal{L}_j x(0), \mathcal{L}_j x(r), \dots, \mathcal{L}_j x(r(\ell-1))\right)$$

has dimension $\ell$.

Therefore, if we consider the vector $\boldsymbol{\alpha} := (\boldsymbol{\alpha}^1, \boldsymbol{\alpha}^2, \dots, \boldsymbol{\alpha}^L)^{\top}$ in $\mathbb{C}^{N_1+N_2+\cdots+N_L}$ associated with $x = \sum_{i=1}^{L}\sum_{k=0}^{N_i-1} \alpha^i_k\, U^k a_i \in \mathcal{V}_{\mathbf{a}}$, the expression (6.8) for the samples can be written in a matrix form as

$$\mathcal{L}_{\text{sam}}x = \mathbf{R}_{\mathbf{a},\mathbf{b}}\,\boldsymbol{\alpha}\,, \tag{6.10}$$

where the $s\ell \times (N_1 + N_2 + \cdots + N_L)$ matrix $\mathbf{R}_{\mathbf{a},\mathbf{b}}$ has the following block structure

$$\mathbf{R}_{\mathbf{a},\mathbf{b}} = \begin{pmatrix} \mathbf{R}_{a_1,b_1} & \mathbf{R}_{a_2,b_1} & \cdots & \mathbf{R}_{a_L,b_1} \\ \mathbf{R}_{a_1,b_2} & \mathbf{R}_{a_2,b_2} & \cdots & \mathbf{R}_{a_L,b_2} \\ \vdots & \vdots & \ddots & \vdots \\ \mathbf{R}_{a_1,b_s} & \mathbf{R}_{a_2,b_s} & \cdots & \mathbf{R}_{a_L,b_s} \end{pmatrix} \tag{6.11}$$

and, for each $1 \le i \le L$ and $1 \le j \le s$, the block $\mathbf{R}_{a_i,b_j}$ is the $\ell \times N_i$ matrix formed by using the cross-covariance $R_{a_i,b_j}(m)$ between $a_i$ and $b_j$ (see (6.9)) as

$$\mathbf{R}_{a_i,b_j} =$$

$$\begin{pmatrix} R_{a_i,b_j}(0) & R_{a_i,b_j}(1) & \cdots & R_{a_i,b_j}(N_i-1) \\ R_{a_i,b_j}(N-r) & R_{a_i,b_j}(N-r+1) & \cdots & R_{a_i,b_j}(N-r+N_i-1) \\ \vdots & \vdots & \ddots & \vdots \\ R_{a_i,b_j}(N-r(\ell-1)) & R_{a_i,b_j}(N-r(\ell-1)+1) & \cdots & R_{a_i,b_j}(N-r(\ell-1)+N_i-1) \end{pmatrix}.$$

Observe that, in sampling terms, the matrix $\mathbf{R}_{\mathbf{a},\mathbf{b}}$ is formed from samples of the generators $a_i$, $i = 1, 2, \ldots, L$, following the $U$-systems $\mathcal{L}_j$, $j = 1, 2, \ldots, s$.

Corollary 6.1 can be now restated in terms of the matrix $\mathbf{R}_{\mathbf{a},\mathbf{b}}$ as:

**Corollary 6.2** *Any element $x \in \mathcal{V}_{\mathbf{a}}$ can be recovered from the sequence of its generalized samples $\big\{\mathcal{L}_j x(rn)\big\}_{\substack{j=1,2,\ldots,s \\ n=0,1,\ldots,\ell-1}}$ if and only if the matrix $\mathbf{R}_{\mathbf{a},\mathbf{b}}$ has rank $N_1 + N_2 + \cdots + N_L$.*

Whenever rank $\mathbf{R}_{\mathbf{a},\mathbf{b}} = N_1 + N_2 + \cdots + N_L$, its Moore-Penrose pseudo-inverse is the $(N_1 + N_2 + \cdots + N_L) \times s\ell$ matrix

$$\mathbf{R}^{\dagger}_{\mathbf{a},\mathbf{b}} = \big[\mathbf{R}^{*}_{\mathbf{a},\mathbf{b}}\,\mathbf{R}_{\mathbf{a},\mathbf{b}}\big]^{-1}\mathbf{R}^{*}_{\mathbf{a},\mathbf{b}}\,.$$

Besides, any left-inverse $\mathbf{H}$ of the matrix $\mathbf{R}_{\mathbf{a},\mathbf{b}}$, i.e., $\mathbf{H}\,\mathbf{R}_{\mathbf{a},\mathbf{b}} = \mathbb{I}_{N_1+N_2+\cdots+N_L}$, can be expressed as

$$\mathbf{H} = \mathbf{R}^{\dagger}_{\mathbf{a},\mathbf{b}} + \mathbf{U}\big[\mathbb{I}_{s\ell} - \mathbf{R}_{\mathbf{a},\mathbf{b}}\,\mathbf{R}^{\dagger}_{\mathbf{a},\mathbf{b}}\big]\,, \tag{6.12}$$

where $\mathbf{U}$ denotes any arbitrary $N \times s\ell$ matrix (see (3.16)).

Let $\mathbf{H}$ be a left-inverse of $\mathbf{R}_{\mathbf{a},\mathbf{b}}$ and denote its columns by

$$\mathbf{H} = \begin{pmatrix} \vdots & & \vdots & \vdots & & \vdots & & \vdots & & \vdots \\ \mathbf{H}_{1,0} & \dots & \mathbf{H}_{1,\ell-1} & \mathbf{H}_{2,0} & \dots & \mathbf{H}_{2,\ell-1} & \dots & \mathbf{H}_{s,0} & \dots & \mathbf{H}_{s,\ell-1} \\ \vdots & & \vdots & \vdots & & \vdots & & \vdots & & \vdots \end{pmatrix}. \tag{6.13}$$

A left multiplication by $\mathbf{H}$ in (6.10) gives

$$\boldsymbol{\alpha} = \mathbf{H}\,\mathcal{L}_{\text{sam}}x = \sum_{j=1}^{s}\sum_{n=0}^{\ell-1} \mathcal{L}_j x(rn)\,\mathbf{H}_{j,n}\,. \tag{6.14}$$

Thus, the columns $\big\{\mathbf{H}_{j,n}\big\}_{\substack{j=1,2,\dots,s\\ n=0,1,\dots,\ell-1}}$ of any left-inverse $\mathbf{H}$ of the matrix $\mathbf{R}_{\mathbf{a},\mathbf{b}}$ give the family of all dual frames of the frame $\big\{\mathbf{G}_{j,n}\big\}_{\substack{j=1,2,\dots,s\\ n=0,1,\dots,\ell-1}}$ in $\mathbb{C}^{N_1+N_2+\cdots+N_L}$.

Applying the isomorphism $\mathcal{T}_{\mathbf{N},\mathbf{a}}$ in formula (6.14), for any $x = \sum_{i=1}^{L}\sum_{k=0}^{N_i-1} \alpha_k^i\, U^k a_i$ in $\mathcal{V}_{\mathbf{a}}$ we finally obtain the sampling formula:

$$x = \mathcal{T}_{\mathbf{N},\mathbf{a}}\big(\boldsymbol{\alpha}\big) = \sum_{j=1}^{s}\sum_{n=0}^{\ell-1} \mathcal{L}_j x(rn)\,\mathcal{T}_{\mathbf{N},\mathbf{a}}\big(\mathbf{H}_{j,n}\big)\,. \tag{6.15}$$

The sampling functions $\mathcal{T}_{\mathbf{N},\mathbf{a}}\big(\mathbf{H}_{j,n}\big)$ in the above formula do not have, in principle, any special $U$-structure as those sampling functions in (6.1) since the columns in $\mathbf{H}$ do not have any particular structure.

However, in next section we find all the left-inverses matrices of $\mathbf{R}_{\mathbf{a},\mathbf{b}}$ compatible with the $U$-structure yielding sampling formulas like (6.1).

### *6.1.1 Structure-Preserving Left-Inverses of Cross-Covariances Matrix*

Assume at least for now that there exists a left-inverse $\mathbf{H}$ with the same structure as the matrix $\mathbf{R}_{\mathbf{a},\mathbf{b}}$ by columns.

In other words, for $j = 1, 2, \dots, s$ and $n = 1, 2, \dots, \ell - 1$, its column $\mathbf{H}_{j,n}$ is obtained from the column $\mathbf{H}_{j,0}$ as follows. Denote the column $\mathbf{H}_{j,0}$ by

$$\mathbf{H}_{j,0} = \big(\mathbf{S}^1_{j,0}, \mathbf{S}^2_{j,0}, \dots, \mathbf{S}^L_{j,0}\big)^\top \in \mathbb{C}^{N_1+N_2+\cdots+N_L}$$

where, for each $i = 1, 2, \dots, L$, $\mathbf{S}^i_{j,0}$ denotes the row vector (of dimension $N_i$):

$$\mathbf{S}^i_{j,0} = \big(S^i_j(0), S^i_j(1), \dots, S^i_j(N_i - 1)\big).$$

Then

$$\mathbf{H}_{j,n} = \big(\mathbf{S}^1_{j,n}, \mathbf{S}^2_{j,n}, \dots, \mathbf{S}^L_{j,n}\big)^\top \in \mathbb{C}^{N_1+N_2+\dots+N_L}$$

where, for each $i = 1, 2, \dots, L$, $\mathbf{S}^i_{j,n}$ denotes the row vector (of dimension $N_i$)

$$\mathbf{S}^i_{j,n} = \big(S^i_j(N - rn), S^i_j(N - rn + 1), \dots, S^i_j(N - rn + N_i - 1)\big).$$

As in Lemma 6.2, we are also extending each $S^i_j(\cdot)$ taking into account $N_i$-periodicity.

Therefore, the shifting property in Lemma 6.2 gives

$$\mathcal{T}_{\mathbf{N},\mathbf{a}}\big(\mathbf{H}_{j,n}\big) = U^{rn}\big(\mathcal{T}_{\mathbf{N},\mathbf{a}}\big(\mathbf{H}_{j,0}\big)\big) \quad \text{for } n = 0, 1, \dots, \ell - 1,$$

and, consequently, formula (6.15) reads

$$\begin{aligned} x = \mathcal{T}_{\mathbf{N},\mathbf{a}}\big(\boldsymbol{\alpha}\big) &= \sum_{j=1}^{s}\sum_{n=0}^{\ell-1} \mathcal{L}_j x(rn)\, \mathcal{T}_{\mathbf{N},\mathbf{a}}\big(\mathbf{H}_{j,n}\big) \\ &= \sum_{j=1}^{s}\sum_{n=0}^{\ell-1} \mathcal{L}_j x(rn)\, U^{rn}\big(\mathcal{T}_{\mathbf{N},\mathbf{a}}\big(\mathbf{H}_{j,0}\big)\big) \\ &= \sum_{j=1}^{s}\sum_{n=0}^{\ell-1} \mathcal{L}_j x(rn)\, U^{rn} c_j, \end{aligned}$$

where $c_j = \mathcal{T}_{\mathbf{N},\mathbf{a}}\big(\mathbf{H}_{j,0}\big) \in \mathcal{V}_\mathbf{a}$, $j = 1, 2, \dots, s$.

We now proceed to construct left-inverses of $\mathbf{R}_{\mathbf{a},\mathbf{b}}$ having the required structure in their columns. To this end we use an algorithm borrowed from Ref. [4], and published for the first time in Ref. [5].

Indeed, let $\mathbf{H}$ be any left-inverse of the matrix $\mathbf{R}_{\mathbf{a},\mathbf{b}}$. This matrix $\mathbf{H}$ has order $(N_1 + N_2 + \dots + N_L) \times s\ell$ and it can be written in blocks as

$$\mathbf{H} = \begin{pmatrix} \mathbf{S}^1_1 & \mathbf{S}^1_2 & \dots & \mathbf{S}^1_s \\ \mathbf{S}^2_1 & \mathbf{S}^2_2 & \dots & \mathbf{S}^2_s \\ \vdots & \vdots & \ddots & \vdots \\ \mathbf{S}^L_1 & \mathbf{S}^L_2 & \dots & \mathbf{S}^L_s \end{pmatrix},$$

where each block $\mathbf{S}^i_j$ denotes an $N_i \times \ell$ matrix, $i = 1, 2, \dots, L$ and $j = 1, 2, \dots, s$.

Following Ref. [5, Section 3] the idea amounts to defining a new left-inverse $\tilde{\mathbf{H}}$ of $\mathbf{R}_{\mathbf{a},\mathbf{b}}$ to be the modification to $\mathbf{H}$ by replacing every block $\mathbf{S}^i_j$ such that $N_i > r$ by a suitable $N_i \times \ell$ block $\tilde{\mathbf{S}}^i_j$.

Namely, the required steps are:

1. If $N_i > r$, denote the first $r$ rows of $\mathbf{S}^i_j$ by

$$\mathbf{S}^{i,r}_j := \begin{pmatrix} S^i_j(0) & S^i_j(N-r) & S^i_j(N-2r) & \dots & S^i_j(N-r(\ell-1)) \\ S^i_j(1) & S^i_j(N-r+1) & S^i_j(N-2r+1) & \dots & S^i_j(N-r(\ell-1)+1) \\ \vdots & \vdots & \vdots & \ddots & \vdots \\ S^i_j(r-1) & S^i_j(N-1) & S^i_j(N-2r+r-1) & \dots & S^i_j(N-r(\ell-1)+r-1) \end{pmatrix}.$$

2. Next we form a $N \times \ell$ matrix (note that $N = r\ell$)

$$\mathbf{S}^{i,N}_j := \begin{pmatrix} S^i_j(0) & S^i_j(N-r) & S^i_j(N-2r) & \dots & S^i_j(r) \\ S^i_j(1) & S^i_j(N-r+1) & S^i_j(N-2r+1) & \dots & S^i_j(r+1) \\ \vdots & \vdots & \vdots & \ddots & \vdots \\ S^i_j(r-1) & S^i_j(N-1) & S^i_j(N-r-1) & \dots & S^i_j(2r-1) \\ S^i_j(r) & S^i_j(0) & S^i_j(N-r) & \dots & S^i_j(2r) \\ S^i_j(r+1) & S^i_j(1) & S^i_j(N-r+1) & \dots & S^i_j(2r+1) \\ \vdots & \vdots & \vdots & \ddots & \vdots \\ S^i_j(N-1) & S^i_j(N-r-1) & S^i_j(N-2r-1) & \dots & S^i_j(r-1) \end{pmatrix}$$

according to the following rules:

- The first column of the matrix $\mathbf{S}^{i,N}_j$ is a concatenation of the columns 1, $\ell$, $\ell - 1, \dots,$ and 2 of $\mathbf{S}^{i,r}_j$;
- The second column of $\mathbf{S}^{i,N}_j$ is a concatenation of the columns 2, 1, $\ell, \dots,$ and 3 of $\mathbf{S}^{i,r}_j$;
- The third column of $\mathbf{S}^{i,N}_j$ is a concatenation of the columns 3, 2, 1, $\dots,$ and 4 of $\mathbf{S}^{i,r}_j$.
  Repeating the process, finally,
- The column $\ell$ of $\mathbf{S}^{i,N}_j$ is a concatenation of the columns $\ell, \ell-1, \ell-2, \dots,$ and 1 of $\mathbf{S}^{i,r}_j$.

3. Take as the block $\tilde{\mathbf{S}}^i_j$ the $N_i$ first rows of the matrix $\mathbf{S}^{i,N}_j$.
4. If $N_i \le r$, take $\tilde{\mathbf{S}}^i_j := \mathbf{S}^i_j$.

5. Finally, consider the new $(N_1 + N_2 + \cdots + N_L) \times s\ell$ matrix $\tilde{\mathbf{H}}$ written in blocks as

$$\tilde{\mathbf{H}} = \begin{pmatrix} \tilde{\mathbf{S}}_1^1 & \tilde{\mathbf{S}}_2^1 & \ldots & \tilde{\mathbf{S}}_s^1 \\ \tilde{\mathbf{S}}_1^2 & \tilde{\mathbf{S}}_2^2 & \ldots & \tilde{\mathbf{S}}_s^2 \\ \vdots & \vdots & \ddots & \vdots \\ \tilde{\mathbf{S}}_1^L & \tilde{\mathbf{S}}_2^L & \ldots & \tilde{\mathbf{S}}_s^L \end{pmatrix}. \tag{6.16}$$

Thus, following the above procedure we have obtained a left-inverse matrix $\tilde{\mathbf{H}}$ for $\mathbf{R_{a,b}}$ having the desired structure:

**Proposition 6.4** *Let* $\mathbf{R_{a,b}}$ *and* $\tilde{\mathbf{H}}$ *be the matrices defined in* (6.11) *and* (6.16) *respectively. Then we have that*

$$\tilde{\mathbf{H}}\,\mathbf{R_{a,b}} = \mathbb{I}_{N_1+N_2+\cdots+N_L}\,.$$

***Proof*** We represent by $\mathbf{H}^p$, $p = 1, 2, \ldots, L$, the $N_p \times s\ell$ matrix extracted from $\mathbf{H}$ by taking the blocks $\mathbf{S}_1^p, \mathbf{S}_2^p, \ldots, \mathbf{S}_s^p$. Similarly, let $\mathbf{R}_{a_q,\mathbf{b}}$, $q = 1, 2, \ldots, L$, the $s\ell \times N_q$ matrix extracted from $\mathbf{R_{a,b}}$ by taking the blocks $\mathbf{R}_{a_q,b_1}, \mathbf{R}_{a_q,b_2}, \ldots, \mathbf{R}_{a_q,b_s}$.

From

$$\mathbf{H}\,\mathbf{R_{a,b}} = \mathbb{I}_{N_1+N_2+\cdots+N_L}$$

we derive that

$$\begin{cases} \mathbf{H}^q\,\mathbf{R}_{a_q,\mathbf{b}} = \mathbb{I}_{N_q}\,, \\ \mathbf{H}^p\,\mathbf{R}_{a_q,\mathbf{b}} = \mathbb{O}_{N_p\times N_q} \quad \text{if } \; p \neq q\,, \end{cases}$$

where $p, q = 1, 2, \ldots, L$, and $\mathbb{I}_{N_p}$ and $\mathbb{O}_{N_p\times N_q}$ denote, respectively, the identity matrix of order $N_p$ and the zero matrix of order $N_p \times N_q$.

Giving to the entries of each block $\mathbf{S}_j^{p,N}$ an $N$-periodic character and having in mind the $N_q$-periodic, and consequently $N$-periodic, character of the entries in matrix $\mathbf{R}_{a_q,\mathbf{b}}$, the product $\alpha_{m,k}^{p,q}$ of row $m+1$ of blocks $\tilde{\mathbf{S}}_1^p, \tilde{\mathbf{S}}_2^p, \ldots, \tilde{\mathbf{S}}_s^p$, $p = 1, 2, \ldots, L$ and $m = 0, 1, \ldots, N_p - 1$, with column $k+1$ of matrix $\mathbf{R}_{a_q,\mathbf{b}}$, $q = 1, 2, \ldots, L$ and $k = 0, 1, \ldots, N_q - 1$, can be written as

$$\alpha_{m,k}^{p,q} = \sum_{j=1}^{s}\sum_{i=0}^{\ell-1} S_j^p(N - ir + m)R_{a_q,b_j}(N - ir + k)\,.$$

We now prove that $\alpha_{m,k}^{p,q}$ can be seen as the product of an appropriate row of $\mathbf{H}^p$ with the corresponding column of $\mathbf{R}_{a_q,\mathbf{b}}$. This is clear if $N_p \le r$ or $m \le r$ since then $\alpha_{m,k}^{p,q}$ is nothing but the product of row $m+1$ of blocks $\mathbf{S}_1^p, \mathbf{S}_2^p, \dots, \mathbf{S}_s^p$, i.e., row $m+1$ of $\mathbf{H}^p$, with column $k+1$ of matrix $\mathbf{R}_{a_q,\mathbf{b}}$. In any case, note that for each $m \in \{0, 1, \dots, N_p - 1\}$ there exist a unique $m' \in \{0, 1, \dots, r-1\}$ and a unique $i'$ such that

$$N - i'r + m = N + m' .$$

In accord with $k$ and $i'$ take the unique $k' \in \{0, 1, \dots, N_q - 1\}$ such that

$$N - i'r + k = k' + zN_q ,$$

with $z \in \mathbb{Z}$. Thus, $\alpha_{m,k}^{p,q} = \alpha_{m',k'}^{p,q}$ is just the product of row $m'+1$ of $\mathbf{H}^p$ with column $k'+1$ of $\mathbf{R}_{a_q,\mathbf{b}}$.

Since $\mathbf{H}^p\,\mathbf{R}_{a_q,\mathbf{b}} = \mathbb{O}_{N_p \times N_q}$ for $p \neq q$, we have that $\alpha_{m,k}^{p,q} = 0$ if $p \neq q$. Whenever $p = q$ we see that $m' = k'$ if and only if $m = k$. In fact, if $m' = k'$ it follows that $m - i'r = N - i'r + k - zN_q$ and consequently, since $N$ is a multiple of $N_q$, we get that $m = N + k - zN_q = k + z''N_q$ with $z'' \in \mathbb{Z}$. The equality $m = k$ follows from the fact that both $m$ and $k$ belong to $\{0, 1, \dots, N_q - 1\}$.

Finally, from $\mathbf{H}^q\,\mathbf{R}_{a_q,\mathbf{b}} = \mathbb{I}_{N_q}$ we obtain that $\alpha_{m,k}^{q,q} = \delta_{m,k}$. □

Having in mind (6.12), we have to compute the pseudo-inverse $\mathbf{R}_{\mathbf{a},\mathbf{b}}^{\dagger}$. In practice this is done by using the singular value decomposition of $\mathbf{R}_{\mathbf{a},\mathbf{b}}$; the singular values are the square root of the eigenvalues of the invertible and positive definite square matrix $\mathbf{R}_{\mathbf{a},\mathbf{b}}^{*}\,\mathbf{R}_{\mathbf{a},\mathbf{b}}$ of order $N_1 + N_2 + \cdots + N_L$ (see, for instance, Refs. [2, 7]).

Note that the singular value decomposition of the matrix $\mathbf{R}_{\mathbf{a},\mathbf{b}}$ is the most reliable method to reveal its rank in practice.

### 6.1.2 The Derived Sampling Results

Collecting the pieces we have obtained until now allows to conclude the main result in this section:

**Theorem 6.1** *Given the $s\ell \times (N_1 + N_2 + \cdots + N_L)$ matrix of cross-covariances $\mathbf{R}_{\mathbf{a},\mathbf{b}}$ defined in* (6.11), *the following statements are equivalent:*

*(a) rank $\mathbf{R}_{\mathbf{a},\mathbf{b}} = N_1 + N_2 + \cdots + N_L$.*
*(b) There exists a structured left-inverse $\widetilde{\mathbf{H}}$ of the matrix $\mathbf{R}_{\mathbf{a},\mathbf{b}}$ as in* (6.16).

(c) *There exist* $c_j \in \mathcal{V}_\mathbf{a}$, $j = 1, 2, \ldots, s$ *such that the sequence* $\{U^{rn} c_j\}_{\substack{j=1,2,\ldots,s \\ n=0,1,\ldots,\ell-1}}$ *is a frame for* $\mathcal{V}_\mathbf{a}$ *and, for any* $x \in \mathcal{V}_\mathbf{a}$, *the expansion*

$$x = \sum_{j=1}^{s} \sum_{n=0}^{\ell-1} \mathcal{L}_j x(rn)\, U^{rn} c_j \tag{6.17}$$

*holds.*

(d) *There exists a frame* $\{C_{j,n}\}_{\substack{j=1,2,\ldots,s \\ n=0,1,\ldots,\ell-1}}$ *for* $\mathcal{V}_\mathbf{a}$ *such that, for each* $x \in \mathcal{V}_\mathbf{a}$, *the expansion*

$$x = \sum_{j=1}^{s} \sum_{n=0}^{\ell-1} \mathcal{L}_j x(rn)\, C_{j,n}$$

*holds.*

***Proof*** That condition ($a$) implies condition ($b$) and that condition ($b$) implies condition ($c$) have been proved above. Obviously, condition ($c$) implies condition ($d$): choose $C_{j,n} = U^{rn} c_j$ for $j = 1, 2, \ldots, s$ and $n = 0, 1, \ldots, \ell - 1$. Finally, condition ($d$) implies condition ($a$) as a consequence of Corollary 6.2. □

In the eventual case where $s\ell = N_1 + N_2 + \cdots + N_L$ we obtain:

**Corollary 6.3** *Assume that* $s\ell = N_1 + N_2 + \cdots + N_L$ *and consider the cross-covariances square matrix* $\mathbf{R}_{\mathbf{a},\mathbf{b}}$ *of order* $N_1 + N_2 + \cdots + N_L$ *defined in* (6.11). *The following statements are equivalent:*

(i) *The square matrix* $\mathbf{R}_{\mathbf{a},\mathbf{b}}$ *is invertible.*

(ii) *There exist* $s$ *unique elements* $c_j \in \mathcal{V}_\mathbf{a}$, $j = 1, 2, \ldots, s$, *such that the sequence* $\{U^{rn} c_j\}_{\substack{j=1,2,\ldots,s \\ n=0,1,\ldots,\ell-1}}$ *is a basis for* $\mathcal{V}_a$, *and the expansion of any* $x \in \mathcal{V}_\mathbf{a}$ *with respect to this basis is*

$$x = \sum_{j=1}^{s} \sum_{n=0}^{\ell-1} \mathcal{L}_j x(rn)\, U^{rn} c_j\,.$$

*In case the equivalent conditions are satisfied, the interpolation property*

$$\mathcal{L}_j c_{j'}(rn) = \delta_{j,j'}\, \delta_{n,0}\,,$$

*whenever* $n = 0, 1, \ldots, \ell - 1$ *and* $j, j' = 1, 2, \ldots, s$, *holds.*

***Proof*** Notice that the inverse matrix $\mathbf{R}_{\mathbf{a},\mathbf{b}}^{-1}$ has necessarily the structure of the matrix $\tilde{\mathbf{H}}$ in (6.16). The uniqueness of the expansion with respect to a basis gives the stated interpolation property. □

**The One Generator Case**

Here we state the sampling result in Theorem 6.1 for the one generator case. For a fixed $a \in \mathcal{H}$, assume that there exists a nonnegative integer $N$ such that $U^N a = a$; let $N$ be the smallest index with this property.

Next, we consider the finite dimensional subspace

$$\mathcal{V}_a := \operatorname{span}\{a, Ua, U^2a, \dots, U^{N-1}a\}$$

in $\mathcal{H}$, where we are assuming that the set $\{a, Ua, U^2a, \dots, U^{N-1}a\}$ is linearly independent.

For each $x \in \mathcal{V}_a$ we consider its samples $\{\mathcal{L}_j x(rn)\}_{\substack{j=1,2,\dots,s \\ n=0,1,\dots,\ell-1}}$ defined in (6.7) where, for a sampling period $r$, the number of samples is $\ell := N/r$.

In this case, the $s\ell \times N$ matrix of cross-covariances $\mathbf{R}_{a,\mathbf{b}}$ reads:

$$\mathbf{R}_{a,\mathbf{b}} := \begin{pmatrix} \mathbf{R}_{a,b_1} \\ \mathbf{R}_{a,b_2} \\ \vdots \\ \mathbf{R}_{a,b_s} \end{pmatrix}, \tag{6.18}$$

where each $\ell \times N$ block is given by $\mathbf{R}_{a,b_j}$, $j = 1, 2, \dots, s$, is given by

$$\mathbf{R}_{a,b_j} = \begin{pmatrix} R_{a,b_j}(0) & R_{a,b_j}(1) & \dots & R_{a,b_j}(N-1) \\ R_{a,b_j}(N-r) & R_{a,b_j}(N-r+1) & \dots & R_{a,b_j}(2N-r-1) \\ \vdots & \vdots & \ddots & \vdots \\ R_{a,b_j}(N-r(\ell-1)) & R_{a,b_j}(N-r(\ell-1)+1) & \dots & R_{a,b_j}(2N-1-r(\ell-1)) \end{pmatrix}.$$

Since $N = r\ell$ we have

$$\mathbf{R}_{a,b_j} = \begin{pmatrix} R_{a,b_j}(0) & R_{a,b_j}(1) & \dots & R_{a,b_j}(N-1) \\ R_{a,b_j}(N-r) & R_{a,b_j}(N-r+1) & \dots & R_{a,b_j}(2N-r-1) \\ \vdots & \vdots & \ddots & \vdots \\ R_{a,b_j}(r) & R_{a,b_j}(r+1) & \dots & R_{a,b_j}(N-1+r) \end{pmatrix}.$$

**Corollary 6.4** *Given the $s\ell \times N$ matrix of cross-covariances $\mathbf{R}_{a,\mathbf{b}}$, the following statements are equivalent:*

*(a) rank $\mathbf{R}_{a,\mathbf{b}} = N$.*
*(b) There exists an $r \times s\ell$ matrix $\mathbf{S}$ such that*

$$\mathbf{S}\,\mathbf{R}_{a,\mathbf{b}} = \left[\mathbb{I}_r, \mathbb{O}_{r\times(N-r)}\right],$$

*where $\mathbb{I}_r$ and $\mathbb{O}_{r\times(N-r)}$ denote, respectively, the identity matrix of order $r$ and the zero matrix of order $r \times (N-r)$.*

(c) *There exist* $c_j \in \mathcal{V}_a$, $j = 1, 2, \ldots, s$ *such that the sequence* $\{U^{rn}c_j\}_{\substack{j=1,2,\ldots,s\\ n=0,1,\ldots,\ell-1}}$ *is a frame for* $\mathcal{V}_a$*, and for any* $x \in \mathcal{V}_a$ *the expansion*

$$x = \sum_{j=1}^{s}\sum_{n=0}^{\ell-1} \mathcal{L}_j x(rn)\, U^{rn} c_j$$

*holds.*

(d) *There exist a frame* $\{C_{j,n}\}_{\substack{j=1,2,\ldots,s\\ n=0,1,\ldots,\ell-1}}$ *for* $\mathcal{V}_a$ *such that, for each* $x \in \mathcal{V}_a$ *the expansion*

$$x = \sum_{j=1}^{s}\sum_{n=0}^{\ell-1} \mathcal{L}_j x(rn)\, C_{j,n}$$

*holds.*

It is worth observing that, as stated in condition $(b)$, we only need to know the first $r$ rows of a left-inverse matrix of $\mathbf{R}_{a,\mathbf{b}}$ to construct the corresponding structured left-inverse $N \times s\ell$ matrix $\widetilde{\mathbf{H}}$ of $\mathbf{R}_{a,\mathbf{b}}$ by following the procedure described in Sect. 6.1.1.

For the particular case where the number of systems $\mathcal{L}_j$ and the sampling period $r$ coincides, i.e., $s = r$, we obtain:

**Corollary 6.5** *Assume that* $s = r$ *and consider the* $N \times N$ *matrix of cross-covariances* $\mathbf{R}_{a,\mathbf{b}}$ *defined in* (6.18). *The following statements are equivalent:*

(i) *The matrix* $\mathbf{R}_{a,\mathbf{b}}$ *is invertible.*

(ii) *There exist* $r$ *unique elements* $c_j \in \mathcal{V}_a$, $j = 1, 2, \ldots, r$*, such that the sequence* $\{U^{rn}c_j\}_{\substack{j=1,2,\ldots,r\\ n=0,1,\ldots,\ell-1}}$ *is a basis for* $\mathcal{V}_a$*, and the expansion of any* $x \in \mathcal{V}_a$ *with respect to this basis is*

$$x = \sum_{j=1}^{r}\sum_{n=0}^{\ell-1} \mathcal{L}_j x(rn)\, U^{rn} c_j\,.$$

*In case the equivalent conditions are satisfied, the interpolation property*

$$\mathcal{L}_j c_{j'}(rn) = \delta_{j,j'}\,\delta_{n,0}\,,$$

*whenever* $n = 0, 1, \ldots, \ell - 1$ *and* $j, j' = 1, 2, \ldots, r$*, holds.*

Note that the basis $\{U^{rn}c_j\}_{\substack{j=1,2,\dots,r\\ n=0,1,\dots,\ell-1}}$ for $\mathcal{V}_a$ in the above corollary has frame (Riesz) bounds

$$\frac{1}{\sigma_N^2}\|\mathcal{T}_{N,a}^{-1}\|^{-2} \quad \text{and} \quad \frac{1}{\sigma_1^2}\|\mathcal{T}_{N,a}\|^2,$$

where $\sigma_1$ (respectively $\sigma_N$) denotes the smallest (respectively the largest) singular value of the matrix $\mathbf{R}_{a,\mathbf{b}}$. Here, $\mathcal{T}_{N,a}$ denotes the isomorphism (6.4) for this particular setting (see finite frames in Sect. 1.2).

**Example 6.1 (A Filter-Bank Interpretation)** Assume that rank of $\mathbf{R}_{\mathbf{a},\mathbf{b}}$ equals $N_1 + N_2 + \dots + N_L$, and let $\tilde{\mathbf{H}}$ be a structured left-inverse of $\mathbf{R}_{\mathbf{a},\mathbf{b}}$ as in (6.16) whose columns $\mathbf{H}_{j,n}$ are denoted as in expression (6.13) for $j = 1, 2, \dots, s$, and $n = 0, 1, \dots, \ell - 1$.

In the sampling formula (6.17) we have $c_j = \mathcal{T}_{N,a}(\mathbf{H}_{j,0})$, $j = 1, 2, \dots, s$, and let denote the coordinates of $\mathbf{H}_{j,0} \in \mathbb{C}^{N_1+N_2+\dots+N_L}$ as

$$\mathbf{H}_{j,0} = \left(\beta_j^1(0), \beta_j^1(1), \dots, \beta_j^1(N_1-1), \dots, \beta_j^L(0), \beta_j^L(1), \dots, \beta_j^L(N_L-1)\right)^\top.$$

Substituting in (6.17), for each $x \in \mathcal{V}_\mathbf{a}$ we get

$$\begin{aligned} x &= \sum_{j=1}^{s}\sum_{n=0}^{\ell-1} \mathcal{L}_j x(rn)\, U^{rn}\Big(\sum_{i=1}^{L}\sum_{k=0}^{N_i-1} \beta_j^i(k)\, U^k a_i\Big) \\ &= \sum_{j=1}^{s}\sum_{n=0}^{\ell-1} \mathcal{L}_j x(rn)\Big(\sum_{i=1}^{L}\sum_{k=0}^{N_i-1} \beta_j^i(m)\, U^{rn+k} a_i\Big). \end{aligned}$$

The change of index $m := rn + k$ and assuming an $(N_1, N_2, \dots, N_L)$-circulant character of each $\mathbf{H}_{j,0}$ gives

$$\begin{aligned} x &= \sum_{j=1}^{s}\sum_{n=0}^{\ell-1} \mathcal{L}_j x(rn)\Big(\sum_{i=1}^{L}\sum_{m=rk}^{rk+N_i-1} \beta_j^i(m-rn)\, U^m a_i\Big) \\ &= \sum_{j=1}^{s}\sum_{n=0}^{\ell-1} \mathcal{L}_j x(rn)\Big(\sum_{i=1}^{L}\sum_{m=0}^{N_i-1} \beta_j^i(m-rn)\, U^m a_i\Big) \\ &= \sum_{i=1}^{L}\sum_{m=0}^{N_i-1}\Big\{\sum_{j=1}^{s}\sum_{n=0}^{\ell-1} \mathcal{L}_j x(rn)\, \beta_j^i(m-rn)\Big\} U^m a_i\,. \end{aligned}$$

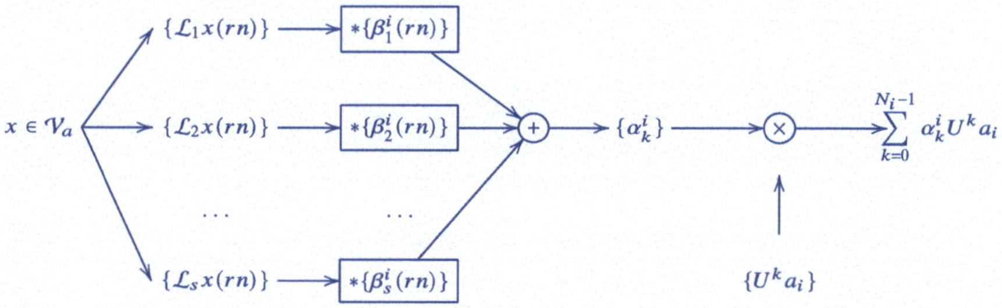


**Fig. 6.2** Sampling and reconstruction scheme for $i = 1, 2, \dots, L$

In other words, for each $i = 1, 2, \dots, L$, the coefficients $\alpha_m^i$ of $x = \sum_{i=1}^{L} \sum_{m=0}^{N_i-1} \alpha_m^i U^m a_i$ are the output of a filter-bank

$$\alpha_m^i = \sum_{j=1}^{s} \sum_{n=0}^{\ell-1} \mathcal{L}_j x(rn)\, \beta_j^i(m - rn)\,, \quad m = 0, 1, \dots, N_i - 1\,,$$

involving the data $\big\{\mathcal{L}_j x(rn)\big\}_{\substack{j=1,2,\dots,s \\ n=0,1,\dots,\ell-1}}$ and the columns $\mathbf{H}_{j,0}$, $j = 1, 2, \dots, s$, of the matrix $\tilde{\mathbf{H}}$ as in expression (6.13).

The above description is outlined in the diagram depicted in Fig. 6.2.

### 6.1.3 An Application: Periodic Extensions of Finite Signals

This section is devoted to show how the usual periodic extensions of finite signals fit as an important example of the developed theory in the previous sections. We are following the terminology used in Ref. [12].

**Example 6.2 (The Wraparound Extension)** In this first example we merely extend a signal of length $M \in \mathbb{N}$ to an $M$-periodic signal (see Fig. 6.3).

**Fig. 6.3** Wraparound extension $M = 4$

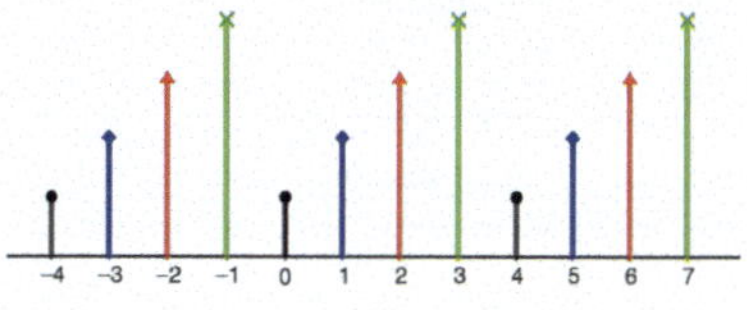

Here we consider the space $\ell^2_M(\mathbb{Z})$ of $M$-periodic signals with inner product

$$\langle \mathbf{x}, \mathbf{y}\rangle_{\ell^2_M} = \sum_{m=0}^{M-1} x(m)\,\overline{y(m)}\,.$$

If we take, for instance, the $M$ periodic signal $\mathbf{a} := (1, 0, \dots, 0)$ and the shift operator

$$U : \mathbf{x} = \{x(m)\} \longmapsto U\mathbf{x} := \{x(m-1)\}$$

in $\ell^2_M(\mathbb{Z})$, then we trivially obtain that $U^M\mathbf{a} = \mathbf{a}$ and $\mathcal{V}_\mathbf{a} = \ell^2_M(\mathbb{Z})$.

Given $\mathbf{b}_j \in \ell^2_M$, $j = 1, 2, \dots, s$, the set of samples $\{\mathcal{L}_j\mathbf{x}(rn)\}_{\substack{j=1,2,\dots,s\\ n=0,1,\dots,\ell-1}}$ of any $\mathbf{x} \in \ell^2_M$ is obtained from the $M$-periodic convolution

$$\mathcal{L}_j\mathbf{x}(rn) = \langle \mathbf{x}, U^{rn}\mathbf{b}_j\rangle_{\ell^2_N} = \sum_{m=0}^{M-1} x(m)\,\overline{b_j(m-rn)} = \big(\mathbf{x} * \mathbf{h}_j\big)(rn)\,,$$

where $h_j(m) := \overline{b_j(-m)}$, $m = 0, 1, \dots, M-1$. Recall that the sampling period $r$ divides $M$, and $\ell = M/r$.

As the cross-covariance $R_{a,b_j}(m) = \langle U^m\mathbf{a}, \mathbf{b}_j\rangle_{\ell^2_N} = \overline{b_j(m)}$, each $\ell \times M$ block $\mathbf{R_{a,b_j}}$, $j = 1, 2, \dots, s$, of the $s\ell \times M$ matrix $\mathbf{R_{a,b}}$ in (6.11) takes the form

$$\mathbf{R_{a,b_j}} = \begin{pmatrix} \overline{b_j(0)} & \overline{b_j(1)} & \dots & \overline{b_j(M-1)} \\ \overline{b_j(M-r)} & \overline{b_j(M-r+1)} & \dots & \overline{b_j(2M-r-1)} \\ \vdots & \vdots & \ddots & \vdots \\ \overline{b_j(M-r(\ell-1))} & \overline{b_j(M-r(\ell-1)+1)} & \dots & \overline{b_j(2M-1-r(\ell-1))} \end{pmatrix}.$$

In case the rank of $\mathbf{R_{a,b}}$ is $M$, which implies $s \ge r$ convolution systems, from Theorem 6.1 we obtain in $\ell^2_M(\mathbb{Z})$ the sampling formula

$$\begin{aligned} \mathbf{x}(m) &= \sum_{j=1}^{s}\sum_{n=0}^{\ell-1} \mathcal{L}_j\mathbf{x}(rn)\,\mathbf{c}_j(m-rn) \\ &= \sum_{j=1}^{s}\sum_{n=0}^{\ell-1} \big(\mathbf{x} * \mathbf{h}_j\big)(rn)\,\mathbf{c}_j(m-rn)\,, \qquad m = 0, 1, \dots, M-1\,. \end{aligned} \tag{6.19}$$

The sampling sequences in $\ell^2_M(\mathbb{Z})$ are $\mathbf{c}_j = \mathcal{T}_{M,a}\big(\mathbf{H}_{j,0}\big)$, $j = 1, 2, \dots, s$, where $\mathbf{H}_{j,0}$ are the corresponding columns (denoted as in expression (6.13)) of a $M \times s\ell$

structured left-inverse $\tilde{\mathbf{H}}$ of $\mathbf{R_{a,b}}$ constructed as (6.16). Note that $\mathbf{c}_j$ is nothing but the $M$-periodic sequence derived from $\mathbf{H}_{j,0} \in \mathbb{C}^M$.

Via Theorem 6.1, the sampling result (6.19) has a straightforward interpretation in terms of *Perfect Reconstruction Filter-banks*. In fact, it gives a full characterization of the perfect reconstruction filter banks in $\ell^2_M(\mathbb{Z})$ in terms of matrix analysis:

*Given, for each* $\mathbf{x} \in \ell^2_M(\mathbb{Z})$, $s \geq r$ *subsampled sequences*

$$\big\{(\mathbf{x} * \mathbf{h}_j)(rn)\big\}_{n=0,1,\ldots,\ell-1} \quad \text{with } j = 1, 2, \ldots, s\,,$$

*then the condition* rank $\mathbf{R_{a,b}} = M$ *on the impulse responses* $\mathbf{h}_j,\ j = 1, 2, \ldots, s$, *is a necessary and sufficient condition for the perfect reconstruction stated in* (6.19).
*Furthermore, if the rank condition is fulfilled, then all the reconstruction filters* $\mathbf{c}_j$, $j = 1, 2, \ldots, s$, *are obtained by means of the structured* $M \times s\ell$ *left-inverse matrices* $\tilde{\mathbf{H}}$ *of the matrix* $\mathbf{R_{a,b}}$.

**Example 6.3 (The Whole-Point Symmetry Extension)** In this case we extend a signal of length $M \in \mathbb{N}$, without repeating the extreme values, to form a $(2M-2)$-periodic signal (see Fig. 6.4).

Here we consider the space $\ell^2_{2M-2}(\mathbb{Z})$ with the inner product

$$\langle \mathbf{x}, \mathbf{y}\rangle_{\ell^2_{2M-2}} = \sum_{m=0}^{2M-3} x(m)\,\overline{y(m)}\,.$$

Now consider the subspace in $\ell^2_{2M-2}(\mathbb{Z})$ spanned by the following $M$ signals:

$$\mathbf{a}_1 := (1, 0, 0, \ldots, 0, 0)\,;\ \mathbf{a}_2 := (0, 1, 0, \ldots, 0, 1)\,;$$
$$\mathbf{a}_3 := (0, 0, 1, \ldots, 1, 0)\,; \ldots;$$
$$\mathbf{a}_i := (0, \ldots, \underset{(i-1)}{1}, \ldots, \underset{(2M-(i+1))}{1}, \ldots, 0)\,; \ldots;$$
$$\mathbf{a}_M := (0, \ldots, 0, \underset{(M-1)}{1}, 0, \ldots, 0)\,,$$

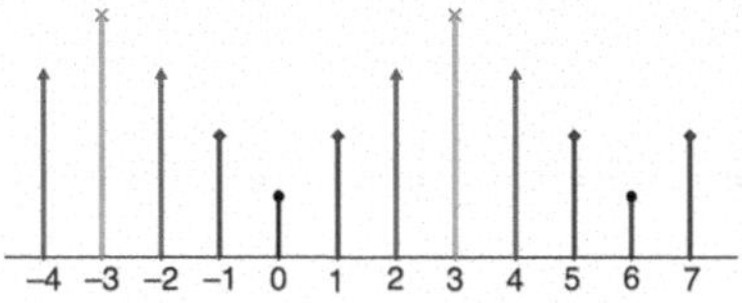


**Fig. 6.4** Whole-point symmetric extension $M = 4$

that is, for $i = 1, 2, \ldots, M$, and $m = 0, 1, \ldots, 2M - 3$, we have

$$\mathbf{a}_i(m) = \begin{cases} 1 & \text{if } m = i - 1 \text{ or } m = 2M - (i+1), \\ 0 & \text{otherwise.} \end{cases}$$

Since the vectors $\{\mathbf{a}_1, \mathbf{a}_2, \ldots, \mathbf{a}_M\}$ are linearly independent, they span an $M$-dimensional subspace $\mathcal{M}$ in $\ell^2_{2M-2}(\mathbb{Z})$.

Now, we consider the shift operator $U_{M-1}$ given by

$$U_{M-1} : \mathbf{x} = \{x(m)\} \longmapsto U_{M-1}\mathbf{x} := \{x(m - M + 1)\}.$$

Under the above circumstances the following result holds:

**Lemma 6.1** *For each $i = 1, 2, \ldots, M$ we have:*

*(i)* $U_{M-1}\,\mathbf{a}_i = \mathbf{a}_{M+1-i}$.
*(ii)* $U^2_{M-1}\,\mathbf{a}_i = \mathbf{a}_i$.
*(iii) For $M$ odd,* $U_{M-1}\big(\mathbf{a}_{\frac{M+1}{2}}\big) = \mathbf{a}_{\frac{M+1}{2}}$.

***Proof*** It suffices to prove statement (i). To this end, note that the vector $U_{M-1}\,\mathbf{a}_i$ has the 1's in positions $i + M - 2$ and $M - i$ which corresponds to the vector $\mathbf{a}_{M+1-i}$. □

As a consequence, we conclude that:

- Whenever $M$ is even, taking $\mathbf{a} := \{\mathbf{a}_1, \mathbf{a}_2, \ldots, \mathbf{a}_{M/2}\}$ we obtain that $\mathcal{M} = \mathcal{V}_\mathbf{a}$, where all the generators have order 2 with respect to $U_{M-1}$.
- Whenever $M$ is odd, taking $\mathbf{a} := \{\mathbf{a}_1, \mathbf{a}_2, \ldots, \mathbf{a}_{\frac{M+1}{2}}\}$ we obtain that $\mathcal{M} = \mathcal{V}_\mathbf{a}$, where the generators have order 2 except for $\mathbf{a}_{\frac{M+1}{2}}$ which has order 1 with respect to $U_{M-1}$.

Concerning sampling we distinguish two cases:

*(a) M Even* In this case we have $L = M/2$ generators of order 2 and $N = 2$. For $r = 1$ and $\ell = 2$, the $2s \times M$ matrix $\mathbf{R}_{\mathbf{a},\mathbf{b}}$ in (6.11) has the blocks:

$$\mathbf{R}_{a_i,b_j} = \begin{pmatrix} R_{a_i,b_j}(0) & R_{a_i,b_j}(1) \\ R_{a_i,b_j}(1) & R_{a_i,b_j}(0) \end{pmatrix}, \qquad 1 \le i \le M/2 \text{ and } 1 \le j \le s.$$

If rank $\mathbf{R_{a,b}} = M$, which implies $s \geq M/2$ systems, the corresponding sampling formula (6.17) for each $\mathbf{x} \in \mathcal{V}_\mathbf{a}$ reads

$$\begin{aligned}\mathbf{x}(m) &= \sum_{j=1}^{s} \big[\mathcal{L}_j\mathbf{x}(0)\,\mathbf{c}_j(m) + \mathcal{L}_j\mathbf{x}(1)\,U_{M-1}\mathbf{c}_j(m)\big] \\ &= \sum_{j=1}^{s} \big[\mathcal{L}_j\mathbf{x}(0)\,\mathbf{c}_j(m) + \mathcal{L}_j\mathbf{x}(1)\,\mathbf{c}_j(m-M+1)\big],\end{aligned} \tag{6.20}$$

where $0 \leq m \leq 2M-3$.

For $r = 2$ and $\ell = 1$, the $s \times M$ matrix $\mathbf{R_{a,b}}$ in (6.11) has the blocks:

$$\mathbf{R}_{a_i,b_j} = \Big(R_{a_i,b_j}(0)\ R_{a_i,b_j}(1)\Big)\,, \quad 1 \leq i \leq M/2 \text{ and } 1 \leq j \leq s\,.$$

If rank $\mathbf{R_{a,b}} = M$, which implies $s \geq M$ systems, the corresponding sampling formula (6.17) for $\mathbf{x} \in \mathcal{V}_\mathbf{a}$ reads

$$\mathbf{x}(m) = \sum_{j=1}^{s} \mathcal{L}_j\mathbf{x}(0)\,\mathbf{c}_j(m)\,, \quad 0 \leq m \leq 2M-3\,. \tag{6.21}$$

*(b) M Odd* In this case we have $L = \frac{M+1}{2}$ generators, $\frac{M-1}{2}$ of order 2 and one generator of order 1; again $N = 2$.

For $r = 1$ and $\ell = 2$, the $2s \times M$ matrix $\mathbf{R_{a,b}}$ in (6.11) has the blocks:

$$\mathbf{R}_{a_i,b_j} = \begin{pmatrix} R_{a_i,b_j}(0) & R_{a_i,b_j}(1) \\ R_{a_i,b_j}(1) & R_{a_i,b_j}(0) \end{pmatrix}, \quad 1 \leq i \leq (M-1)/2 \text{ and } 1 \leq j \leq s\,,$$

and

$$\mathbf{R}_{a_{\frac{M+1}{2}},b_j} = \begin{pmatrix} R_{a_{\frac{M+1}{2}},b_j}(0) \\ R_{a_{\frac{M+1}{2}},b_j}(0) \end{pmatrix}, \quad 1 \leq j \leq s\,.$$

If rank $\mathbf{R_{a,b}} = M$, which implies $s \geq M/2$ systems, then we obtain in $\mathcal{V}_\mathbf{a}$ a sampling formula as (6.20).

For $r = 2$ and $\ell = 1$, the $s \times M$ matrix $\mathbf{R_{a,b}}$ in (6.11) has the blocks:

$$\mathbf{R}_{a_i,b_j} = \Big(R_{a_i,b_j}(0)\ R_{a_i,b_j}(1)\Big), \quad 1 \leq i \leq (M-1)/2 \text{ and } 1 \leq j \leq s\,,$$

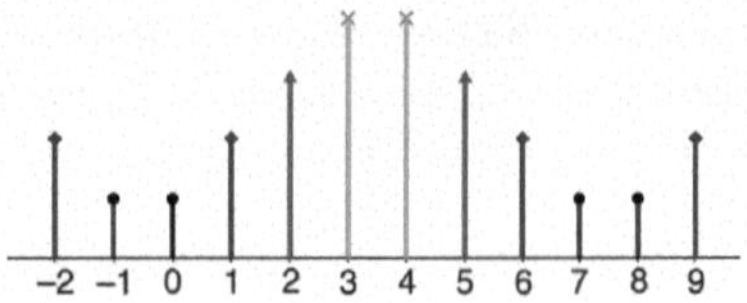


**Fig. 6.5** Half-point symmetric extension $M = 4$

and

$$\mathbf{R}_{a\frac{M+1}{2},b_j} = \left(R_{a\frac{M+1}{2},b_j}(0)\right), \quad 1 \le j \le s\,.$$

If rank $\mathbf{R}_{\mathbf{a},\mathbf{b}} = M$, which implies $s \ge M$ systems, we obtain in $\mathcal{V}_{\mathbf{a}}$ a sampling formula as (6.21).

**Example 6.4 (The Half-Point Symmetry Extension)** In this case we extend a signal of length $M \in \mathbb{N}$, repeating the extreme values, to form a $2M$-periodic signal (see Fig. 6.5).

Here we consider the space $\ell^2_{2M}(\mathbb{Z})$ with the inner product

$$\langle \mathbf{x}, \mathbf{y} \rangle_{\ell^2_{2M}} = \sum_{m=0}^{2M-1} x(m)\,\overline{y(m)}\,.$$

Now consider the subspace in $\ell^2_{2M}(\mathbb{Z})$ spanned by the following $M$ signals:

$$\mathbf{a}_1 := (1, 0, 0, \ldots, 0, 1)\,;\ \mathbf{a}_2 := (0, 1, 0, \ldots, 1, 0)\,; \ldots;$$
$$\mathbf{a}_M := (0, 0, \ldots, 1, 1, \ldots, 0, 0)\,,$$

that is, for $i = 1, 2, \ldots, M$, and $m = 0, 1, \ldots, 2M - 3$, we have

$$\mathbf{a}_i(m) = \begin{cases} 1 & \text{if } m = i - 1 \text{ or } m = 2M - i, \\ 0 & \text{otherwise.} \end{cases}$$

Since the vectors $\{\mathbf{a}_1, \mathbf{a}_2, \ldots, \mathbf{a}_M\}$ are linearly independent, they span an $M$-dimensional subspace $\mathcal{M}$ in $\ell^2_{2M}(\mathbb{Z})$. Now, we consider the shift operator $U_M$ given by

$$U_M : \mathbf{x} = \{x(m)\} \longmapsto U_M\,\mathbf{x} := \{x(m - M)\}\,.$$

Under the above circumstances the following result holds:

**Lemma 6.2** *For each* $i = 1, 2, \ldots, M$ *we have:*

*(i)* $U_M\, \mathbf{a}_i = \mathbf{a}_{M+1-i}$ .
*(ii)* $U_M^2\, \mathbf{a}_i = \mathbf{a}_i$ .
*(iii)* *For $M$ odd,* $U_M\big(\mathbf{a}_{\frac{M+1}{2}}\big) = \mathbf{a}_{\frac{M+1}{2}}$ .

***Proof*** It suffices to prove statement (i). To this end, note that the vector $U_M\, \mathbf{a}_i$ has the 1's in positions $i + M - 1$ and $M - i$ which corresponds to the vector $\mathbf{a}_{M+1-i}$. □

As a consequence, we conclude that:

- Whenever $M$ is even, taking $\mathbf{a} := \{\mathbf{a}_1, \mathbf{a}_2, \ldots, \mathbf{a}_{M/2}\}$ we obtain that $\mathcal{M} = \mathcal{V}_{\mathbf{a}}$, where all the generators have order 2 with respect to $U_M$.
- Whenever $M$ is odd, taking $\mathbf{a} := \{\mathbf{a}_1, \mathbf{a}_2, \ldots, \mathbf{a}_{\frac{M+1}{2}}\}$ we obtain that $\mathcal{M} = \mathcal{V}_{\mathbf{a}}$, where the generators have order 2 except for $\mathbf{a}_{\frac{M+1}{2}}$ which has order 1 with respect to $U_M$.

For the present case, analogous sampling results to those in the whole-point symmetry case can be derived in a similar way; we omit the details.

There is another extension of a finite signal which is not considered here: the *zero-padding extension*, i.e., extending the signal by zeros to get a signal in $\ell^2(\mathbb{Z})$. Thus, we are interested in sampling formulas in $\ell^2(\mathbb{Z})$ involving only signals of finite support, i.e., with a finite number of nonzero terms.

The corresponding sampling formulas can be easily deduced by using the technique in Example 5.12. When the corresponding matrix $A$ had entries with finite support, the entries of the transfer matrix $\widehat{A}$ are *Laurent polynomials*. Under the same hypotheses involved there, the matrix $\widehat{A}$ has a left-inverse whose entries are also Laurent polynomials (see Ref. [3, Theorem 3.4]). Thus we obtain the suitable sampling sequences, i.e., with only a finite number of nonzero terms.

**Remark 6.1 (A Comment on the Sampling Period** $r$**)** Once we have handled some examples, a comment on the sampling period $r$ is pertinent. Along this section we have assumed that the sampling period $r$ divides $N$ where $N = \text{l.c.m.}\,(N_1, N_2, \ldots N_L)$.

Actually, the case where the sampling period does not divide $N$ can be reduced to the case considered in this section: we have taken $N$ to be the least common multiple of $N_1, N_2, \ldots N_L$ but everything works with any common multiple of $N_1, N_2, \ldots N_L$. Thus, denoting as $r'$ any sampling period, we could take

$$N' = \text{l.c.m.}\,(r', N_1, N_2, \ldots N_L)\,.$$

The used techniques in Sect. 6.1 could be applied for $N'$ and $r'$ instead of $N$ and $r$. Moreover, this case reduces to the situation where $r$ is the greatest common divisor of $N$ and $r'$ since $\ell = N/r = N'/r'$ and for each $x \in \mathcal{V}_{\mathbf{a}}$ and each $j = 1, 2, \ldots, s$, the set of samples $\{\mathcal{L}_j x(r'n') : n' = 0, 1, \ldots, \ell - 1\}$ with sampling period $r'$ is nothing but the set of samples $\{\mathcal{L}_j x(rn) : n = 0, 1, \ldots, \ell - 1\}$ with sampling period $r$. That is, the corresponding $s\ell \times (N_1 + N_2 + \cdots + N_L)$ matrix $\mathbf{R}_{\mathbf{a},\mathbf{b}}$ in (6.11) for $r'$ and $N'$ is the same (up to a permutation of its rows) than those obtained for $r$ and $N$.

## 6.2 A Generalization Involving the Knit Product of Groups

In this section we obtain an abstract sampling theory associated with a unitary representation of a non-abelian finite group $\mathbf{G}$, which is the *knit product* $\mathbf{G} = \mathbf{N} \bowtie \mathbf{H}$ of two finite subgroups, $\mathbf{N}$ and $\mathbf{H}$, on a Hilbert space $\mathcal{H}$.

Specifically, given $\mathbf{G} \ni g \longmapsto U(g) \in \mathcal{U}(\mathcal{H})$ a unitary representation of the group $\mathbf{G}$ on a Hilbert space $\mathcal{H}$, for a fixed vector $\mathsf{a} \in \mathcal{H}$ we consider the subspace $\mathcal{V}_\mathsf{a} := \operatorname{span}\{U(g)\mathsf{a}\}_{g \in \mathbf{G}}$ of $\mathcal{H}$.

The aim is to obtain sampling formulas in $\mathcal{V}_\mathsf{a}$ having a compatible $U$-structure and involving samples indexed by either $\mathbf{N}$ or $\mathbf{H}$. Namely, fixing $\kappa$ vectors $\mathsf{b}_k$ in $\mathcal{H}$, which do not necessarily belong to $\mathcal{V}_\mathsf{a}$, for each $\mathsf{f} \in \mathcal{V}_\mathsf{a}$ we consider the data samples

$$\begin{gathered} \mathcal{L}_k \mathsf{f}(\nu) := \langle \mathsf{f}, U(\nu)\mathsf{b}_k \rangle_{\mathcal{H}}, \quad \nu \in \mathbf{N}, \\ \text{or} \\ \mathcal{L}_k \mathsf{f}(\tau) := \langle \mathsf{f}, U(\tau)\mathsf{b}_k \rangle_{\mathcal{H}}, \quad \tau \in \mathbf{H}, \end{gathered} \tag{6.22}$$

for $k = 1, 2, \ldots, \kappa$.

The aim is to get sampling formulas in $\mathcal{V}_\mathsf{a}$ having, for each $\mathsf{f} \in \mathcal{V}_\mathsf{a}$, the form

$$\begin{gathered} \mathsf{f} = \sum_{k=1}^{\kappa} \sum_{\nu \in \mathbf{N}} \mathcal{L}_k \mathsf{f}(\nu)\, U(\nu)\mathsf{c}_k, \\ \text{or} \\ \mathsf{f} = \sum_{k=1}^{\kappa} \sum_{\tau \in \mathbf{H}} \mathcal{L}_k \mathsf{f}(\tau)\, U(\tau)\mathsf{d}_k, \end{gathered} \tag{6.23}$$

for some vectors $\mathsf{c}_k$ or $\mathsf{d}_k$, $k = 1, 2, \ldots, \kappa$, in $\mathcal{V}_\mathsf{a}$, depending on whether $\mathbf{N}$ or $\mathbf{H}$ is abelian. This section is broadly in line with Ref. [6].

### 6.2.1 *A Brief on the Knit Product* $\mathbf{G} = \mathbf{N} \bowtie \mathbf{H}$ *of Groups*

Let $\mathbf{G}$ be a group with identity $1_{\mathbf{G}}$, and let $\mathbf{N}$ and $\mathbf{H}$ be two subgroups of $\mathbf{G}$ such that $\mathbf{G} = \mathbf{NH}$ and $\mathbf{N} \cap \mathbf{H} = \{1_{\mathbf{G}}\}$, or equivalently, for each $g \in \mathbf{G}$ there exists a unique $h \in \mathbf{H}$ and a unique $n \in \mathbf{N}$ such that $g = nh$. In this case, $\mathbf{G}$ is said to be the *internal knit product* (or *Zappa-Szép product*) of $\mathbf{N}$ and $\mathbf{H}$, and it is denoted by $\mathbf{G} = \mathbf{N} \bowtie \mathbf{H}$.

For each $h \in \mathbf{H}$ and $n \in \mathbf{N}$ there exist $\alpha(n, h) \in \mathbf{N}$ and $\beta(n, h) \in \mathbf{H}$ such that $hn = \alpha(n, h)\beta(n, h)$. This defines mappings

$$\alpha^* : \mathbf{H} \to Aut(\mathbf{N}) \quad \text{and} \quad \beta^* : \mathbf{N} \to Aut(\mathbf{H}),$$

where $\alpha^*(h)(n) := \alpha(n, h)$ and $\beta^*(n)(h) := \beta(n, h)$ for $n \in \mathbf{N}$ and $h \in \mathbf{H}$, and satisfying:

1. $\alpha^*$ is a group isomorphism and $\beta^*$ is a group anti-isomorphism (i.e., $\beta_{n_1n_2} = \beta_{n_2}\beta_{n_1}$).
2. $\alpha_h(n_1n_2) = \alpha_h(n_1)\alpha_{\beta_{n_1}(h)}(n_2)$ for each $h \in \mathbf{H}$ and $n_1, n_2 \in \mathbf{N}$.
3. $\beta_n(h_1h_2) = \beta_{\alpha_{h_2}(n)}(h_1)\beta_n(h_2)$ for each $n \in \mathbf{N}$ and $h_1, h_2 \in \mathbf{H}$,

where we have used the notation $\alpha_h(n) := \alpha^*(h)(n)$ and $\beta_n(h) := \beta^*(n)(h)$ (see Refs. [1, 13, 14]).

From now on, we denote the order of the involved groups as $\mathfrak{g} := |\mathbf{G}|$, $\mathfrak{n} := |\mathbf{N}|$ and $\mathfrak{h} := |\mathbf{H}|$ respectively.

In case $\mathbf{N}$ is a normal subgroup of $\mathbf{G}$ then $\alpha_h(n) := hnh^{-1}$ and $\beta_n := \mathrm{id}_H$, the corresponding knit product $\mathbf{N} \bowtie \mathbf{H}$ coincides with the *internal semidirect product* of $\mathbf{N}$ and $\mathbf{H}$, denoted by $\mathbf{G} = \mathbf{N} \rtimes \mathbf{H}$.

Whenever $\mathbf{G} = \mathbf{N} \bowtie \mathbf{H}$ we can choose exactly one element of $\mathbf{N}$ in each left coset of the quotient set $\mathbf{G}/\mathbf{H}$. Thus, in case $\mathbf{N} = \{\nu_1 = 1_{\mathbf{G}}, \nu_2, \ldots, \nu_{\mathfrak{n}}\}$ we can describe the quotient set $\mathbf{G}/\mathbf{H}$ as

$$\mathbf{G}/\mathbf{H} = \big\{[1_{\mathbf{G}} = \nu_1], [\nu_2], \cdots, [\nu_{\mathfrak{n}}]\big\}. \tag{6.24}$$

There is an external version of the knit product for groups. In this case, we have two groups $\mathbf{N}$ and $\mathbf{H}$ which are not known to be subgroups of a given group and mappings $\alpha : \mathbf{H} \times \mathbf{N} \to \mathbf{N}$ and $\beta : \mathbf{H} \times \mathbf{N} \to \mathbf{H}$ satisfying the above properties $1 - 3$. On the product $\mathbf{N} \times \mathbf{H}$ we define a product law

$$(n_1, h_1)(n_2, h_2) = \big(n_1\alpha_{h_1}(n_2), \beta_{n_2}(h_1)h_2\big) \quad \text{for any } n_1,\ n_2 \in \mathbf{N} \text{ and } h_1,\ h_2 \in \mathbf{H}.$$

Thus, $(1_{\mathbf{N}}, 1_{\mathbf{H}})$ is the identity, and for any $(n, h) \in \mathbf{N} \times \mathbf{H}$

$$(n, h)^{-1} = \big(\alpha_{h^{-1}}(n^{-1}),\ \beta_{n^{-1}}(h^{-1})\big).$$

With this product the set $\mathbf{N} \times \mathbf{H}$ is a group called *external knit product* of the groups $\mathbf{N}$ and $\mathbf{H}$, and it is denoted again as $\mathbf{N} \bowtie \mathbf{H}$. The subsets $\mathbf{N} \times \{1_\mathbf{H}\}$ and $\{1_\mathbf{N}\} \times \mathbf{H}$ are subgroups of $\mathbf{N} \bowtie \mathbf{H}$ isomorphic to $\mathbf{N}$ and $\mathbf{H}$, respectively.

Clearly, the external knit product of the groups $\mathbf{N}$ and $\mathbf{H}$ coincides with the internal knit product of subgroups $\mathbf{N} \times \{1_\mathbf{H}\}$ and $\{1_\mathbf{N}\} \times \mathbf{H}$.

### 6.2.2 Invariant Subspaces Associated with a Unitary Representation

Let $\mathbf{G}$ be a finite, not necessarily abelian, group of order $\mathfrak{g}$ with identity element $1_\mathbf{G}$. Let $\mathbf{G} \ni g \longmapsto U(g) \in \mathcal{U}(\mathcal{H})$ be a *unitary representation* of $\mathbf{G}$ on a Hilbert space $\mathcal{H}$, i.e., a homomorphism from the group $\mathbf{G}$ into the group $\mathcal{U}(\mathcal{H})$ of unitary operators on $\mathcal{H}$, i.e., a map satisfying $U(gg') = U(g)U(g')$ for all $g, g' \in G$; in particular, $U(1_\mathbf{G}) = I_\mathcal{H}$.

From now on, for a fixed vector $\mathsf{a} \in \mathcal{H}$ we consider the subspace $\mathcal{V}_\mathsf{a}$ of $\mathcal{H}$ spanned by the vectors $U(g)\mathsf{a}$, $g \in \mathbf{G}$, i.e.,

$$\mathcal{V}_\mathsf{a} := \operatorname{span}\{U(g)\mathsf{a}\}_{g\in\mathbf{G}} \,.$$

In case this set is linearly independent in $\mathcal{H}$, each $\mathsf{f} \in \mathcal{V}_\mathsf{a}$ can be expressed uniquely as the expansion $\mathsf{f} = \sum_{g\in\mathbf{G}} \alpha_g \, U(g)\mathsf{a}$, with scalars $\alpha_g \in \mathbb{C}$.

There is a close relationship between the finite sequence $\big\{U(g)\mathsf{a}\big\}_{g\in\mathbf{G}}$ in the Hilbert space $\mathcal{H}$ and the so-called *stationary sequences* (see Ref. [8]).

We say that the finite sequence $\{\mathsf{a}_g\}_{g\in\mathbf{G}}$ in $\mathcal{H}$ is (left) $\mathbf{G}$-*stationary* if

$$\langle \mathsf{a}_g, \mathsf{a}_{g'}\rangle_\mathcal{H} = \langle \mathsf{a}_{hg}, \mathsf{a}_{hg'}\rangle_\mathcal{H}\,, \quad \text{for all } g, g', h \in \mathbf{G}\,.$$

In other words, the inner product $\langle \mathsf{a}_g, \mathsf{a}_h\rangle_\mathcal{H}$ only depends on $h^{-1}g$. Then, it is easy to deduce that there exists a unitary representation $g \mapsto U(g)$ of the group $\mathbf{G}$ on $\mathcal{H}$ and $\mathsf{a} \in \mathcal{H}$ such that $\mathsf{a}_g = U(g)\mathsf{a}$, $g \in \mathbf{G}$.

We define the *auto-covariance* of the finite sequence $\big\{U(g)\mathsf{a}\big\}_{g\in\mathbf{G}}$ as the positive semidefinite function:

$$r_\mathsf{a}(g) := \big\langle U(g)\mathsf{a}, \mathsf{a}\big\rangle_\mathcal{H}\,, \quad g \in \mathbf{G}\,.$$

Similarly, we define the *cross-covariance* between the finite sequences $\big\{U(g)\mathsf{a}\big\}_{g\in\mathbf{G}}$ and $\big\{U(g)\mathsf{b}\big\}_{g\in\mathbf{G}}$, where $\mathsf{a}, \mathsf{b} \in \mathcal{H}$, as

$$r_{\mathsf{a},\mathsf{b}}(g) := \big\langle U(g)\mathsf{a}, \mathsf{b}\big\rangle_\mathcal{H}\,, \quad g \in \mathbf{G}\,.$$

Note that $r_{\mathsf{a},\mathsf{b}}(g) = \overline{r_{\mathsf{b},\mathsf{a}}(g^{-1})}$ for $\mathsf{a}, \mathsf{b} \in \mathcal{H}$ and $g \in \mathbf{G}$.

**Proposition 6.5** *Let* $\mathbb{R}_{\mathsf{a}}$ *denote the square matrix of order* $\mathfrak{g}$ *defined from the auto-covariance* $r_{\mathsf{a}}$ *as* $\mathbb{R}_{\mathsf{a}} := [r_{\mathsf{a}}(t^{-1}g)]_{t,g\in\mathbf{G}}$. *Then, the set of vectors* $\{U(g)\mathsf{a}\}_{g\in\mathbf{G}}$ *is linearly independent in* $\mathcal{H}$ *if and only if* $\det \mathbb{R}_{\mathsf{a}} \neq 0$.

***Proof*** If $\det \mathbb{R}_{\mathsf{a}} = 0$, then there exists a vector $\boldsymbol{\lambda} = (\lambda_g)_{g\in\mathbf{G}} \in \mathbb{C}^{\mathfrak{g}}$ such that $\boldsymbol{\lambda} \neq \mathbf{0}$ and $\mathbb{R}_{\mathsf{a}}\boldsymbol{\lambda} = 0$. Thus $\sum_{g\in\mathbf{G}} \lambda_g\, U(g)\mathsf{a}$ is orthogonal to $U(g)\mathsf{a}$ for all $g \in \mathbf{G}$ so that $\sum_g \lambda_g\, U(g)\mathsf{a} = 0$.

Conversely, if $\sum_{g\in G} \lambda_g\, U(g)\mathsf{a} = 0$ for some $\boldsymbol{\lambda} \neq \mathbf{0}$, then the inner product in the above expression with each $U(t)\mathsf{a}$, $t \in \mathbf{G}$, yields $\mathbb{R}_{\mathsf{a}}\boldsymbol{\lambda} = 0$. □

**The Isomorphism $\mathcal{T}_{\mathsf{a}}^{\mathbf{G}}$ connecting $\ell^2(\mathbf{G})$ with $\mathcal{V}_{\mathsf{a}}$**

The space of functions $\boldsymbol{\alpha}\colon \mathbf{G} \to \mathbb{C}$, $g \mapsto \boldsymbol{\alpha}(g)$, in short $\boldsymbol{\alpha} = \big(\boldsymbol{\alpha}(g)\big)_{g\in\mathbf{G}}$, can be identified with $\ell^2(\mathbf{G})$ endowed with its natural inner product $\langle \boldsymbol{\alpha}, \boldsymbol{\beta}\rangle = \sum_{g\in G} \alpha(g)\,\overline{\beta(g)}$, which becomes a Hilbert space isomorphic to $\mathbb{C}^{\mathfrak{g}}$.

The Hilbert space $\ell^2(\mathbf{G})$ supports a natural unitary representation of $\mathbf{G}$ called the *left regular representation* $L_s$, $s \in \mathbf{G}$, defined by

$$L_s\boldsymbol{\alpha}(g) = \boldsymbol{\alpha}(s^{-1}g) \quad \text{for } s, g \in \mathbf{G}\,.$$

Next we define the natural surjective linear map $\mathcal{T}_{\mathsf{a}}^{\mathbf{G}}$ between $\ell^2(\mathbf{G})$ and $\mathcal{V}_{\mathsf{a}}$:

$$\begin{aligned} \mathcal{T}_{\mathsf{a}}^{\mathbf{G}} : \ell^2(\mathbf{G}) &\longrightarrow \mathcal{V}_{\mathsf{a}} \\ \boldsymbol{\alpha} &\longmapsto \mathsf{f} = \sum_{g\in\mathbf{G}} \boldsymbol{\alpha}(g)\, U(g)\mathsf{a}\,. \end{aligned} \tag{6.25}$$

If the matrix $\mathbb{R}_{\mathsf{a}}$ is invertible, due to Proposition 6.5, the above map $\mathcal{T}_{\mathsf{a}}^{\mathbf{G}}$ becomes an isomorphism and it has the following *shifting property* with respect to the left representation $L_s$:

**Proposition 6.6** *For any* $s \in \mathbf{G}$ *and* $\boldsymbol{\alpha} \in \ell^2(\mathbf{G})$ *we have that*

$$\mathcal{T}_{\mathsf{a}}^{\mathbf{G}}\big(L_s\boldsymbol{\alpha}\big) = U(s)\,\mathcal{T}_{\mathsf{a}}^{\mathbf{G}}(\boldsymbol{\alpha})\,. \tag{6.26}$$

***Proof*** Indeed, denoting $g' = s^{-1}g$ we have

$$\begin{aligned} \mathcal{T}_{\mathsf{a}}^{\mathbf{G}}\big(L_s\boldsymbol{\alpha}\big) &= \sum_{g\in\mathbf{G}} \boldsymbol{\alpha}(s^{-1}g)U(g)\mathsf{a} = \sum_{g'\in\mathbf{G}} \boldsymbol{\alpha}(g')U(sg')\mathsf{a} \\ &= \sum_{g'\in\mathbf{G}} \boldsymbol{\alpha}(g')U(s)U(g')\mathsf{a} = U(s)\,\mathcal{T}_{\mathsf{a}}^{\mathbf{G}}(\boldsymbol{\alpha})\,. \end{aligned}$$

□

### 6.2.3 Sampling Indexed by N when N is an Abelian Subgroup

Having in mind the description of $\mathbf{G}/\mathbf{H}$ in (6.24), we write the group $\mathbf{G}$ as

$$\mathbf{G} = \bigcup_{n=1}^{\mathfrak{n}} \nu_n^{-1}\mathbf{H}\,.$$

In the sequel we fix the way of writing the elements of the group $\mathbf{H}$; this will be important for maintaining the structure of the matrices $\mathbb{R}_{\mathsf{a},\mathsf{b}_k}$ introduced below. We also need the abelian character of subgroup $\mathbf{N}$ to get a block symmetry for the matrices $\mathbb{R}_{\mathsf{a},\mathsf{b}_k}$.

Given $\kappa$ elements $\mathsf{b}_k \in \mathcal{H}$, $k = 1, 2, \ldots, \kappa$, not necessarily in $\mathcal{V}_\mathsf{a}$, for each $\mathsf{f} \in \mathcal{V}_\mathsf{a}$ we define its *generalized samples*, indexed by the elements in $\mathbf{N}$, as

$$\mathcal{L}_k\mathsf{f}(\nu_n) := \big\langle \mathsf{f}, U(\nu_n)\mathsf{b}_k \big\rangle_{\mathcal{H}}, \quad n = 1, 2, \ldots, \mathfrak{n}, \text{ and } k = 1, 2, \ldots, \kappa\,. \tag{6.27}$$

Since, in general, to recover any $\mathsf{f} \in \mathcal{V}_\mathsf{a}$ we need at least $\mathfrak{g}$ samples, if the samples are indexed by elements in $\mathbf{N}$, we will need at least $\mathfrak{n}\kappa \geq \mathfrak{g} = \mathfrak{n}\mathfrak{h}$ samples, i.e., $\kappa \geq \mathfrak{h}$.

The main aim of this section is to recover any $\mathsf{f} \in \mathcal{V}_\mathsf{a}$ through its generalized samples in (6.27) by means of a sampling formula which takes care of the unitary structure of $\mathcal{V}_\mathsf{a}$.

To this end, we first obtain an alternative expression for $\mathcal{L}_k\mathsf{f}(\nu_n)$ with $n = 1, 2, \ldots, \mathfrak{n}$ and $k = 1, 2, \ldots, \kappa$.

Namely, introducing the expression of $\mathsf{f} \in \mathcal{V}_\mathsf{a}$ in (6.27) we get

$$\begin{aligned}
\mathcal{L}_k\mathsf{f}(\nu_n) &= \Big\langle \sum_{g\in\mathbf{G}} \alpha_g\, U(g)\mathsf{a}, U(\nu_n)\mathsf{b}_k \Big\rangle_{\mathcal{H}} \\
&= \sum_{g\in\mathbf{G}} \alpha_g \big\langle U(g)\mathsf{a}, U(\nu_n)\mathsf{b}_k \big\rangle_{\mathcal{H}} \\
&= \big\langle (\alpha_g)_{g\in\mathbf{G}}, (\overline{\langle U(g)\mathsf{a}, U(\nu_n)\mathsf{b}_k\rangle})_{g\in\mathbf{G}} \big\rangle_{\ell^2(\mathbf{G})} \\
&= \big\langle \boldsymbol{\alpha}, \mathbf{g}_{k,\nu_n} \big\rangle_{\ell^2(\mathbf{G})}\,,
\end{aligned} \tag{6.28}$$

where $\boldsymbol{\alpha} = (\alpha_g)_{g\in\mathbf{G}}$ and $\mathbf{g}_{k,\nu_n} = (\overline{\langle U(g)\mathsf{a}, U(\nu_n)\mathsf{b}_k\rangle})_{g\in\mathbf{G}}$ belong to $\ell^2(\mathbf{G})$.

The vectors $\mathbf{g}_{k,\nu_n} \in \ell^2(\mathbf{G})$, $k = 1, 2, \ldots, \kappa$ and $n = 1, 2, \ldots, \mathfrak{n}$, can be expressed in terms of the cross-covariances $r_{\mathsf{a},\mathsf{b}_k}$ as

$$\begin{aligned}\mathbf{g}_{k,\nu_n} &= \big(\overline{\langle U(g)\mathsf{a}, U(\nu_n)\mathsf{b}_k\rangle}\big)_{g\in\mathbf{G}} \\ &= \big(\overline{\langle U(\nu_n^{-1}g)\mathsf{a}, \mathsf{b}_k\rangle}\big)_{g\in\mathbf{G}} \\ &= \big(\overline{r_{\mathsf{a},\mathsf{b}_k}}(\nu_n^{-1}g)\big)_{g\in\mathbf{G}}\,.\end{aligned}$$

Having in mind the expression (6.28) for the samples and the isomorphism $\mathcal{T}_{\mathsf{a}}^{\mathbf{G}}$ defined in (6.25) we deduce the following result (see finite frames in Sect. 1.2):

**Proposition 6.7** *Any $\mathsf{f} \in \mathcal{V}_{\mathsf{a}}$ can be recovered from its samples $\big\{\mathcal{L}_k\mathsf{f}(\nu_n)\big\}_{\substack{k=1,2,\ldots,\kappa\\ n=1,2,\ldots,\mathfrak{n}}}$ if and only if the set of vectors $\big\{\mathbf{g}_{k,\nu_n}\big\}_{\substack{k=1,2,\ldots,\kappa\\ n=1,2,\ldots,\mathfrak{n}}}$ in $\ell^2(\mathbf{G})$ is a spanning set (a frame) for $\ell^2(\mathbf{G})$.*

Equivalently, the $\mathfrak{g} \times \mathfrak{n}\kappa$ matrix having columns $\mathbf{g}_{k,\nu_n}$, $k = 1, 2, \ldots, \kappa$ and $n = 1, 2, \ldots, \mathfrak{n}$, i.e.,

$$\Big(\mathbf{g}_{1,\nu_1}\cdots\mathbf{g}_{1,\nu_\mathfrak{n}}\ \mathbf{g}_{2,\nu_1}\cdots\mathbf{g}_{2,\nu_\mathfrak{n}}\cdots\mathbf{g}_{\kappa,\nu_1}\cdots\mathbf{g}_{\kappa,\nu_\mathfrak{n}}\Big) \tag{6.29}$$

has rank $\mathfrak{g}$. Hence, we deduce that $\mathfrak{g} \le \mathfrak{n}\kappa$, that is, $\kappa \ge \mathfrak{h}$.

The vector $\mathbf{g}_{k,\nu_n}$ can be written as

$$\mathbf{g}_{k,\nu_n} = \Big(\overline{r_{\mathsf{a},\mathsf{b}_k}}(\nu_n^{-1}\nu_1^{-1}\mathbf{H}), \overline{r_{\mathsf{a},\mathsf{b}_k}}(\nu_n^{-1}\nu_2^{-1}\mathbf{H}), \ldots, \overline{r_{\mathsf{a},\mathsf{b}_k}}(\nu_n^{-1}\nu_\mathfrak{n}^{-1}\mathbf{H})\Big)^\top,$$

where $\nu_n \in \mathbf{N}$ and $\overline{r_{\mathsf{a},\mathsf{b}_k}}(\nu_n^{-1}\nu_r^{-1}\mathbf{H})$ is the row vector

$$\overline{r_{\mathsf{a},\mathsf{b}_k}}(\nu_n^{-1}\nu_r^{-1}\mathbf{H}) = \Big(\overline{r_{\mathsf{a},\mathsf{b}_k}}(\nu_n^{-1}\nu_r^{-1}\tau_1), \overline{r_{\mathsf{a},\mathsf{b}_k}}(\nu_n^{-1}\nu_r^{-1}\tau_2), \ldots, \overline{r_{\mathsf{a},\mathsf{b}_k}}(\nu_n^{-1}\nu_r^{-1}\tau_\mathfrak{h})\Big),$$

being $\mathbf{H} = \{\tau_1 = 1_{\mathbf{G}}, \tau_2, \ldots, \tau_\mathfrak{h}\}$.

For each $k = 1, 2, \ldots, \kappa$, let $\mathbb{R}_{\mathsf{a},\mathsf{b}_k}$ be the $\mathfrak{n} \times \mathfrak{g}$ matrix

$$\mathbb{R}_{\mathsf{a},\mathsf{b}_k} = \begin{pmatrix} r_{\mathsf{a},\mathsf{b}_k}(\nu_1^{-1}\nu_1^{-1}\mathbf{H}) & r_{\mathsf{a},\mathsf{b}_k}(\nu_1^{-1}\nu_2^{-1}\mathbf{H}) & \ldots & r_{\mathsf{a},\mathsf{b}_k}(\nu_1^{-1}\nu_\mathfrak{n}^{-1}\mathbf{H}) \\ r_{\mathsf{a},\mathsf{b}_k}(\nu_2^{-1}\nu_1^{-1}\mathbf{H}) & r_{\mathsf{a},\mathsf{b}_k}(\nu_2^{-1}\nu_2^{-1}\mathbf{H}) & \ldots & r_{\mathsf{a},\mathsf{b}_k}(\nu_2^{-1}\nu_\mathfrak{n}^{-1}\mathbf{H}) \\ \vdots & \vdots & \ldots & \vdots \\ r_{\mathsf{a},\mathsf{b}_k}(\nu_\mathfrak{n}^{-1}\nu_1^{-1}\mathbf{H}) & r_{\mathsf{a},\mathsf{b}_k}(\nu_\mathfrak{n}^{-1}\nu_2^{-1}\mathbf{H}) & \ldots & r_{\mathsf{a},\mathsf{b}_k}(\nu_\mathfrak{n}^{-1}\nu_\mathfrak{n}^{-1}\mathbf{H}) \end{pmatrix}.$$

Since $\mathbf{N}$ is an abelian subgroup of $\mathbf{G}$, the left cosets $v_n^{-1}v_r^{-1}\mathbf{H}$ and $v_r^{-1}v_n^{-1}\mathbf{H}$ coincide. As a consequence, $\mathbb{R}_{\mathsf{a},\mathsf{b}_k}$ is the block symmetric matrix

$$\mathbb{R}_{\mathsf{a},\mathsf{b}_k}=\begin{pmatrix} r_{\mathsf{a},\mathsf{b}_k}(v_1^{-1}v_1^{-1}\mathbf{H}) & r_{\mathsf{a},\mathsf{b}_k}(v_2^{-1}v_1^{-1}\mathbf{H}) & \dots & r_{\mathsf{a},\mathsf{b}_k}(v_{\mathfrak{n}}^{-1}v_1^{-1}\mathbf{H}) \\ r_{\mathsf{a},\mathsf{b}_k}(v_1^{-1}v_2^{-1}\mathbf{H}) & r_{\mathsf{a},\mathsf{b}_k}(v_2^{-1}v_2^{-1}\mathbf{H}) & \dots & r_{\mathsf{a},\mathsf{b}_k}(v_{\mathfrak{n}}^{-1}v_2^{-1}\mathbf{H}) \\ \vdots & \vdots & \dots & \vdots \\ r_{\mathsf{a},\mathsf{b}_k}(v_1^{-1}v_{\mathfrak{n}}^{-1}\mathbf{H}) & r_{\mathsf{a},\mathsf{b}_k}(v_2^{-1}v_{\mathfrak{n}}^{-1}\mathbf{H}) & \dots & r_{\mathsf{a},\mathsf{b}_k}(v_{\mathfrak{n}}^{-1}v_{\mathfrak{n}}^{-1}\mathbf{H}) \end{pmatrix}. \tag{6.30}$$

The matrix given in (6.29) can be written as $\big(\mathbb{R}^*_{\mathsf{a},\mathsf{b}_1}\ \mathbb{R}^*_{\mathsf{a},\mathsf{b}_2}\ \cdots\ \mathbb{R}^*_{\mathsf{a},\mathsf{b}_\kappa}\big)$, where, as usual, the symbol $*$ denotes the transpose conjugate matrix.

Thus, Proposition 6.7 can be restated in terms of the $\mathfrak{n}\kappa\times\mathfrak{g}$ *cross-covariance matrix* $\mathbb{R}_{\mathsf{a},\mathbf{b}}$ defined by

$$\mathbb{R}_{\mathsf{a},\mathbf{b}}:=\begin{pmatrix}\mathbb{R}_{\mathsf{a},\mathsf{b}_1}\\ \mathbb{R}_{\mathsf{a},\mathsf{b}_2}\\ \vdots\\ \mathbb{R}_{\mathsf{a},\mathsf{b}_\kappa}\end{pmatrix}. \tag{6.31}$$

**Corollary 6.6** *Any* $\mathsf{f}\in\mathcal{V}_{\mathsf{a}}$ *can be recovered from its samples* $\big\{\mathcal{L}_k\mathsf{f}(v_n)\big\}_{\substack{k=1,2,\dots,\kappa\\ n=1,2,\dots,\mathfrak{n}}}$ *if and only* rank $\mathbb{R}_{\mathsf{a},\mathbf{b}}=\mathfrak{g}$.

Besides, for any $\mathsf{f}=\sum_{g\in\mathbf{G}}\alpha_g\,U(g)\mathsf{a}$ in $\mathcal{V}_{\mathsf{a}}$ Eq. (6.28) can be expressed as

$$\big(\mathcal{L}_k\mathsf{f}(v_1),\ \mathcal{L}_k\mathsf{f}(v_2),\ \cdots,\ \mathcal{L}_k\mathsf{f}(v_{\mathfrak{n}})\big)^\top=\mathbb{R}_{\mathsf{a},\mathsf{b}_k}\,\boldsymbol{\alpha}\,,$$

where $\boldsymbol{\alpha}=(\alpha_g)_{g\in\mathbf{G}}\in\mathbb{C}^{\mathfrak{g}}$. As a consequence, we deduce the expression:

**Proposition 6.8** *For any* $\mathsf{f}=\sum_{g\in\mathbf{G}}\alpha_g\,U(g)\mathsf{a}$ *in* $\mathcal{V}_{\mathsf{a}}$, *consider its samples vector*

$$\mathcal{L}_{\text{samp}}\mathsf{f}:=\big(\mathcal{L}_1\mathsf{f}(v_1),\dots,\mathcal{L}_1\mathsf{f}(v_p),\cdots,\mathcal{L}_\kappa\mathsf{f}(v_1),\dots\mathcal{L}_\kappa\mathsf{f}(v_{\mathfrak{n}})\big)^\top. \tag{6.32}$$

*Then, the matrix relationship*

$$\mathcal{L}_{\text{samp}}\mathsf{f}=\mathbb{R}_{\mathsf{a},\mathbf{b}}\,\boldsymbol{\alpha} \tag{6.33}$$

*holds, where* $\boldsymbol{\alpha}=(\alpha_g)_{g\in\mathbf{G}}$ *and* $\mathbb{R}_{\mathsf{a},\mathbf{b}}$ *is the* $\mathfrak{n}\kappa\times\mathfrak{g}$ *cross-covariances matrix defined in* (6.11).

Assuming that $\big\{\mathbf{g}_{k,v_n}\big\}_{\substack{k=1,2,\dots,\kappa\\ n=1,2,\dots,\mathfrak{n}}}$ is a frame for $\ell^2(\mathbf{G})$ we have that the rank of the matrix $\mathbb{R}_{\mathsf{a},\mathbf{b}}$ is $\mathfrak{g}$. Let $\mathbb{M}$ be a left-inverse of the matrix $\mathbb{R}_{\mathsf{a},\mathbf{b}}$ whose columns are denoted by $\mathbf{m}_{k,v_n}$, $k=1,2,\dots,\kappa$ and $n=1,2,\dots,\mathfrak{n}$, as in matrix (6.29).

All these matrices are expressed as

$$\mathbb{M} = \mathbb{R}^{\dagger}_{\mathsf{a},\mathbf{b}} + \mathbb{U}[\mathbb{I}_{\mathfrak{n}\kappa} - \mathbb{R}_{\mathsf{a},\mathbf{b}}\mathbb{R}^{\dagger}_{\mathsf{a},\mathbf{b}}], \tag{6.34}$$

where the $\mathfrak{g} \times \mathfrak{n}\kappa$ matrix $\mathbb{R}^{\dagger}_{\mathsf{a},\mathbf{b}} = [\mathbb{R}^{*}_{\mathsf{a},\mathbf{b}}\mathbb{R}_{\mathsf{a},\mathbf{b}}]^{-1}\mathbb{R}^{*}_{\mathsf{a},\mathbf{b}}$ is the Moore-Penrose pseudoinverse of $\mathbb{R}_{\mathsf{a},\mathbf{b}}$, and $\mathbb{U}$ denotes an arbitrary $\mathfrak{g} \times \mathfrak{n}\kappa$ matrix (see (3.16)).

From (6.33) we obtain the frame expansion

$$\begin{aligned}\alpha = \mathbb{M}\,\mathcal{L}_{\text{samp}}\mathsf{f} &= \sum_{k=1}^{\kappa}\sum_{n=1}^{\mathfrak{n}} \mathcal{L}_k\mathsf{f}(v_n)\mathbf{m}_{k,v_n} \\ &= \sum_{k=1}^{\kappa}\sum_{n=1}^{\mathfrak{n}} \langle \alpha, \mathbf{g}_{k,v_n}\rangle_{\ell^2(\mathbf{G})}\, \mathbf{m}_{k,v_n}\,. \end{aligned} \tag{6.35}$$

**The Sampling Result**

Let $\mathsf{f} = \sum_{g\in\mathbf{G}} \alpha_g\, U(g)\mathsf{a}$ be a vector of $\mathcal{V}_\mathsf{a}$; applying the isomorphism (6.25) in expansion (6.35) we get

$$\mathsf{f} = \mathcal{T}^{\mathbf{G}}_{\mathsf{a}}(\alpha) = \sum_{k=1}^{\kappa}\sum_{n=1}^{\mathfrak{n}} \mathcal{L}_k\mathsf{f}(v_n)\,\mathcal{T}^{\mathbf{G}}_{\mathsf{a}}(\mathbf{m}_{k,v_n})\,. \tag{6.36}$$

The columns $\mathbf{m}_{k,v_n}$ in the formula above do not have, in principle, a suitable structure for applying the shifting property (6.26).

Although we will see that the columns of the Moore-Penrose pseudo-inverse $\mathbb{R}^{\dagger}_{\mathsf{a},\mathbf{b}}$ fulfil the required attribute, we will construct in the next section all the left-inverses of $\mathbb{R}_{\mathsf{a},\mathbf{b}}$ allowing it.

In order to prove that $\mathbb{R}^{\dagger}_{\mathsf{a},\mathbf{b}}$ has the suitable structure, note that each $\mathfrak{n} \times \mathfrak{g}$ block $\mathbb{R}_{\mathsf{a},\mathsf{b}_k}$ has an $\mathfrak{h}$-circulant character in the sense that each row of $\mathbb{R}_{\mathsf{a},\mathsf{b}_k}$ is the previous row moved to the right $\mathfrak{h}$ places and wrapped around.

Let $P_i$ denote the 1-circulant square matrix of order $i \in \mathbb{N}$ with first row $(0, 1, 0, \cdots, 0)$. In general, in terms of a matrix $C$ of order $\kappa\mathfrak{n} \times \mathfrak{g}$ partitioned into $\kappa$ submatrices of order $\mathfrak{n} \times \mathfrak{g}$, each block has a $\mathfrak{h}$-circulant character if and only if $CP^{\mathfrak{h}}_{\mathfrak{g}} = \mathbb{P}C$, or equivalently,

$$C = \mathbb{P}^{*}CP^{\mathfrak{h}}_{\mathfrak{g}}\,,$$

where $\mathbb{P}$ is the square matrix of order $\kappa\mathfrak{n}$ given by $\mathbb{P} = \operatorname{diag}(P_{\mathfrak{n}}, \ldots, P_{\mathfrak{n}})$, the direct sum of $\kappa$ times the matrix $P_{\mathfrak{n}}$.

The above characterization allows to conclude easily that $(C^{\dagger})^{*}$ inherits, and consequently $(C^{\dagger})^{\top}$, the $\mathfrak{h}$-circulant character from $C$. Indeed,

$$(C^{\dagger})^{*} = ((\mathbb{P}^{*}CP^{\mathfrak{h}}_{\mathfrak{g}})^{\dagger})^{*} = \big((P^{\mathfrak{h}}_{\mathfrak{g}})^{*}C^{\dagger}\mathbb{P}\big)^{*} = \mathbb{P}^{*}(C^{\dagger})^{*}P^{\mathfrak{h}}_{\mathfrak{g}}\,.$$

For more details on pseudoinverses of circulant matrices see Refs. [10, 11]. In these sources the above results are to be found although for a square matrix $C$.

**G-Compatible Left-Inverses**

Now, we proceed to construct a specific left-inverse $\mathbb{M}_{\mathbb{S}}$ of $\mathbb{R}_{\mathsf{a},\mathbf{b}}$ from any left-inverse $\mathbb{M}$ given by (6.34) in the following way: We denote by $\mathbb{S}$ the first $\mathfrak{h}$ rows of $\mathbb{M}$; i.e,

$$\mathbb{S}\mathbb{R}_{\mathsf{a},\mathbf{b}} = \left(\mathbb{I}_{\mathfrak{h}}\ \mathbb{O}_{\mathfrak{h}\times(\mathfrak{g}-\mathfrak{h})}\right). \tag{6.37}$$

Having in mind the structure of $\mathbb{R}_{\mathsf{a},\mathbf{b}}$ we write the $\mathfrak{h} \times \mathfrak{n}\kappa$ matrix $\mathbb{S}$ as

$$\mathbb{S} = \left(\mathbb{S}_1\, \mathbb{S}_2 \cdots \mathbb{S}_\kappa\right),$$

where each block $\mathbb{S}_k$ is a $\mathfrak{h} \times \mathfrak{n}$ matrix denoted by $\mathbb{S}_k = \left(S_k^1\ S_k^2\ \cdots\ S_k^{\mathfrak{n}}\right)$, where $S_k^n \in \mathbb{C}^{\mathfrak{h}}$ for each $n = 1, 2, \ldots, \mathfrak{n}$, and $k = 1, 2, \ldots, \kappa$.

From (6.30) and (6.37) we have

$$\sum_{k=1}^{\kappa}\sum_{n=1}^{\mathfrak{n}} S_k^n\, r_{\mathsf{a},\mathsf{b}_k}(\nu_1^{-1}\nu_n^{-1}\mathbf{H}) = \mathbb{I}_{\mathfrak{h}}\,,$$

$$\sum_{k=1}^{\kappa}\sum_{n=1}^{\mathfrak{n}} S_k^n\, r_{\mathsf{a},\mathsf{b}_k}(\nu_l^{-1}\nu_n^{-1}\mathbf{H}) = \mathbb{O}_{\mathfrak{h}}, \quad l = 2, \ldots, \mathfrak{n}\,,$$

or equivalently

$$\sum_{k=1}^{\kappa}\sum_{n=1}^{\mathfrak{n}} S_k^n\, r_{\mathsf{a},\mathsf{b}_k}(\nu_n^{-1}\mathbf{H}) = \mathbb{I}_{\mathfrak{h}}\,, \tag{6.38}$$

$$\sum_{k=1}^{\kappa}\sum_{n=1}^{\mathfrak{n}} S_k^n\, r_{\mathsf{a},\mathsf{b}_k}(\nu_n^{-1}\nu_l^{-1}\mathbf{H}) = \mathbb{O}_{\mathfrak{h}}, \quad l = 2, \ldots, \mathfrak{n}\,. \tag{6.39}$$

Now, we form the $\mathfrak{g} \times \mathfrak{n}\kappa$ matrix $\mathbb{M}_{\mathbb{S}} = \left(\widetilde{\mathbb{S}}_1\ \widetilde{\mathbb{S}}_2\ \cdots\ \widetilde{\mathbb{S}}_\kappa\right)$, where each $\mathfrak{g} \times \mathfrak{n}$ block $\widetilde{\mathbb{S}}_k$, $k = 1, 2, \ldots, \kappa$, is formed from the columns of $\mathbb{S}_k$ in the following manner:

$$\widetilde{\mathbb{S}}_k := \begin{pmatrix} S_k^1 & S_k^2 & \cdots & S_k^{\mathfrak{n}} \\ S_k^{1,2} & S_k^{2,2} & \cdots & S_k^{\mathfrak{n},2} \\ \vdots & \vdots & \cdots & \vdots \\ S_k^{1,\mathfrak{n}} & S_k^{2,\mathfrak{n}} & \cdots & S_k^{\mathfrak{n},\mathfrak{n}} \end{pmatrix},$$

where, for $i = 2, \ldots, \mathfrak{n}$, and $n = 1, 2, \ldots, \mathfrak{n}$, we set

$$s_k^{n,i} := s_k^m , \tag{6.40}$$

where $\nu_m \in \mathbf{N}$ is the unique element such that $\nu_m^{-1} = \nu_n^{-1} \nu_i^{-1}$.

**Proposition 6.9** *The above* $\mathfrak{g} \times \mathfrak{n}\kappa$ *matrix* $\mathbb{M}_{\mathbb{S}}$ *is a left-inverse of* $\mathbb{R}_{\mathsf{a},\mathbf{b}}$, *i.e.,*

$$\mathbb{M}_{\mathbb{S}}\, \mathbb{R}_{\mathsf{a},\mathbf{b}} = \mathbb{I}_{\mathfrak{g}} .$$

***Proof*** From (6.38), for each $i = 2, 3, \ldots, \mathfrak{n}$, we have

$$\sum_{k=1}^{\kappa} \sum_{n=1}^{\mathfrak{n}} s_k^{n,i}\, r_{a,\mathsf{b}_k}(\nu_n^{-1} \nu_i^{-1} \mathbf{H}) = \sum_{k=1}^{\kappa} \sum_{m=1}^{\mathfrak{n}} s_k^m\, r_{\mathsf{a},\mathsf{b}_k}(\nu_m^{-1} \mathbf{H}) = \mathbb{I}_{\mathfrak{h}}$$

and, from (6.39)

$$\sum_{k=1}^{\kappa} \sum_{n=1}^{\mathfrak{n}} s_k^{n,i}\, r_{\mathsf{a},\mathsf{b}_k}(\nu_n^{-1} \nu_l^{-1} \mathbf{H}) = \sum_{k=1}^{\kappa} \sum_{m=1}^{\mathfrak{n}} s_k^m\, r_{\mathsf{a},\mathsf{b}_k}\big(\nu_m^{-1}(\nu_i \nu_l^{-1})\mathbf{H}\big) = \mathbb{O}_{\mathfrak{h}}, \quad l \neq i ,$$

since $\nu_i \nu_l^{-1} \neq 1_{\mathbf{G}}$ for $l \neq i$. As a consequence, we deduce that $\mathbb{M}_{\mathbb{S}}\, \mathbb{R}_{\mathsf{a},\mathbf{b}} = \mathbb{I}_{\mathfrak{g}}$. □

Denoting the columns of the matrix $\mathbb{M}_{\mathbf{S}}$ as $\widetilde{\mathbf{s}}_{k,n}$, where $k = 1, 2, \ldots, \kappa$, and $n = 1, 2, \ldots, \mathfrak{n}$, we have

$$\mathbb{M}_{\mathbf{S}} = \left(\widetilde{\mathbf{s}}_{1,1} \ldots \widetilde{\mathbf{s}}_{1,\mathfrak{n}}\ \widetilde{\mathbf{s}}_{2,1} \ldots \widetilde{\mathbf{s}}_{2,\mathfrak{n}} \ldots \widetilde{\mathbf{s}}_{\kappa,1} \ldots \widetilde{\mathbf{s}}_{\kappa,\mathfrak{n}}\right) . \tag{6.41}$$

Using the left-inverse $\mathbb{M}_{\mathbb{S}}$ of $\mathbb{R}_{\mathsf{a},\mathbf{b}}$ instead of $\mathbb{M}$ in (6.35), for each $\mathsf{f} = \sum_{g \in \mathbf{G}} \alpha_g\, U(g)\mathsf{a}$ in $\mathcal{V}_{\mathsf{a}}$ we obtain

$$\boldsymbol{\alpha} = \mathbb{M}_{\mathbb{S}}\, \mathcal{L}_{\text{samp}} = \sum_{k=1}^{\kappa} \sum_{n=1}^{\mathfrak{n}} \mathcal{L}_k \mathsf{f}(\nu_n)\, \widetilde{\mathbf{s}}_{k,n} .$$

On the other hand, the columns $\widetilde{\mathbf{s}}_{k,n}$, $k = 1, 2, \ldots, \kappa$, and $n = 1, 2, \ldots, \mathfrak{n}$, as vectors of $\ell^2(\mathbf{G})$ satisfy, by construction (see Eq. (6.40)) the crucial property

$$\widetilde{\mathbf{s}}_{k,n} = L_{\nu_n} \widetilde{\mathbf{s}}_{k,1}, \quad k = 1, 2, \ldots, \kappa, \text{ and } n = 1, 2, \ldots, \mathfrak{n} .$$

Hence, the shifting property (6.26) gives

$$\mathsf{f} = \sum_{k=1}^{\kappa}\sum_{n=1}^{\mathfrak{n}} \mathcal{L}_k\mathsf{f}(\nu_n)\,\mathcal{T}_{\mathsf{a}}^{\mathbf{G}}(\widetilde{\mathsf{s}}_{k,n}) = \sum_{k=1}^{\kappa}\sum_{n=1}^{\mathfrak{n}} \mathcal{L}_k\mathsf{f}(\nu_n)\,\mathcal{T}_{\mathsf{a}}^{\mathbf{G}}(L_{\nu_n}\widetilde{\mathsf{s}}_{k,1})$$

$$= \sum_{k=1}^{\kappa}\sum_{n=1}^{\mathfrak{n}} \mathcal{L}_k\mathsf{f}(\nu_n)\,U(\nu_n)\,\mathcal{T}_{\mathsf{a}}^{\mathbf{G}}(\widetilde{\mathsf{s}}_{k,1})\,.$$

Therefore, we have proved that, for any $\mathsf{f} \in \mathcal{V}_{\mathsf{a}}$ the sampling expansion

$$\mathsf{f} = \sum_{k=1}^{\kappa}\sum_{n=1}^{\mathfrak{n}} \mathcal{L}_k\mathsf{f}(\nu_n)\,U(\nu_n)\,\mathsf{c}_k$$

holds, where $\mathsf{c}_k = \mathcal{T}_{\mathsf{a}}^{\mathbf{G}}(\widetilde{\mathsf{s}}_{k,1}) \in \mathcal{V}_{\mathsf{a}}$, $k = 1, 2, \ldots, \kappa$.

Actually, collecting all the pieces we have obtained until now we can state the following sampling result:

**Theorem 6.2** *Consider the* $\mathfrak{n}\kappa \times \mathfrak{g}$ *matrix* $\mathbb{R}_{\mathsf{a},\mathbf{b}}$ *defined in* (6.31). *The following statements are equivalent:*

1. rank $\mathbb{R}_{\mathsf{a},\mathbf{b}} = \mathfrak{g}$.
2. *There exists a* $\mathfrak{h} \times \mathfrak{n}\kappa$ *matrix* $\mathbb{S}$ *such that*

$$\mathbb{S}\,\mathbb{R}_{\mathsf{a},\mathbf{b}} = \big(\mathbb{I}_{\mathfrak{h}}\ \mathbb{O}_{\mathfrak{h}\times(\mathfrak{g}-\mathfrak{h})}\big)\,.$$

3. *There exist vectors* $\mathsf{c}_k \in \mathcal{V}_{\mathsf{a}}$, $k = 1, 2, \ldots, \kappa$, *such that* $\big\{U(\nu_n)\mathsf{c}_k\big\}_{\substack{k=1,2,\ldots,\kappa\\ n=1,2,\ldots,\mathfrak{n}}}$ *is a frame (spanning set) for* $\mathcal{V}_{\mathsf{a}}$, *and for any* $\mathsf{f} \in \mathcal{V}_{\mathsf{a}}$ *we have the expansion*

$$\mathsf{f} = \sum_{k=1}^{\kappa}\sum_{n=1}^{\mathfrak{n}} \mathcal{L}_k\mathsf{f}(\nu_n)\,U(\nu_n)\mathsf{c}_k\,.$$

4. *There exists a frame* $\{\mathsf{C}_{k,n}\}_{\substack{k=1,2,\ldots,\kappa\\ n=1,2,\ldots,\mathfrak{n}}}$ *for* $\mathcal{V}_{\mathsf{a}}$ *such that, for each* $\mathsf{f} \in \mathcal{V}_{\mathsf{a}}$ *we have the expansion*

$$\mathsf{f} = \sum_{k=1}^{\kappa}\sum_{n=1}^{\mathfrak{n}} \mathcal{L}_k\mathsf{f}(\nu_n)\,\mathsf{C}_{k,n}\,.$$

***Proof*** That condition (1) implies condition (2) and condition (2) implies condition (3) have been proved above. Obviously, condition (3) implies condition (4): take $\mathsf{C}_{k,n} = U(\nu_n)\mathsf{c}_k$ for $k = 1, 2, \ldots, \kappa$, and $n = 1, 2, \ldots, \mathfrak{n}$. Finally, as a consequence of Corollary 6.1, condition (4) implies condition (1). □

For the particular case where $\kappa = \mathfrak{h}$ we obtain:

**Corollary 6.7** *Assume that* $\kappa = \mathfrak{h}$ *and consider the* $\mathfrak{g} \times \mathfrak{g}$ *cross-covariance matrix* $\mathbb{R}_{\mathsf{a},\mathbf{b}}$ *defined in* (6.31). *The following statements are equivalent:*

*(i) The matrix* $\mathbb{R}_{\mathsf{a},\mathbf{b}}$ *is invertible.*
*(ii) There exist* $\mathfrak{h}$ *unique elements* $\mathsf{c}_k \in \mathcal{V}_\mathsf{a}$, $k = 1, 2, \dots, \mathfrak{h}$, *such that the sequence* $\{U(\nu_n)\mathsf{c}_k\}_{\substack{k=1,2,\dots,\mathfrak{h}\\ n=1,2,\dots,\mathfrak{n}}}$ *is a basis for* $\mathcal{V}_\mathsf{a}$, *and the expansion of any* $\mathsf{f} \in \mathcal{V}_\mathsf{a}$ *with respect to this basis is*

$$\mathsf{f} = \sum_{k=1}^{\mathfrak{h}} \sum_{n=1}^{\mathfrak{n}} \mathcal{L}_k \mathsf{f}(\nu_n)\, U(\nu_n)\mathsf{c}_k\,.$$

*In case the equivalent conditions are satisfied, the interpolation property*

$$\mathcal{L}_k \mathsf{c}_{k'}(\nu_n) = \delta_{k,k'}\, \delta_{n,1}\,,$$

*holds for* $n = 1, 2, \dots, \mathfrak{n}$, *and* $k, k' = 1, 2, \dots, \mathfrak{h}$.

***Proof*** Notice that the inverse matrix $\mathbb{R}^{-1}_{\mathsf{a},\mathbf{b}}$ has necessarily the structure of the matrix $\mathbb{M}_\mathbb{S}$ in (6.41). The uniqueness of the expansion with respect to a basis gives the interpolation property. □

### 6.2.4 *Sampling Indexed by* H *when* H *is an Abelian Subgroup*

In case $\mathbf{G} = \mathbf{N} \bowtie \mathbf{H}$, there is just one element of $\mathbf{H}$ in each left coset of $\mathbf{G}/\mathbf{N}$. Indeed, suppose that there exist $\tau, \tau^* \in \mathbf{H}$ such that $\tau = \tau^* \nu$ for some $\nu \in \mathbf{N}$. Then,

$$\tau = \tau^* \nu = \alpha_{\tau^*}(\nu)\beta_\nu(\tau^*),$$

and, as a consequence, $\alpha_{\tau^*}(\nu) = 1_\mathbf{G}$, which implies $\nu = 1_\mathbf{G}$, since $\alpha_{\tau^*} \in Aut(\mathbf{N})$, and $\tau = \tau^*$.

Hence, we can choose an element of $\mathbf{H}$ in each left coset of the quotient set $\mathbf{G}/\mathbf{N}$. As $\mathfrak{h} = |\mathbf{H}| = |\mathbf{G}/\mathbf{N}|$, in case $\mathbf{H} = \{\tau_1 = 1_\mathbf{G}, \tau_2, \dots, \tau_\mathfrak{h}\}$ we can describe the quotient set $\mathbf{G}/\mathbf{N}$ as

$$\mathbf{G}/\mathbf{N} = \big\{[1_\mathbf{G} = \tau_1], [\tau_2], \cdots, [\tau_\mathfrak{h}]\big\}\,. \tag{6.42}$$

Hence, for a fixed order of the elements in $\mathbf{N}$ we can write

$$\mathbf{G} = \bigcup_{n=1}^{\mathfrak{h}} \tau_n^{-1}\mathbf{N}\,.$$

Therefore, in case the subgroup $\mathbf{H}$ is abelian the above partition of $\mathbf{G}$ allow us to proceed as in the previous section in order to obtain a sampling formula for $\mathcal{V}_\mathsf{a}$ by indexing the data samples in $\mathbf{H}$.

Indeed, for fixed $\kappa$ elements $\mathsf{b}_k \in \mathcal{H}$, $k = 1, 2, \dots, \kappa$, for each $\mathsf{f} = \sum_{g\in\mathbf{G}} \alpha_g\, U(g)\mathsf{a}$ in $\mathcal{V}_\mathsf{a}$ we define again its generalized samples, now indexed in $\mathbf{H}$, by

$$\mathcal{L}_k\mathsf{f}(\tau_n) := \big\langle \mathsf{f}, U(\tau_n)\mathsf{b}_k\big\rangle_{\mathcal{H}}, \quad n = 1, 2, \dots, \mathfrak{h}, \text{ and } k = 1, 2, \dots, \kappa\,. \tag{6.43}$$

Roughly speaking, to recover any $\mathsf{f} \in \mathcal{V}_\mathsf{a}$ we need at least $\mathfrak{g}$ samples; if we are sampling at $\mathbf{H}$, we will need at least $\kappa\mathfrak{h} \geq \mathfrak{g} = \mathfrak{n}\mathfrak{h}$ samples, i.e., $\kappa \geq \mathfrak{n}$.

Let $\mathsf{f}$ be in $\mathcal{V}_\mathsf{a}$, in this case the expression for $\mathcal{L}_k\mathsf{f}(\tau_n)$ with $n = 1, 2, \dots, \mathfrak{h}$, similar to (6.28), is

$$\mathcal{L}_k\mathsf{f}(\tau_n) = \big\langle \boldsymbol{\alpha}, \mathbf{g}_{k,\tau_n}\big\rangle_{\ell^2(\mathbf{G})}\,, \tag{6.44}$$

where $\boldsymbol{\alpha} = (\alpha_g)_{g\in\mathbf{G}}$ and $\mathbf{g}_{k,\tau_n} = \big(\overline{\langle U(g)\mathsf{a}, U(\tau_n)\mathsf{b}_k\rangle}\big)_{g\in\mathbf{G}}$ belong to $\ell^2(\mathbf{G})$.

The vectors $\mathbf{g}_{k,\tau_n} \in \ell^2(\mathbf{G})$, $k = 1, 2, \dots, \kappa$, $n = 1, 2, \dots, \mathfrak{h}$, can be expressed in terms of the cross-covariance $r_{\mathsf{a},\mathsf{b}_k}$ as $\mathbf{g}_{k,\tau_n} = \big(\overline{r_{\mathsf{a},\mathsf{b}_k}}(\tau_n^{-1}g)\big)_{g\in\mathbf{G}}$.

Now, proceeding as before, since $\mathbf{H}$ is abelian, for each $k = 1, 2, \dots, \kappa$, we get the $\mathfrak{h} \times \mathfrak{g}$ matrix $\mathbb{R}_{\mathsf{a},\mathsf{b}_k}$

$$\mathbb{R}_{\mathsf{a},\mathsf{b}_k} = \begin{pmatrix} r_{\mathsf{a},\mathsf{b}_k}(\tau_1^{-1}\tau_1^{-1}\mathbf{N}) & r_{\mathsf{a},\mathsf{b}_k}(\tau_2^{-1}\tau_1^{-1}\mathbf{N}) & \dots & r_{\mathsf{a},\mathsf{b}_k}(\tau_\mathfrak{h}^{-1}\tau_1^{-1}\mathbf{N}) \\ r_{\mathsf{a},\mathsf{b}_k}(\tau_1^{-1}\tau_2^{-1}\mathbf{N}) & r_{\mathsf{a},\mathsf{b}_k}(\tau_2^{-1}\tau_2^{-1}\mathbf{N}) & \dots & r_{\mathsf{a},\mathsf{b}_k}(\tau_\mathfrak{h}^{-1}\tau_2^{-1}\mathbf{N}) \\ \vdots & \vdots & \dots & \vdots \\ r_{\mathsf{a},\mathsf{b}_k}(\tau_1^{-1}\tau_\mathfrak{h}^{-1}\mathbf{N}) & r_{\mathsf{a},\mathsf{b}_k}(\tau_2^{-1}\tau_\mathfrak{h}^{-1}\mathbf{N}) & \dots & r_{\mathsf{a},\mathsf{b}_k}(\tau_\mathfrak{h}^{-1}\tau_\mathfrak{h}^{-1}\mathbf{N}) \end{pmatrix}, \tag{6.45}$$

where $\tau_n \in \mathbf{H}$ and $\overline{r_{\mathsf{a},\mathsf{b}_k}}(\tau_n^{-1}\tau_m^{-1}\mathbf{N})$ is the row vector

$$\overline{r_{\mathsf{a},\mathsf{b}_k}}(\tau_n^{-1}\tau_m^{-1}\mathbf{N}) = \Big(\overline{r_{\mathsf{a},\mathsf{b}_k}}(\tau_n^{-1}\tau_m^{-1}\nu_1), \overline{r_{\mathsf{a},\mathsf{b}_k}}(\tau_n^{-1}\tau_m^{-1}\nu_2), \dots, \overline{r_{\mathsf{a},\mathsf{b}_k}}(\tau_n^{-1}\tau_m^{-1}\nu_\mathfrak{n})\Big)\,,$$

where $\mathbf{N} = \{\nu_1 = 1_\mathbf{G}, \nu_2, \dots, \nu_\mathfrak{n}\}$.

From matrices $\mathbb{R}_{\mathsf{a},\mathsf{b}_k}$ in (6.45), $k = 1, 2, \dots, \kappa$, we form now the new $\mathfrak{h}\kappa \times \mathfrak{g}$ cross-covariance matrix $\mathbb{R}_{\mathsf{a},\mathbf{b}}$ as in (6.31):

$$\mathbb{R}_{\mathsf{a},\mathbf{b}} := \begin{pmatrix} \mathbb{R}_{\mathsf{a},\mathsf{b}_1} \\ \mathbb{R}_{\mathsf{a},\mathsf{b}_2} \\ \vdots \\ \mathbb{R}_{\mathsf{a},\mathsf{b}_\kappa} \end{pmatrix}. \tag{6.46}$$

Now, the sampling theory goes in the same manner as in section above. In fact we have the following result, completely analogous to Theorem 6.2:

**Theorem 6.3** *Consider the* $\mathfrak{h}\kappa \times \mathfrak{g}$ *matrix* $\mathbb{R}_{\mathsf{a},\mathbf{b}}$ *defined in* (6.46). *The following statements are equivalent:*

1. rank $\mathbb{R}_{\mathsf{a},\mathbf{b}} = \mathfrak{g}$.
2. *There exists a* $\mathfrak{n} \times \mathfrak{h}\kappa$ *matrix* $\mathbb{S}$ *such that*

$$\mathbb{S}\,\mathbb{R}_{\mathsf{a},\mathbf{b}} = \big(\mathbb{I}_{\mathfrak{n}}\ \mathbb{O}_{\mathfrak{n}\times(\mathfrak{g}-\mathfrak{n})}\big)\,.$$

3. *There exist vectors* $\mathsf{d}_k \in \mathcal{V}_{\mathsf{a}}$, $k = 1, 2, \ldots, \kappa$, *such that* $\big\{U(\tau_n)\mathsf{d}_k\big\}_{\substack{k=1,2,\ldots,\kappa\\ n=1,2,\ldots,\mathfrak{h}}}$ *is a frame for* $\mathcal{V}_{\mathsf{a}}$, *and for any* $\mathsf{f} \in \mathcal{V}_{\mathsf{a}}$ *we have the expansion*

$$\mathsf{f} = \sum_{k=1}^{\kappa}\sum_{n=1}^{\mathfrak{h}} \mathcal{L}_k\mathsf{f}(\tau_n)\, U(\tau_n)\mathsf{d}_k\,.$$

4. *There exists a frame* $\{\mathsf{D}_{k,n}\}_{\substack{k=1,2,\ldots,\kappa\\ n=1,2,\ldots,\mathfrak{h}}}$ *for* $\mathcal{V}_{\mathsf{a}}$ *such that, for each* $\mathsf{f} \in \mathcal{V}_{\mathsf{a}}$ *we have the expansion*

$$\mathsf{f} = \sum_{k=1}^{\kappa}\sum_{n=1}^{\mathfrak{h}} \mathcal{L}_k\mathsf{f}(\tau_n)\, \mathsf{D}_{k,n}\,.$$

Note that an analogous of Corollary 6.7 can be stated in case $\kappa = \mathfrak{n}$.

**Example 6.5 (A Toy Example: Sampling Associated with the Dihedral Group)** In geometry, the *dihedral group* $\mathcal{D}_{2N}$ refers to the symmetries of the $N$-gon, which has order $2N$.

Indeed, the group $\mathcal{D}_{2N} = \mathbf{NH}$, where $\mathbf{N} = \{e, r, r^2, \ldots, r^{N-1}\}$ is the group generated by $r$, the rotation of $2\pi/N$ radians and $\mathbf{H} = \{e, s\}$ being $s$ the axial reflection.

The subgroup $\mathbf{N}$ is a normal subgroup of $\mathcal{D}_{2N}$ and, as a consequence, $\mathcal{D}_{2N}$ is commonly referred as the internal semi-direct product of $\mathbf{N}$ and $\mathbf{H}$.

Since the subgroups $\mathbf{N}$ and $\mathbf{H}$ of $D_{2N}$ are both abelians, the sampling theorems in Theorems 6.2 and 6.3 apply.

Let $\mathcal{D}_{2N} \ni g \mapsto U(g) \in \mathcal{U}(\mathcal{H})$ be a unitary representation of $\mathcal{D}_{2N}$ on a Hilbert space $\mathcal{H}$.

Given $\mathsf{a} \in \mathcal{H}$, we consider the subspace $\mathcal{V}_{\mathsf{a}}$ of $\mathcal{H}$ spanned by

$$\big\{U(g)\mathsf{a}\,:\, g \in \mathcal{D}_{2N}\big\}\,.$$

In case this set is linearly independent in $\mathcal{H}$ the subspace $\mathcal{V}_{\mathsf{a}}$ can be described as

$$\mathcal{V}_{\mathsf{a}} = \Big\{ \sum_{g \in \mathcal{D}_{2N}} \alpha_g U(g)\mathsf{a} \,:\, \alpha_g \in \mathbb{C} \Big\} \subset \mathcal{H}.$$

Assume that $\kappa$ $\mathcal{L}_k$-systems are defined on $\mathcal{V}_{\mathsf{a}}$ from $\kappa$ vectors $\mathsf{b}_k \in \mathcal{H}$, $k = 1, 2, \ldots, \kappa$, as in (6.43). In this case, each $2 \times 2N$ block $\mathbb{R}_{\mathsf{a},\mathsf{b}_k}$ of the $2\kappa \times 2N$ matrix $\mathbb{R}_{\mathsf{a},\mathbf{b}}$ in (6.46) is given by

$$\mathbb{R}_{\mathsf{a},\mathsf{b}_k} = \begin{pmatrix} r_{\mathsf{a},\mathsf{b}_k}(\mathbf{N}) & r_{\mathsf{a},\mathsf{b}_k}(s\mathbf{N}) \\ r_{\mathsf{a},\mathsf{b}_k}(s\mathbf{N}) & r_{\mathsf{a},\mathsf{b}_k}(\mathbf{N}) \end{pmatrix}.$$

If $\operatorname{rank} \mathbb{R}_{\mathsf{a},\mathbf{b}} = 2N$, according to Theorem 6.3 there exist $\kappa$ vectors $\mathsf{d}_k \in \mathcal{V}_{\mathsf{a}}$, $k = 1, 2, \ldots, \kappa$, with $\kappa \geq N$ such that the sequence $\big\{U(h)\mathsf{d}_k\big\}_{h\in\mathbf{H};\, k=1,2,\ldots,\kappa}$ is a frame for $\mathcal{V}_{\mathsf{a}}$ and, for any $\mathsf{f} \in \mathcal{V}_{\mathsf{a}}$ we obtain the sampling expansion

$$\mathsf{f} = \sum_{k=1}^{\kappa} \sum_{h\in\mathbf{H}} \mathcal{L}_k\mathsf{f}(h)\, U(h)\mathsf{d}_k\,.$$

Moreover, $\mathsf{d}_k = \mathcal{T}_{\mathsf{a}}^{\mathcal{D}_{2N}}(\widetilde{\mathbf{s}}_{k,e})$, where $\widetilde{\mathbf{s}}_{k,e}$, $k = 1, 2, \ldots, \kappa$, denote the corresponding column of a $2N \times 2\kappa$ left-inverse $\mathbb{M}_{\mathbb{S}}$ of the matrix $\mathbb{R}_{\mathsf{a},\mathbf{b}}$.

Analogously, we can take generalized samples indexed on $\mathbf{N}$ as in (6.27) from $\kappa$ vectors $\mathsf{b}_k \in \mathcal{H}$, $k = 1, 2, \ldots, \kappa$. In this case each $2 \times 2N$ block $\mathbb{R}_{\mathsf{a},\mathsf{b}_k}$ of the $2\kappa \times 2N$ matrix $\mathbb{R}_{\mathsf{a},\mathbf{b}}$ in (6.31) is given by

$$\mathbb{R}_{\mathsf{a},\mathsf{b}_k} = \begin{pmatrix} r_{\mathsf{a},\mathsf{b}_k}(\mathbf{H}) & r_{\mathsf{a},\mathsf{b}_k}(r\mathbf{H}) & \cdots & r_{\mathsf{a},\mathsf{b}_k}(r^{n-1}\mathbf{H}) \\ r_{\mathsf{a},\mathsf{b}_k}(r^{n-1}\mathbf{H}) & r_{\mathsf{a},\mathsf{b}_k}(r^{n-2}\mathbf{H}) & \cdots & r_{\mathsf{a},\mathsf{b}_k}(\mathbf{H}) \\ \vdots & \vdots & \cdots & \vdots \\ r_{\mathsf{a},\mathsf{b}_k}(r\mathbf{H}) & r_{\mathsf{a},\mathsf{b}_k}(\mathbf{H}) & \cdots & r_{\mathsf{a},\mathsf{b}_k}(r^2\mathbf{H}) \end{pmatrix}.$$

If $\operatorname{rank} \mathbb{R}_{\mathsf{a},\mathbf{b}} = 2N$, according to Theorem 6.2 there exist $\kappa$ vectors $\mathsf{c}_k \in \mathcal{V}_{\mathsf{a}}$, $k = 1, 2, \ldots, \kappa$, with $\kappa \geq 2$, such that the sequence $\big\{U(n)\mathsf{c}_k\big\}_{n\in\mathbf{N};\, k=1,2,\ldots,\kappa}$ is a frame for $\mathcal{V}_{\mathsf{a}}$ and, for any $\mathsf{f} \in \mathcal{V}_{\mathsf{a}}$ we have the sampling expansion

$$\mathsf{f} = \sum_{k=1}^{\kappa} \sum_{n\in\mathbf{N}} \mathcal{L}_k\mathsf{f}(n)\, U(n)\mathsf{c}_k\,.$$

Moreover, $\mathsf{c}_k = \mathcal{T}_{\mathsf{a}}^{\mathcal{D}_{2N}}(\widetilde{\mathbf{s}}_{k,e})$, where $\widetilde{\mathbf{s}}_{k,e}$, $k = 1, 2, \ldots, \kappa$, denote the corresponding column of a $2N \times 2\kappa$ left-inverse matrix $\mathbb{M}_{\mathbb{S}}$ of the cross-covariance matrix $\mathbb{R}_{\mathsf{a},\mathbf{b}}$.

# References


1. Brin, M.G.: On the Zappa-Szép product. Commun. Algebra **33**(2), 393–424 (2005)
2. Christensen, O.: An Introduction to Frames and Riesz Bases, 2nd edn. Birkhäuser, Boston (2016)
3. Fernández-Morales, H.R., García, A.G., Hernández-Medina, M.A., Muñoz-Bouzo, M.J.: Generalized sampling: from shift-invariant to $U$-invariant spaces. Anal. Appl. **13**(3), 303–329 (2015)
4. Fernández-Morales, H.R., García, A.G., Muñoz-Bouzo, M.J., Ortega, A.: Finite sampling in multiple generated U-invariant subspaces. IEEE Trans. Inform. Theory **62**(4), 2203–2212 (2016)
5. García, A.G., Muñoz-Bouzo, M.J.: Sampling-related frames in finite $U$-invariant subspaces. Appl. Comput. Harmon. Anal. **39**, 173–184 (2015)
6. García, A.G., Hernández-Medina, M.A., Ibort, A.: Knit product of finite groups and sampling. Mediterr. J. Math. **16**(6), 146 (2019)
7. Horn, R.A., Johnson, C.R.: Matrix Analysis. Cambridge University Press, Cambridge UK (1999)
8. Kolmogorov, A.N.: Stationary sequences in Hilbert space. In: Shiryayev, A.N. (ed.) Selected Works of A.N. Kolmogorov, Vol. II, pp. 228–271. Kluwer Academic Publ., Dordrecht (1992)
9. Pohl, V., Boche, H.: $U$-invariant sampling and reconstruction in atomic spaces with multiple generators. IEEE Trans. Signal Process. **60**(7), 3506–3519 (2012)
10. Pye, W.C., Boullion, T.L., Atchison, T.A.: The pseudoinverse of a composite matrix of circulants. SIAM J. Appl. Math. **24**, 552–555 (1973)
11. Stallings, W.T., Boullion, T.L.: The pseudoinverse of an $r$-circulant matrix Proc. Amer. Math. Soc. **34**, 385–388 (1972)
12. Strang, G., Nguyen, T.: Wavelets and Filter Banks. Wellesley-Cambridge Press, Cambridge MA (1996)
13. Szép, J.: On the structure of groups which can be represented as the product of two subgroups. Acta Sci. Math. Szeged **12**, 57–61 (1950)
14. Zappa, G.: Sulla costruzione dei gruppi prodotto di due dati sottogruppi permutabili traloro. Atti Secondo Congresso Un. Mat. Ital., Bologna; Edizioni Cremonense, Rome, pp. 119–125 (1942)

# Chapter 7
# Sampling in Shift-Invariant-Like Subspaces of Hilbert-Schmidt Operators

In this chapter we obtain sampling results for shift-invariant-like subspaces of the class $\mathcal{HS}(\mathbb{R}^d)$ of Hilbert-Schmidt operators on $L^2(\mathbb{R}^d)$. To be more precise, these subspaces are obtained by translation in a lattice $\Lambda \subset \mathbb{R}^{2d}$ of a fixed set of *Hilbert-Schmidt* operators $S_1, S_2, \dots, S_N$. At this point we can consider a Hilbert-Schmidt operator $T \in \mathcal{HS}(\mathbb{R}^d)$ as an integral operator $T : L^2(\mathbb{R}^d) \longrightarrow L^2(\mathbb{R}^d)$

$$Tf(t) = \int_{\mathbb{R}^d} \kappa(t,x) f(x) dx \quad \text{a.e. } t \in \mathbb{R}^d ,$$

having a kernel $\kappa$ in $L^2(\mathbb{R}^{2d})$.

The *translation of an operator* $S$ by $z \in \mathbb{R}^{2d}$ is defined by using conjugation with the *time-frequency shift* $\pi(z)$, where $z = (x, \omega)$ belongs to the *phase space* $\mathbb{R}^d \times \widehat{\mathbb{R}}^d$ (which in the sequel will be identified with $\mathbb{R}^{2d}$) by

$$\alpha_z(S) := \pi(z) S \pi(z)^* , \quad z \in \mathbb{R}^{2d} .$$

The time-frequency shift $\pi(z)$ acts on $f \in L^2(\mathbb{R}^d)$ as

$$\pi(z) f(t) = e^{2\pi i \omega \cdot t} f(t-x) , \quad z = (x, \omega) \in \mathbb{R}^{2d} ,$$

and $\pi(z)^*$ denotes the adjoint operator of $\pi(z)$.

Moreover, the set of translations $\{\alpha_z\}_{z \in \mathbb{R}^{2d}}$ is a *unitary representation* of the group $\mathbb{R}^{2d}$ on the Hilbert space $\mathcal{HS}(\mathbb{R}^d)$.

If we take a *full rank lattice* $\Lambda$ in $\mathbb{R}^{2d}$, i.e., $\Lambda = A\mathbb{Z}^{2d}$, where $A$ is a $2d \times 2d$ real invertible matrix, such that the sequence $\{\alpha_\lambda(S_n)\}_{\lambda \in \Lambda;\, n=1,2,\dots,N}$ is a *Riesz sequence* in the Hilbert space $\mathcal{HS}(\mathbb{R}^d)$, we consider the closed subspace of $\mathcal{HS}(\mathbb{R}^d)$ given by

$$V_{\mathbf{S}}^2 = \Big\{ \sum_{n=1}^{N} \sum_{\lambda \in \Lambda} c_n(\lambda)\, \alpha_\lambda(S_n) \ : \ \{c_n(\lambda)\}_{\lambda \in \Lambda} \in \ell^2(\Lambda) \, ; \, n = 1, 2, \dots, N \Big\} .$$

 
A. García García, *The Use of Frames in Sampling Theory*,
RSME Springer Series 14, https://doi.org/10.1007/978-3-031-63242-6_7

From now on, the subspaces $V_{\mathbf{S}}^2$ obtained in this way will be called $\Lambda$-*shift-invariant subspaces* in $\mathcal{HS}(\mathbb{R}^d)$.

These spaces are similar in structure as the shift-invariant subspaces in $L^2(\mathbb{R}^d)$ that have appeared in this book:

$$V_\Phi^2 := \Big\{ \sum_{n=1}^{N} \sum_{\alpha \in \mathbb{Z}^d} c_n(\alpha)\, \varphi_n(t-\alpha) \ : \ \{c_n(\alpha)\}_{\alpha \in \mathbb{Z}^d} \in \ell^2(\mathbb{Z}^d)\,;\ n = 1, 2, \dots, N \Big\},$$

where $\Phi = \{\varphi_1, \varphi_2, \dots, \varphi_N\}$ denotes a set of generators of $V_\Phi^2$.

As we have seen in Chap. 5, sampling in the classical shift-invariant subspace $V_\Phi^2$ usually involves, for each $f \in V_\Phi^2$, pointwise samples $\{f(\alpha+\beta_m)\}_{\alpha \in \mathbb{Z}^d}$ and/or average samples $\{\langle f, \psi_m(\cdot - \alpha)\rangle\}_{\alpha \in \mathbb{Z}^d}$, where $\psi_m$ is an *average function* in $L^2(\mathbb{R}^d)$, which does not necessarily belong to $V_\Phi^2$. Any stable sampling in $V_\Phi^2$ will require, necessarily, $M \geq N$ sequences of samples. See also Refs. [1, 15].

A challenging problem here is to choose an appropriate set of samples that should be used for the stable reconstruction of the operators in $V_{\mathbf{S}}^2$. Inspired in Ref. [18] and motivated by the problem of channel estimation in wireless communications, in this chapter for any $T \in V_{\mathbf{S}}^2$ we propose its *diagonal channel samples* at a lattice $\Lambda \subset \mathbb{R}^{2d}$ defined by

$$s_{T,m}(\lambda) := \langle \alpha_{-\lambda}(T) g_m, \widetilde{g}_m \rangle_{L^2(\mathbb{R}^d)}, \quad \lambda \in \Lambda\,, \quad m = 1, 2, \dots, M\,, \tag{7.1}$$

where $g_m, \widetilde{g}_m$, $m = 1, 2, \dots, M$, are $2M$ fixed functions in $L^2(\mathbb{R}^d)$. It will be proved that, necessarily, $M \geq N$.

The name *diagonal channel samples* coined for these samples will be clarified later on, where a little explanation will be done for both, the choice of Hilbert-Schmidt operators (in $V_{\mathbf{S}}^2$) to be sampled, and the choice of the above samples for any $T \in V_{\mathbf{S}}^2$.

The samples defined in (7.1) can also be considered as a particular case of the *average sampling* in $V_{\mathbf{S}}^2$ (see Sect. 7.3.4). For any operator $T \in V_{\mathbf{S}}^2$, its *average samples* at $\Lambda$ are defined by

$$\big\langle T, \alpha_\lambda(Q_m) \big\rangle_{\mathcal{HS}}, \quad \lambda \in \Lambda\,, \quad m = 1, 2, \dots, M\,, \tag{7.2}$$

from $M$ fixed operators $Q_1, Q_2, \dots, Q_M$ in $\mathcal{HS}(\mathbb{R}^d)$, not necessarily in $V_{\mathbf{S}}^2$.

Besides, as we will see in Sect. 7.3.4, the samples defined in Eq. (7.1) are nothing but the *lower symbol of the operator* $T$ with respect $g_m, \widetilde{g}_m \in L^2(\mathbb{R}^d)$ and lattice $\Lambda$, i.e.,

$$\big\langle T\pi(\lambda) g_m, \pi(\lambda)\widetilde{g}_m \big\rangle_{L^2(\mathbb{R}^d)}, \quad \lambda \in \Lambda\,,$$

used in time-frequency analysis (see Ref. [6]), or the samples of the *Berezin transform*

$$\mathcal{B}^{g_m,\widetilde{g}_m} T(z) := \langle T\pi(z)g_m, \pi(z)\widetilde{g}_m \rangle_{L^2(\mathbb{R}^d)}, \quad z \in \mathbb{R}^{2d},$$

at the lattice $\Lambda$ used in quantum physics (see Ref. [21]).

As announced before, these samples are also a particular case of the *average samples* $\langle T, \alpha_\lambda(Q_m)\rangle_{\mathcal{HS}}$ where the *average operator* $Q_m$ is precisely the rank-one operator $\widetilde{g}_m \otimes g_m$. Average sampling has been previously used in Refs. [6, 13].

The main aim here is the stable recovery of any $T \in V_{\mathbf{S}}^2$ from its samples in Eq. (7.1) by means of a sampling formula in $V_{\mathbf{S}}^2$ having the form

$$T = \sum_{m=1}^{M} \sum_{\lambda \in \Lambda} s_{T,m}(\lambda)\, \alpha_\lambda(H_m) \quad \text{in } \mathcal{HS}\text{-norm},$$

for each $T \in V_{\mathbf{S}}^2$. The operators $H_m$, $m = 1, 2, \ldots, M$, in the above formula belong to $V_{\mathbf{S}}^2$ and satisfy that the sequence $\{\alpha_\lambda(H_m)\}_{\lambda\in\Lambda;\, m=1,2,\ldots,M}$ is a *frame* for the Hilbert space $V_{\mathbf{S}}^2$.

For sampling in classical shift-invariant spaces see, for instance, Chap. 5 and references therein. For the less known topic on sampling operators, see Refs. [6, 13, 14, 18, 20, 22, 23].

The used sampling techniques in this chapter are those of the frame theory in a separable Hilbert space. To be precise, the samples used in this chapter will be expressed as the output of a discrete convolution system in the product Hilbert space $\ell_N^2(\Lambda) := \ell^2(\Lambda) \times \cdots \times \ell^2(\Lambda)$ ($N$ times), and then the close relationship between discrete convolution systems and sequences of translates in $\ell_N^2(\Lambda)$ will be used to find suitable sampling formulas. In other words, the techniques used in Chap. 5.

The other tools involved are the Kohn-Nirenberg transform, or the Weyl transform for Hilbert-Schmidt operators: both are unitary operators from $L^2(\mathbb{R}^{2d})$ onto $\mathcal{HS}(\mathbb{R}^d)$ which preserve the translations in the following sense: If we denote either of both by $\mathcal{L}$, we have

$$\mathcal{L}(T_z f) = \alpha_z(\mathcal{L}f) \quad \text{for } f \in L^2(\mathbb{R}^{2d}) \text{ and } z \in \mathbb{R}^{2d}.$$

Most of the necessary mathematical background to follow this approach is provided by Gröchenig's book [16] and Ref. [25]. In any case, as far as possible, the chapter is intended to be self-contained. The chapter is broadly in line with Refs. [13, 14].

## 7.1 A Motivation: Time-Varying Systems

Let us first briefly explain a practical motivation for considering the samples defined in Eq. (7.1) for the Hilbert-Schmidt operators in $V_S^2$.

Namely, it is a well-known fact in mobile wireless channels that the relative location between transmitter and receiver is varying with time and, consequently, the input-output relation is modeled by a *time-varying system* $x \longmapsto Hx$ that can be expressed as the integral operator

$$Hx(t) = \int_{\mathbb{R}^d} h_t(s)\, x(t-s)\, ds = \int_{\mathbb{R}^d} \sigma(t,\omega)\, \widehat{x}(\omega)\, e^{2\pi i\, \omega\cdot t}\, d\omega\,,$$

where $\sigma(t,\omega) = \mathcal{F}(h_t)(\omega)$, i.e., the Fourier transform with respect to the last $d$ variables in $h(t,s) := h_t(s)$.

In the last formulation, the operator $H$ becomes a *pseudodifferential operator* with *Kohn-Nirenberg symbol* $\sigma$ (see, for instance, Refs. [16, 26]).

As pointed out in Ref. [18], in *orthogonal frequency-division multiplexing* (OFDM) the digital information, i.e., a sequence of numbers $\{c_\lambda\}$, $\lambda$ in the lattice $\Lambda = a\mathbb{Z}^d \times b\mathbb{Z}^d$ $(a, b > 0)$, is used as the coefficients of the input signal

$$x(t) = \sum_{\mu\in\Lambda} c_\mu\, \pi(\mu) g(t)$$

of a time-varying system $H$ producing the output $y(t) = Hx(t)$. Then, the sequence of numbers

$$d_\lambda = \big\langle y, \pi(\lambda)\widetilde{g}\big\rangle_{L^2(\mathbb{R}^d)} = \sum_{\mu\in\Lambda} c_\mu \big\langle H\pi(\mu)g, \pi(\lambda)\widetilde{g}\big\rangle_{L^2(\mathbb{R}^d)}\,, \quad \lambda\in\Lambda\,, \tag{7.3}$$

is considered.

The main task of the engineer is to recover the original data $\{c_\lambda\}_{\lambda\in\Lambda}$ from the received data $\{d_\lambda\}_{\lambda\in\Lambda}$.

The matrix $A = [a_{\lambda,\mu}]$, where $a_{\lambda,\mu} = \big\langle H\pi(\mu)g, \pi(\lambda)\widetilde{g}\big\rangle_{L^2(\mathbb{R}^d)}$, which appears in Eq. (7.3), involving $H$ and the time-frequency shifts of a pair of fixed functions $g, \widetilde{g} \in L^2(\mathbb{R}^d)$, is the so-called *channel matrix* associated with $H$ and the functions (windows) $g, \widetilde{g}$ in $L^2(\mathbb{R}^d)$.

As it will be proved in Sect. 7.3.4 (see Eq. (7.14) below), we have that

$$\big\langle H\pi(\lambda)g, \pi(\lambda)\widetilde{g}\big\rangle_{L^2(\mathbb{R}^d)} = \big\langle \alpha_{-\lambda}(H)g, \widetilde{g}\big\rangle_{L^2(\mathbb{R}^d)}\,, \quad \lambda\in\Lambda\,,$$

i.e., the samples $\big\langle \alpha_{-\lambda}(H)g, \widetilde{g}\big\rangle_{L^2(\mathbb{R}^d)}$, $\lambda\in\Lambda$, coincide with the diagonal entries of the channel matrix associated with $H$ and the windows $g, \widetilde{g}$.

This is the reason to consider the samples defined in Eq. (7.1) and to name them as the *diagonal channel samples* of the operator $H$ with respect to the fixed functions $g, \widetilde{g} \in L^2(\mathbb{R}^d)$ and lattice $\Lambda$.

Besides, a simple class of operators $H$ describing time-varying systems, and allowing to live in the Hilbert space setting, is given by the class of Hilbert-Schmidt operators on $L^2(\mathbb{R}^d)$.

A Hilbert-Schmidt operator $H$ on $L^2(\mathbb{R}^d)$ is a *compact operator* on $L^2(\mathbb{R}^d)$ having an integral representation with kernel $\kappa \in L^2(\mathbb{R}^{2d})$. That is,

$$\begin{aligned} Hx(t) &= \int_{\mathbb{R}^d} \kappa(t,s)\,x(s)\,ds = \int_{\mathbb{R}^d} \kappa(t,t-s)\,x(t-s)\,ds \\ &= \int_{\mathbb{R}^d} h_t(s)\,x(t-s)\,ds\,, \qquad \text{a.e. } t \in \mathbb{R}^d\,, \end{aligned}$$

with $h_t(s) = \kappa(t, t-s)$.

Although only Hilbert-Schmidt operators on $L^2(\mathbb{R}^d)$ can be described as integral operators with kernel in $L^2(\mathbb{R}^{2d})$, every bounded operator on $L^2(\mathbb{R}^d)$ can be uniquely described, via the *Schwartz kernel theorem*, by a distributional kernel in the space of tempered distributions $\mathcal{S}'(\mathbb{R}^{2d})$ (see, for instance, Ref. [16]).

## 7.2 The Kohn-Nirenberg and Weyl Transforms

Before to define the Kohn-Nirenberg and Weyl transforms in $L^2(\mathbb{R}^{2d})$ we introduce the class of Hilbert-Schmidt operators on the Hilbert space $L^2(\mathbb{R}^d)$:

### *7.2.1 The Class of Hilbert-Schmidt Operators*

The class of Hilbert-Schmidt operators on a Hilbert space, $L^2(\mathbb{R}^d)$ in our case, can be introduced by using the *Schmidt decomposition* (singular value decomposition) of a compact operator on $L^2(\mathbb{R}^d)$ (see, for instance, Ref. [4]). Namely:

*For a compact operator $S$ on $L^2(\mathbb{R}^d)$ there exist two orthonormal sequences $\{x_n\}_{n\in\mathbb{N}}$ and $\{y_n\}_{n\in\mathbb{N}}$ in $L^2(\mathbb{R}^d)$ and a bounded sequence of positive numbers $\{s_n(S)\}_{n\in\mathbb{N}}$ (the singular values of $S$) such that*

$$S = \sum_{n\in\mathbb{N}} s_n(S)\, x_n \otimes y_n\,,$$

*with convergence of the series in the operator norm.*

Here, $x_n \otimes y_n$ denotes the *rank-one operator* $L^2(\mathbb{R}^d) \longrightarrow L^2(\mathbb{R}^d)$ defined by

$$\big(x_n \otimes y_n\big)(e) = \langle e, y_n\rangle_{L^2} x_n \quad \text{for } e \in L^2(\mathbb{R}^d)\,.$$

The sequence $\big\{s_n(S)\big\}_{n\in\mathbb{N}}$ consists of the eigenvalues of the operator $|T| = \big(S^*S\big)^{1/2}$ in a decreasing order and repeated according to their multiplicities.

For $1 \leq p < \infty$ the *Schatten p-class* $\mathcal{T}^p$ is defined by

$$\mathcal{T}^p := \big\{S \text{ compact on } L^2(\mathbb{R}^d) \;:\; \{s_n(S)\}_{n\in\mathbb{N}} \in \ell^p(\mathbb{N})\big\}\,.$$

The Schatten $p$-class $\mathcal{T}^p$ becomes a Banach space endowed with the norm

$$\|S\|^p_{\mathcal{T}^p} = \sum_{n\in\mathbb{N}} s_n^p(S)\,, \quad S \in \mathcal{T}^p\,.$$

For $1 \leq p \leq q < \infty$ we have the continuous inclusions

$$\mathcal{T}^1 \hookrightarrow \mathcal{T}^2 \hookrightarrow \ldots \hookrightarrow \mathcal{T}^p \hookrightarrow \mathcal{T}^q \hookrightarrow \ldots \hookrightarrow B\big(L^2(\mathbb{R}^d)\big)\,.$$

where $B\big(L^2(\mathbb{R}^d)\big)$ denotes the Banach space of bounded operators from $L^2(\mathbb{R}^d)$ into itself. Thus, convergence in $\mathcal{T}^p$-norm implies, in particular, convergence in operator norm.

In particular, for $p = 1$ we obtain the so-called *trace class operators* $\mathcal{T}^1$. The *trace* of an operator $S \in \mathcal{T}^1$, defined by

$$\operatorname{tr}(S) = \sum_{n\in\mathbb{N}} \langle Se_n, e_n\rangle_{L^2}\,,$$

is a well-defined bounded linear functional on $\mathcal{T}^1$, and independent of the used orthonormal basis $\{e_n\}_{n\in\mathbb{N}}$ in $L^2(\mathbb{R}^d)$.

For $p = 2$ we obtain the class of *Hilbert-Schmidt operators* $\mathcal{HS}(\mathbb{R}^d) := \mathcal{T}^2$. The space $\mathcal{HS}(\mathbb{R}^d)$ endowed with the inner product

$$\big\langle S, T\big\rangle_{\mathcal{HS}} = \operatorname{tr}(ST^*)\,, \quad S, T \in \mathcal{HS}(\mathbb{R}^d)\,,$$

becomes a Hilbert space. For the norm of $S \in \mathcal{HS}(\mathbb{R}^d)$ we have the expressions:

$$\|S\|^2_{\mathcal{HS}} = \operatorname{tr}(SS^*) = \sum_{n\in\mathbb{N}} \|S^*(e_n)\|^2_{L^2} = \sum_{n\in\mathbb{N}} \|S(e_n)\|^2_{L^2} = \sum_{n\in\mathbb{N}} s_n^2(S)\,,$$

where $\{e_n\}_{n\in\mathbb{N}}$ denotes any orthonormal basis in $L^2(\mathbb{R}^d)$.

Equivalently, a Hilbert-Schmidt operator $S \in \mathcal{HS}(\mathbb{R}^d)$ can be seen as a compact operator on $L^2(\mathbb{R}^d)$ defined for each $f \in L^2(\mathbb{R}^d)$ by

$$Sf(t) = \int_{\mathbb{R}^d} \kappa_S(t, x) f(x) dx \quad \text{a.e. } t \in \mathbb{R}^d ,$$

with kernel $\kappa_S \in L^2(\mathbb{R}^{2d})$. Furthermore, for $S, T \in \mathcal{HS}(\mathbb{R}^d)$:

$$\langle S, T\rangle_{\mathcal{HS}} = \langle \kappa_S, \kappa_T\rangle_{L^2(\mathbb{R}^{2d})} \quad \text{and} \quad \|S\|_{\mathcal{HS}} = \|\kappa_S\|_{L^2(\mathbb{R}^{2d})} .$$

### 7.2.2 *The Kohn-Nirenberg and Weyl Transforms*

Now, we introduce the Kohn-Nirenberg and Weyl transforms in $L^2(\mathbb{R}^{2d})$, the setting where they will be used in this chapter (see Ref. [16, Theorem 14.6.1]). More information and details about these transforms, also valid in more general settings, can be found in Refs. [9, 10, 16, 25, 27].

(*a*) *The Weyl transform in* $L^2(\mathbb{R}^{2d})$ is defined in terms of the so-called cross-Wigner distribution. The *cross-Wigner distribution* of the functions $\psi, \phi \in L^2(\mathbb{R}^d)$ is defined by

$$W(\psi, \phi)(x, \omega) = \int_{\mathbb{R}^d} \psi\big(x + \frac{t}{2}\big) \overline{\phi\big(x - \frac{t}{2}\big)} \, e^{-2\pi i\, \omega\cdot t} dt , \quad (x, \omega) \in \mathbb{R}^{2d} .$$

The function $W(\psi, \phi)$ is well-defined for $\psi, \phi \in L^2(\mathbb{R}^d)$ and, as we will deduce from Lemma 7.1 below, belongs to $L^2(\mathbb{R}^{2d})$.

**Theorem 7.1 (The Weyl Transform in $L^2(\mathbb{R}^{2d})$)** *The Weyl transform*

$$L^2(\mathbb{R}^{2d}) \ni f \longmapsto L_f \in \mathcal{HS}(\mathbb{R}^d)$$

*is a unitary operator where* $L_f : L^2(\mathbb{R}^d) \to L^2(\mathbb{R}^d)$ *is the Hilbert-Schmidt operator defined in weak sense by*

$$\langle L_f \phi, \psi\rangle_{L^2(\mathbb{R}^d)} = \langle f, W(\psi, \phi)\rangle_{L^2(\mathbb{R}^{2d})} , \quad \phi, \psi \in L^2(\mathbb{R}^d) . \tag{7.4}$$

Thus, for each operator $S \in \mathcal{HS}(\mathbb{R}^d)$ there exists a unique function $a_S \in L^2(\mathbb{R}^{2d})$, called its *Weyl symbol*, that is, $S = L_{a_S}$, and such that

$$\langle S, T\rangle_{\mathcal{HS}} = \langle a_S, a_T\rangle_{L^2(\mathbb{R}^{2d})} \quad \text{for each } S, T \in \mathcal{HS}(\mathbb{R}^d) .$$

For the sake of completeness, we include the proofs concerning the above Weyl transform in $L^2(\mathbb{R}^{2d})$ theorem. The idea is to help the readers who are not familiar with these mathematical techniques.

We introduce $\mathcal{T}_s$, the *symmetric coordinate transformation* defined for $F$ in $L^2(\mathbb{R}^{2d})$ as

$$\mathcal{T}_s F(x, y) = F\big(x + \frac{y}{2}, x - \frac{y}{2}\big), \quad (x, y) \in \mathbb{R}^{2d}.$$

$\mathcal{T}_s$ defines a unitary operator on $L^2(\mathbb{R}^{2d})$ with inverse $\mathcal{T}_s^{-1} F(x, y) = F(\frac{x+y}{2}, x - y)$.

**Lemma 7.1** *For the cross-Wigner distribution we have the expression:*

$$W(\phi, \psi) = \mathcal{F}_2 \mathcal{T}_s(\phi \otimes \psi), \quad \phi, \psi \in L^2(\mathbb{R}^d),$$

*where $\mathcal{F}_2$ denotes the Fourier transform with respect to the last d-variables, and $\mathcal{T}_s$ denotes the symmetric coordinate transformation.*

***Proof*** In line with the definition of a rank-one operator, the tensor product $\phi \otimes \psi$ of the functions $\phi, \psi \in L^2(\mathbb{R}^d)$ means

$$(\phi \otimes \psi)(x, y) = \phi(x)\, \overline{\psi(y)}.$$

Observe that

$$\int_{\mathbb{R}^d} (\phi \otimes \psi)(x, y)\, e(y)\, dy = \langle e, \psi \rangle\, \phi(x).$$

In particular, for $\phi, \psi \in L^2(\mathbb{R}^d)$, the operator $\phi \otimes \psi$ belongs to $\mathcal{HS}(\mathbb{R}^d)$.

Thus,

$$\begin{aligned} W(\phi, \psi)(x, \omega) &= \int_{\mathbb{R}^d} \phi\big(x + \frac{t}{2}\big)\, \overline{\psi\big(x - \frac{t}{2}\big)}\, e^{-2\pi i\, \omega \cdot t}\, dt \\ &= \int_{\mathbb{R}^d} \mathcal{T}_s(\phi \otimes \psi)(x, t)\, e^{-2\pi i\, \omega \cdot t}\, dt \\ &= \mathcal{F}_2 \mathcal{T}_s(\phi \otimes \psi)(x, \omega), \qquad (x, \omega) \in \mathbb{R}^{2d}. \end{aligned}$$

□

Next, for any fixed $f \in L^2(\mathbb{R}^{2d})$ we define the *sesquilinear form*

$$Q : L^2(\mathbb{R}^d) \times L^2(\mathbb{R}^d) \longrightarrow \mathbb{C} \quad \text{with} \quad Q(\phi, \psi) = \langle f, W(\psi, \phi) \rangle, \quad \phi, \psi \in L^2(\mathbb{R}^d),$$

that is, $Q(\phi, \psi)$ is linear in the first component $\phi$ and conjugate linear in the second component $\psi$.

The sesquilinear form $Q(\phi, \psi)$ is also bounded since it is easy to check that

$$\left|Q(\phi, \psi)\right| \le \|f\|_{L^2(\mathbb{R}^{2d})}\|\phi\|_{L^2(\mathbb{R}^d)}\|\psi\|_{L^2(\mathbb{R}^d)} \quad \text{for all } \phi, \psi \in L^2(\mathbb{R}^d).$$

By using the *representation theorem for bounded sesquilinear forms* (see Ref. [3]), there exists a unique operator $A : L^2(\mathbb{R}^d) \to L^2(\mathbb{R}^d)$ such that

$$\langle A\phi, \psi\rangle_{L^2(\mathbb{R}^d)} = \langle f, W(\psi, \phi)\rangle_{L^2(\mathbb{R}^{2d})}, \quad \phi, \psi \in L^2(\mathbb{R}^d). \tag{7.5}$$

**Lemma 7.2** *Let $A$ be the operator defined by* (7.5). *Then, $A \in \mathcal{HS}(\mathbb{R}^d)$ and*

$$\|A\|_{\mathcal{HS}} = \|f\|_{L^2(\mathbb{R}^{2d})}.$$

***Proof*** Indeed, we have that

$$\begin{aligned}\langle A\phi, \psi\rangle_{L^2(\mathbb{R}^d)} &= \langle f, W(\psi, \phi)\rangle_{L^2(\mathbb{R}^{2d})} = \langle f, \mathcal{F}_2\mathcal{T}_s(\psi \otimes \phi)\rangle_{L^2(\mathbb{R}^{2d})} \\ &= \langle \mathcal{T}_s^{-1}\mathcal{F}_2^{-1} f, \psi \otimes \phi\rangle_{L^2(\mathbb{R}^{2d})} = \langle k, \psi \otimes \phi\rangle_{L^2(\mathbb{R}^{2d})},\end{aligned}$$

where $k = \mathcal{T}_s^{-1}\mathcal{F}_2^{-1} f$ belongs to $L^2(\mathbb{R}^{2d})$. Since $\mathcal{F}_2$ and $\mathcal{T}_s$ are unitary operators on $L^2(\mathbb{R}^{2d})$ we obtain that $\|k\|_{L^2(\mathbb{R}^{2d})} = \|f\|_{L^2(\mathbb{R}^{2d})}$.

Moreover, we get

$$\begin{aligned}\left\langle k, \psi \otimes \phi\right\rangle_{L^2(\mathbb{R}^{2d})} &= \iint_{\mathbb{R}^{2d}} k(x, \omega)\, \overline{\psi(x)}\, \phi(\omega)\, dx d\omega \\ &= \int_{\mathbb{R}^d} \Big(\int_{\mathbb{R}^d} k(x, \omega)\phi(\omega) d\omega\Big)\overline{\psi(x)}\, dx,\end{aligned}$$

that is, the operator $A$ is an integral operator with kernel $k \in L^2(\mathbb{R}^{2d})$:

$$A\phi(x) = \int_{\mathbb{R}^d} k(x, \omega)\phi(\omega) d\omega \quad \text{a.e. } x \in \mathbb{R}^d.$$

Finally, we prove that $\|A\|_{\mathcal{HS}} = \|k\|_{L^2(\mathbb{R}^{2d})}$.

To this end, let $\{e_n\}_{n\in\mathbb{N}}$ be any orthonormal basis in $L^2(\mathbb{R}^d)$. We have that

$$\begin{aligned}\|A\|^2_{\mathcal{HS}} &= \sum_{n\in\mathbb{N}} \|Ae_n\|^2 = \sum_{n,m} |\langle Ae_n, e_m\rangle|^2 \\ &= \sum_{n,m\in\mathbb{N}} \left|\iint_{\mathbb{R}^{2d}} k(x, y)\, e_n(x)\, \overline{e_m(y)}\, dx dy\right|^2 \\ &= \sum_{n,m\in\mathbb{N}} \left|\left\langle k, \overline{e_n(x)}\, e_m(y)\right\rangle_{L^2(\mathbb{R}^{2d})}\right|^2 = \|k\|_{L^2(\mathbb{R}^{2d})},\end{aligned}$$

since the sequence $\left\{\overline{e_n(x)}\, e_m(y)\right\}_{n,m\in\mathbb{N}}$ is an orthonormal basis for $L^2(\mathbb{R}^{2d})$.

Observe that the map $f \mapsto A$ defined by (7.5) is also surjective. Indeed, let $A$ be an operator in $\mathcal{HS}(\mathbb{R}^d)$ with kernel $k \in L^2(\mathbb{R}^{2d})$. Then the corresponding function $f \in L^2(\mathbb{R}^{2d})$ is given by $f = \mathcal{F}_2\mathcal{T}_s k$. □

In conclusion, by using the above two lemmas, the linear Weyl mapping

$$L^2(\mathbb{R}^{2d}) \ni f \longmapsto A \in \mathcal{HS}(\mathbb{R}^d)$$

is bijective and satisfies

$$\|A\|_{\mathcal{HS}} = \|k\|_{L^2(\mathbb{R}^{2d})} = \|f\|_{L^2(\mathbb{R}^{2d})}\,.$$

(*b*) *The Kohn-Nirenberg transform in* $L^2(\mathbb{R}^{2d})$ can be defined in an analogously way by using the the Rihaczek distribution in $L^2(\mathbb{R}^d)$. The *Rihaczek distribution* of the functions $\psi, \phi \in L^2(\mathbb{R}^d)$ is defined as

$$R(\psi,\phi)(x,\omega) = \psi(x)\,\overline{\widehat{\phi}(\omega)}\,e^{-2\pi i\, x\cdot\omega}\,, \quad (x,\omega) \in \mathbb{R}^{2d}\,.$$

**Theorem 7.2 (The Kohn-Nirenberg Transform in $L^2(\mathbb{R}^{2d})$)** *The Kohn-Nirenberg transform*

$$L^2(\mathbb{R}^{2d}) \ni \sigma \longmapsto K_\sigma \in \mathcal{HS}(\mathbb{R}^d)$$

*is a unitary operator where* $K_\sigma : L^2(\mathbb{R}^d) \to L^2(\mathbb{R}^d)$ *is the Hilbert-Schmidt operator defined in weak sense by*

$$\big\langle K_\sigma\phi, \psi\big\rangle_{L^2(\mathbb{R}^d)} = \big\langle \sigma, R(\psi,\phi)\big\rangle_{L^2(\mathbb{R}^{2d})}\,, \quad \phi, \psi \in L^2(\mathbb{R}^d)\,. \tag{7.6}$$

Thus, for each operator $S \in \mathcal{HS}(\mathbb{R}^d)$ there exists a unique function $\sigma_S \in L^2(\mathbb{R}^{2d})$, called its *Kohn-Nirenberg symbol*, that is, $S = K_{\sigma_S}$, and such that

$$\langle S, T\rangle_{\mathcal{HS}} = \langle \sigma_S, \sigma_T\rangle_{L^2(\mathbb{R}^{2d})} \quad \text{for each } S, T \in \mathcal{HS}(\mathbb{R}^d)\,.$$

The proofs for the Kohn-Nirenberg transform are completely analogous. Both, Kohn-Nirenberg and Weyl symbols are related by means of a unitary operator on $L^2(\mathbb{R}^{2d})$. Namely, if $a_S$ denotes the Weyl symbol of $S$, its Kohn-Nirenberg symbol $\sigma_S$ is given by $Ua_S$ where $U$ is the unitary operator on $L^2(\mathbb{R}^{2d})$ such that

$$\widehat{Ua_S}(\xi,u) = e^{\pi i u\cdot\xi}\,\widehat{a}_S(\xi,u)\,, \quad (\xi,u) \in \mathbb{R}^{2d}\,,$$

that is, $K_{Ua_S} = L_{a_S}$.

This relationship comes out by comparing the expressions of $K_\sigma$ and $L_\sigma$ as pseudodifferential operators; see the details in Ref. [16].

The Kohn-Nirenberg and Weyl transforms can be defined for $\sigma$ (or $f$) in $\mathcal{S}'(\mathbb{R}^{2d})$, i.e., for tempered distributions by using the dualities

$$\big(\mathcal{S}(\mathbb{R}^d), \mathcal{S}'(\mathbb{R}^d)\big) \quad \text{and} \quad \big(\mathcal{S}(\mathbb{R}^{2d}), \mathcal{S}'(\mathbb{R}^{2d})\big)$$

in Eq. (7.6) (or Eq. (7.4)); see, for instance, Refs. [16, 25].

## 7.3 Riesz Sequences of Translated Operators in $\mathcal{HS}(\mathbb{R}^d)$

As it was announced in the introduction section, given a fixed operator $S \in \mathcal{HS}(\mathbb{R}^d)$, we are interested in that the sequence of its translates $\{\alpha_\lambda(S)\}_{\lambda\in\Lambda}$ forms a Riesz sequence in $\mathcal{HS}(\mathbb{R}^d)$, where $\Lambda \subset \mathbb{R}^{2d}$ is a full rank lattice in $\mathbb{R}^{2d}$. To this end, we firstly remind the meaning of the translation of an operator, as well as the Fourier transform in $\ell^2(\Lambda)$ describing the dual group $\widehat{\Lambda}$.

**The Translation of an Operator**

For $z = (x, \omega) \in \mathbb{R}^{2d}$, the *time-frequency shift* operator $\pi(z) : L^2(\mathbb{R}^d) \to L^2(\mathbb{R}^d)$ is defined as

$$\pi(z)\varphi(t) := e^{2\pi i\omega\cdot t}\varphi(t-x) \quad \text{for } \varphi \in L^2(\mathbb{R}^d)\,.$$

In other words, the time-frequency shift is the composition of the *translation* $T_x$ which maps $f(t) \mapsto f(t-x)$, and the *modulation* $M_\omega$ which maps $f(t) \mapsto e^{2\pi i\omega\cdot t} f(t)$.

That is,

$$\pi(z) = M_\omega T_x\,, \quad z = (x, \omega) \in \mathbb{R}^{2d}\,.$$

Thus we get a unitary operator on $L^2(\mathbb{R}^d)$.

The time-frequency shift is used to define the *short-time Fourier transform* (or Gabor transform) $V_\psi\varphi$ of $\varphi$ with window $\psi$, both in $L^2(\mathbb{R}^d)$, by

$$V_\psi\varphi(z) = \langle\varphi, \pi(z)\psi\rangle_{L^2(\mathbb{R}^d)}\,, \quad z \in \mathbb{R}^{2d}\,.$$

The adjoint operator of $\pi(z)$ is given by

$$\pi(z)^* = e^{-2\pi i x\cdot\omega}\,\pi(-z)\,, \quad z = (x, \omega) \in \mathbb{R}^{2d}\,.$$

**Definition 7.1** *The translation $\alpha_z(S)$ by $z \in \mathbb{R}^{2d}$ of an operator $S \in \mathcal{HS}(\mathbb{R}^d)$ is defined as*

$$\alpha_z(S) := \pi(z)\, S\, \pi(z)^*\,, \quad z \in \mathbb{R}^{2d}\,.$$

**Example 7.1 (The Translation of a Rank-One Operator)** For $\varphi, \psi \in L^2(\mathbb{R}^d)$ the translation of the rank-one operator $\varphi \otimes \psi$ is given by

$$\alpha_z(\varphi \otimes \psi) = [\pi(z)\varphi] \otimes [\pi(z)\psi]\,, \quad z \in \mathbb{R}^{2d}\,.$$

Indeed, for each function $e \in L^2(\mathbb{R}^d)$ we have

$$\begin{aligned}\alpha_z(\varphi \otimes \psi)(e) &= \pi(z)(\varphi \otimes \psi)\pi(z)^*(e) = \pi(z)\,\langle \pi(z)^*(e), \psi\rangle\, \varphi \\ &= \langle e, \pi(z)\psi\rangle\, \pi(z)\varphi = \big[\pi(z)\varphi \otimes \pi(z)\psi\big](e)\,.\end{aligned}$$

**Theorem 7.3** *The map $z \longmapsto \alpha_z$ is a unitary representation of the additive group $\mathbb{R}^{2d}$ on the Hilbert space $\mathcal{HS}(\mathbb{R}^d)$ of Hilbert-Schmidt operators in $L^2(\mathbb{R}^d)$.*

***Proof*** The translation $\alpha_z : S \longmapsto \alpha_z(S)$ defines a unitary operator on $\mathcal{HS}(\mathbb{R}^d)$. Indeed,

$$\begin{aligned}\|\alpha_z(S)\|^2_{\mathcal{HS}} &= \operatorname{tr}\left(\pi(z)S\pi(z)^*\pi(z)S^*\pi(z)^*\right) = \operatorname{tr}\left(\pi(z)SS^*\pi(z)^*\right) \\ &= \sum_{n\in\mathbb{N}} \langle \pi(z)SS^*\pi(z)^* e_n, e_n\rangle = \sum_{n\in\mathbb{N}} \langle SS^*\pi(z)^* e_n, \pi(z)^* e_n\rangle \\ &= \operatorname{tr}(SS^*) = \|S\|^2_{\mathcal{HS}}\,,\end{aligned}$$

where $\{e_n\}_{n\in\mathbb{N}}$ denotes any orthonormal basis for $L^2(\mathbb{R}^d)$ and, as a consequence, the sequence $\{\pi(z)^* e_n\}_{n\in\mathbb{N}}$ is also an orthonormal basis for $L^2(\mathbb{R}^d)$ since $\pi(z)$ is a unitary operator on $L^2(\mathbb{R}^d)$.

Furthermore, the map $z \mapsto \alpha_z$ should satisfy (see Definition 5.1):

- $\alpha_z \alpha_{z'} = \alpha_{z+z'}$ for $z, z' \in \mathbb{R}^{2d}$, and
- the map $z \mapsto \alpha_z(S)$ is continuous for each $S \in \mathcal{HS}(\mathbb{R}^d)$.

To prove the first statement we use the relationship

$$\pi(z)\pi(z') = e^{-2\pi i \omega' \cdot x}\, \pi(z + z')\,, \quad (x, \omega),\, (x', \omega') \in \mathbb{R}^{2d}\,.$$

Indeed, since $T_x M_{\omega'} = e^{-2\pi i \omega' \cdot x} M_{\omega'} T_x$ we obtain

$$\begin{aligned}\pi(z)\pi(z') &= M_\omega T_x M_{\omega'} T_{x'} = e^{-2\pi i \omega' \cdot x} M_\omega M_{\omega'} T_x T_{x'} \\ &= e^{-2\pi i \omega' \cdot x} M_{\omega+\omega'} T_{x+x'} = e^{-2\pi i \omega' \cdot x}\, \pi(z + z')\,.\end{aligned}$$

Hence,

$$\alpha_{z+z'}(S) = \pi(z+z')\, S\, \pi(z+z')^* = e^{2\pi i\omega'\cdot x}\, \pi(z)\pi(z')\, S\, e^{-2\pi i\omega'\cdot x}\, \pi(z')^*\pi(z)^*$$
$$= \pi(z)\alpha_{z'}(S)\pi(z)^* = \alpha_z\big(\alpha_{z'}(S)\big)\,.$$

For the second statement we refer to the *Grümm convergence theorem* (see Ref. [24, Theorem 2.19]).

In short, this theorem says that if $A_n f \to Af$, $A_n^* f \to A^* f$ in $L^2(\mathbb{R}^d)$ for each $f \in L^2(\mathbb{R}^d)$, and $\|A_n\|_{\mathcal{HS}} \to \|A\|_{\mathcal{HS}}$, then $\|A_n - A\|_{\mathcal{HS}} \to 0$.

In our case, having in mind that $\pi(z_n) f \to \pi(z_0) f$ in $L^2(\mathbb{R}^d)$ whenever $z_n \to z_0$ in $\mathbb{R}^{2d}$, we deduce that $\|\alpha_{z_n}(S) - \alpha_{z_0}(S)\|_{\mathcal{HS}} \to 0$ as $z_n \to z_0$. □

Notice that for a discrete countable subgroup of the additive group $\mathbb{R}^{2d}$, for instance a full rank lattice $\Lambda$ in $\mathbb{R}^{2d}$, the map $\lambda \in \Lambda \mapsto \alpha_\lambda$ defines a unitary representation of $\Lambda$ on the Hilbert space $\mathcal{HS}(\mathbb{R}^d)$ of Hilbert-Schmidt operators in $L^2(\mathbb{R}^d)$. In this case, the second condition always holds.

Properties and applications of the translation of an operator $\alpha_z$ can be found, for instance, in Refs. [21, 25, 27].

**The Fourier Transform in $\ell^2(\Lambda)$**

Let $\Lambda$ be a *full rank lattice* in $\mathbb{R}^{2d}$, i.e., $\Lambda = A\mathbb{Z}^{2d}$ with $A \in GL(2d, \mathbb{R})$ and volume $|\Lambda| = \det A$. Its dual group $\widehat{\Lambda}$ is identified with $\mathbb{R}^{2d}/\Lambda^\circ$, where $\Lambda^\circ$ is the *annihilator group*

$$\Lambda^\circ = \big\{\lambda^\circ \in \mathbb{R}^{2d} \;:\; e^{2\pi i\, \sigma(\lambda^\circ, \lambda)} = 1 \text{ for all } \lambda \in \Lambda\big\}\,,$$

where $\sigma$ denotes here the *standard symplectic form* in $\mathbb{R}^{2d}$ defined as

$$\sigma(z, z') := \omega \cdot x' - \omega' \cdot x \quad \text{for } z = (x, \omega) \text{ and } z' = (x', \omega') \text{ in } \mathbb{R}^{2d}\,. \tag{7.7}$$

Notice that, since $\Lambda$ is discrete its dual group $\widehat{\Lambda}$ is compact (see Ref. [11]). The group $\Lambda^\circ$ is itself a lattice: the so-called *adjoint lattice* of $\Lambda$.

The *symplectic characters*

$$\chi_z(z') := e^{2\pi i\, \sigma(z,z')} \quad \text{with } z = (x, \omega) \text{ and } z' = (x', \omega') \text{ in } \mathbb{R}^{2d}\,,$$

are the natural way of identifying the group $\mathbb{R}^{2d}$ with its dual group via the bijection $z \longmapsto \chi_z$.

The *Fourier transform of the sequence* $c \in \ell^1(\Lambda)$ is defined as

$$\mathcal{F}_s^\Lambda(c)(\dot{z}) := \sum_{\lambda\in\Lambda} c(\lambda)\, e^{2\pi i\, \sigma(\lambda, z)}\,, \quad \dot{z} \in \mathbb{R}^{2d}/\Lambda^\circ\,,$$

where $\dot{z}$ denotes the image of $z$ under the natural quotient map $\mathbb{R}^{2d} \longrightarrow \mathbb{R}^{2d}/\Lambda^\circ$.

Since $\mathcal{F}_s^\Lambda$ is a Fourier transform it extends to a unitary mapping

$$\mathcal{F}_s^\Lambda : \ell^2(\Lambda) \longrightarrow L^2(\widehat{\Lambda})$$

which is the *Fourier transform in* $\ell^2(\Lambda)$.

When $\Lambda = \mathbb{Z}^{2d}$, the Fourier transform $\mathcal{F}_s^\Lambda(c)$ coincides with the *symplectic Fourier series* of the sequence $c \in \ell^2(\Lambda)$.

As usual, the convolution $*_\Lambda$ of two sequences $c$ and $d$ is defined by

$$(c *_\Lambda d)(\lambda) = \sum_{\mu \in \Lambda} c(\mu)\, d(\lambda - \mu), \quad \lambda \in \Lambda\,.$$

For $c$ and $d$ in $\ell^1(\Lambda)$, the convolution $c *_\Lambda d$ belongs to $\ell^1(\Lambda)$. For $c \in \ell^1(\Lambda)$ and $d \in \ell^2(\Lambda)$, the Fourier transform $\mathcal{F}_s^\Lambda$ satisfies

$$\mathcal{F}_s^\Lambda(c *_\Lambda d) = \mathcal{F}_s^\Lambda(c)\,\mathcal{F}_s^\Lambda(d)\,.$$

Moreover, if $c, d \in \ell^2(\Lambda)$ with $c *_\Lambda d \in \ell^2(\Lambda)$, then also $\mathcal{F}_s^\Lambda(c *_\Lambda d) = \mathcal{F}_s^\Lambda(c)\mathcal{F}_s^\Lambda(d)$. See the details, for instance, in Refs. [5, 11, 12, 25].

### 7.3.1 Riesz Sequences of Translated Operators

For a fixed set $\mathbf{S} = \{S_1, S_2, \ldots, S_N\} \subset \mathcal{HS}(\mathbb{R}^d)$, we are interested in that the sequence of translates $\{\alpha_\lambda(S_n)\}_{\lambda\in\Lambda;\, n=1,2,\ldots,N}$ forms a Riesz sequence for $\mathcal{HS}(\mathbb{R}^d)$, where $\Lambda \subset \mathbb{R}^{2d}$ is a full rank lattice with dual group $\widehat{\Lambda}$.

As was said before, the Weyl transform $f \mapsto L_f$ is a unitary operator from $L^2(\mathbb{R}^{2d})$ onto $\mathcal{HS}(\mathbb{R}^d)$ which respects translations in the sense that

$$L_{T_z f} = \alpha_z(L_f) \quad \text{for } f \in L^2(\mathbb{R}^{2d}) \text{ and } z \in \mathbb{R}^{2d}\,.$$

Consequently, as pointed out in Refs. [6, 25], for a fixed $S \in \mathcal{HS}(\mathbb{R}^d)$ with Weyl symbol $a_S \in L^2(\mathbb{R}^{2d})$ and lattice $\Lambda$ in $\mathbb{R}^{2d}$, the sequence $\{\alpha_\lambda(S)\}_{\lambda\in\Lambda}$ is a Riesz sequence in $\mathcal{HS}(\mathbb{R}^d)$, i.e., a Riesz basis for $V_S^2 := \overline{\text{span}}_{\mathcal{HS}}\{\alpha_\lambda(S)\}_{\lambda\in\Lambda}$, if and only if the sequence $\{T_\lambda(a_S)\}_{\lambda\in\Lambda}$ is a Riesz sequence in $L^2(\mathbb{R}^{2d})$, i.e., a Riesz basis for the shift-invariant subspace $V_{a_S}^2$ in $L^2(\mathbb{R}^{2d})$ generated by $a_S$.

A necessary and sufficient condition for $\{\alpha_\lambda(S)\}_{\lambda\in\Lambda}$ to be a Riesz sequence in $\mathcal{HS}(\mathbb{R}^d)$ is given in Ref. [25]. There, it is assumed that $S \in \mathcal{B}$, a Banach space of continuous operators with Weyl symbol $a_S$ in the *Feichtinger's algebra* $S_0(\mathbb{R}^{2d})$. In essence, $\mathcal{B}$ consists of trace class operators on $L^2(\mathbb{R}^d)$ with a norm-continuous inclusion $\iota : \mathcal{B} \hookrightarrow \mathcal{T}^1$ (see the details in Refs. [17, 25]).

Recall that the *Feichtinger's algebra* $\mathcal{S}_0(\mathbb{R}^d)$ is the space of all tempered distributions $\psi$ in $\mathbb{R}^d$ such that

$$\|\psi\|_{\mathcal{S}_0} := \int_{\mathbb{R}^{2d}} |V_{\varphi_0}\psi(z)|dz < \infty ,$$

where $\varphi_0$ denotes the $L^2$-normalized gaussian $\varphi_0(x) = 2^{d/4}e^{-\pi x\cdot x}$ for $x \in \mathbb{R}^d$. With this norm, $\mathcal{S}_0(\mathbb{R}^d)$ is a Banach space of continuous functions and an algebra under multiplication and convolution; see the details in Refs. [16, 19, 25].

The simplest examples of operators in the Banach space $\mathcal{B}$ are the rank-one operators $\phi \otimes \psi$ with $\phi, \psi \in \mathcal{S}_0(\mathbb{R}^d)$.

**Theorem 7.4** *Let $\Lambda$ be a lattice and $S \in \mathcal{B}$. The sequence $\{\alpha_\lambda(S)\}_{\lambda\in\Lambda}$ is a Riesz sequence in $\mathcal{HS}(\mathbb{R}^d)$ if and only if the function*

$$P_{\Lambda^\circ}\big(|\mathcal{F}_W(S)|^2\big)(\dot{z}) := \frac{1}{|\Lambda|}\sum_{\lambda^\circ\in\Lambda^\circ} |\mathcal{F}_W(S)(z+\lambda^\circ)|^2 , \quad z \in \mathbb{R}^{2d} ,$$

*has no zeros in $\widehat{\Lambda}$.*

This is Theorem 6.1 in Ref. [25]. It involves the *periodization operator* $P_{\Lambda^\circ}$ in $\Lambda^\circ$ and the *Fourier-Wigner transform* $\mathcal{F}_W$ of an operator $S$. In this case, we have that $\mathcal{F}_W(S) = \mathcal{F}_s(a_S)$, where $\mathcal{F}_s$ denotes the *symplectic Fourier transform* of $a_S$ defined by

$$\mathcal{F}_s(a_S)(z) := \int_{\mathbb{R}^{2d}} a_S(z')\,e^{-2\pi i\,\sigma(z,z')}dz' , \quad z \in \mathbb{R}^{2d} ,$$

where $\sigma$ denotes the standard symplectic form in $\mathbb{R}^{2d}$ (see (7.7)). This fact will be justified in Sect. 7.5($f$).

The symplectic Fourier transform $\mathcal{F}_s$, defined in $L^1(\mathbb{R}^{2d})\cap L^2(\mathbb{R}^{2d})$ as the above integral, extends to a unitary operator $L^2(\mathbb{R}^{2d}) \longrightarrow L^2(\mathbb{R}^{2d})$. Moreover, $\mathcal{F}_s$ is its own inverse, that is, $\mathcal{F}_s(\mathcal{F}_s f) = f$ for each $f \in L^2(\mathbb{R}^{2d})$. The relationship between the symplectic Fourier transform and the classical Fourier transform of $f \in L^2(\mathbb{R}^{2d})$ is $\mathcal{F}_s f(x,\omega) = \mathcal{F}f(\omega,-x)$.

The Fourier-Wigner transform of an operator $S$ is defined as the function

$$\mathcal{F}_W(S)(z) := e^{-\pi i\,x\cdot\omega}\,\mathrm{tr}[\pi(-z)S] , \quad z = (x,\omega) \in \mathbb{R}^{2d} . \tag{7.8}$$

See the details and proof of Theorem 7.4 in Ref. [25].

A similar result just for a rank-one operator $S = \psi \otimes \phi$, with $\psi, \phi \in L^2(\mathbb{R}^d)$, can be found in Refs. [2, 6].

### 7.3.2 Λ-Shift-Invariant Subspaces

In case $\{\alpha_\lambda(S)\}_{\lambda\in\Lambda}$ is a Riesz sequence for $\mathcal{HS}(\mathbb{R}^d)$, the operator $S$ is the *generator* of the Λ-*shift-invariant subspace* $V_S^2$ which can be described by

$$V_S^2 := \overline{\text{span}}_{\mathcal{HS}}\big\{\alpha_\lambda(S)\big\}_{\lambda\in\Lambda} = \Big\{\sum_{\lambda\in\Lambda} c(\lambda)\,\alpha_\lambda(S) \;:\; \{c(\lambda)\}_{\lambda\in\Lambda} \in \ell^2(\Lambda)\Big\}.$$

**Example 7.2 (The Case Where $S = \varphi \otimes \psi$)** Observe that, in case $S = \varphi \otimes \psi$, the operators in $V_S^2$ are nothing but the so-called Gabor multipliers used in time-frequency analysis (see, for instance, Ref. [25]).

**Definition 7.2** *The Gabor multiplier $\mathcal{G}_{\mathbf{c}}^{\psi,\varphi}$ with windows $\psi$, $\varphi$ and mask $\mathbf{c}$ in $\ell^2(\Lambda)$ is the operator $\mathcal{G}_{\mathbf{c}}^{\psi,\varphi} : L^2(\mathbb{R}^d) \longrightarrow L^2(\mathbb{R}^d)$ defined as*

$$\mathcal{G}_{\mathbf{c}}^{\psi,\varphi}(\eta) := \sum_{\lambda\in\Lambda} c(\lambda)\, V_\psi\,\eta(\lambda)\pi(\lambda)\varphi\,, \quad \eta \in L^2(\mathbb{R}^d)\,. \tag{7.9}$$

Indeed, for $\eta \in L^2(\mathbb{R}^d)$ we have

$$\sum_{\lambda\in\Lambda} c(\lambda)\,\alpha_\lambda(S)(\eta) = \sum_{\lambda\in\Lambda} c(\lambda)\,\big(\pi(\lambda)\varphi \otimes \pi(\lambda)\psi\big)(\eta) = \sum_{\lambda\in\Lambda} c(\lambda)\, V_\psi\,\eta(\lambda)\pi(\lambda)\varphi\,,$$

that is,

$$\sum_{\lambda\in\Lambda} c(\lambda)\,\alpha_\lambda(S) = \mathcal{G}_{\mathbf{c}}^{\psi,\varphi}\,.$$

Analogously, a necessary and sufficient condition can be obtained for the multiply generated case. Indeed, let $\mathbf{S} = \{S_1, S_2, \ldots, S_N\}$ be a fixed subset of $\mathcal{HS}(\mathbb{R}^d)$ and let Λ be a lattice in $\mathbb{R}^{2d}$. We are searching for a necessary and sufficient condition such that $\{\alpha_\lambda(S_n)\}_{\lambda\in\Lambda;\, n=1,2,\ldots,N}$ is a Riesz sequence for $\mathcal{HS}(\mathbb{R}^d)$, i.e., a Riesz basis for the closed subspace

$$V_{\mathbf{S}}^2 := \overline{\text{span}}_{\mathcal{HS}}\big\{\alpha_\lambda(S_n)\big\}_{\lambda\in\Lambda;\, n=1,2,\ldots,N} \subset \mathcal{HS}(\mathbb{R}^d)\,.$$

For the multiply generated case we have the following result:

**Theorem 7.5** *Let Λ be a lattice and $S_n \in \mathcal{B}$, $n = 1, 2, \ldots, N$. Then, the sequence $\{\alpha_\lambda(S_n)\}_{\lambda\in\Lambda;\, n=1,2,\ldots,N}$ is a Riesz sequence for $\mathcal{HS}(\mathbb{R}^d)$ if and only if there exist two constants $0 < m \le M$ such that*

$$m\,\mathbb{I}_N \le G_{\mathbf{S}}^W(z) \le M\,\mathbb{I}_N \quad \textit{for all } z \in \mathbb{R}^{2d}\,,$$

*where* $G_{\mathbf{S}}^{W}(z)$ *denotes the* $N \times N$ *matrix-valued function*

$$G_{\mathbf{S}}^{W}(z) := \sum_{\lambda^\circ \in \Lambda^\circ} \mathcal{F}_W(\mathbf{S})(z+\lambda^\circ)\, \overline{\mathcal{F}_W(\mathbf{S})(z+\lambda^\circ)}^{\top}, \quad z \in \mathbb{R}^{2d},$$

*and*

$$\mathcal{F}_W(\mathbf{S}) = \big(\mathcal{F}_W(S_1), \mathcal{F}_W(S_2), \dots, \mathcal{F}_W(S_N)\big)^{\top}.$$

***Proof*** The sequence $\{\alpha_\lambda(S_n)\}_{\lambda \in \Lambda;\, n=1,2,\dots,N}$ is a Riesz sequence in $\mathcal{HS}(\mathbb{R}^d)$ if and only if the sequence $\{T_\lambda(a_{S_n})\}_{\lambda \in \Lambda;\, n=1,2,\dots,N}$ is a Riesz sequence in $L^2(\mathbb{R}^{2d})$.

To check this, we introduce the $N \times N$ matrix-valued function

$$G_{\mathbf{S}}^{s}(z) := \sum_{\lambda^\circ \in \Lambda^\circ} \mathcal{F}_s(a_{\mathbf{S}})(z+\lambda^\circ)\, \overline{\mathcal{F}_s(a_{\mathbf{S}})(z+\lambda^\circ)}^{\top}, \quad z \in \mathbb{R}^{2d},$$

where $\mathcal{F}_s(a_{\mathbf{S}}) = \big(\mathcal{F}_s(a_{S_1}), \mathcal{F}_s(a_{S_2}), \dots, \mathcal{F}_s(a_{S_N})\big)^{\top}$. It is known (see Example 5.8 in Chap. 5) that the sequence $\{T_\lambda(a_{S_n})\}_{\lambda \in \Lambda;\, n=1,2,\dots,N}$ is a Riesz sequence in $L^2(\mathbb{R}^{2d})$ if and only if there exist two constants $0 < m \le M$ such that

$$m\, \mathbb{I}_N \le G_{\mathbf{S}}^{s}(z) \le M\, \mathbb{I}_N, \quad \text{a.e. } z \in \mathbb{R}^{2d}.$$

Assuming as before that $S_n \in \mathcal{B}$, $n = 1, 2, \dots, N$, the functions $\mathcal{F}_s(a_{S_n})$ are continuous and $\mathcal{F}_W(S_n) = \mathcal{F}_s(a_{S_n})$ for $n = 1, 2, \dots, N$.

Hence, the above necessary and sufficient condition can be expressed in terms of the hermitian matrix $G_{\mathbf{S}}^{W}(z)$ as in the statement of the theorem. □

Observe that, proceeding as in the above theorem, the existence of two constants $0 < m \le M$ such that

$$m\, \mathbb{I}_N \le G_{\mathbf{S}}^{W}(z) \le M\, \mathbb{I}_N, \quad \text{a.e. } z \in \mathbb{R}^{2d},$$

is a necessary and sufficient condition to be the sequence $\{\alpha_\lambda(S_n)\}_{\lambda \in \Lambda;\, n=1,2,\dots,N}$ a Riesz sequence for $\mathcal{HS}(\mathbb{R}^d)$, for the operators $S_n \in \mathcal{HS}(\mathbb{R}^d)$, $n = 1, 2, \dots, N$.

**Example 7.3 (The Fourier-Wigner Transform of a Rank-One Operator)** For a rank-one operator $\varphi \otimes \psi$, where $\varphi, \psi \in L^2(\mathbb{R}^d)$, an easy computation gives:

$$\mathcal{F}_W(\varphi \otimes \psi)(z) = e^{\pi i x \cdot \omega}\, V_\psi \varphi(z), \quad z = (x, \omega) \in L^2(\mathbb{R}^{2d}). \tag{7.10}$$

Indeed, let $\{e_n\}_{n\in\mathbb{N}}$ be an orthonormal basis for $L^2(\mathbb{R}^d)$; by using the definition (7.8) of the Fourier-Wigner transform we get

$$\begin{aligned}\mathcal{F}_W(\varphi\otimes\psi)(z) &= \mathrm{e}^{-\pi i x\cdot\omega}\,\mathrm{tr}\big[\pi(-z)\varphi\otimes\psi\big]\\ &= \mathrm{e}^{-\pi i x\cdot\omega}\sum_{n\in\mathbb{N}}\big\langle(\pi(-z)\varphi\otimes\psi)e_n, e_n\big\rangle\\ &= \mathrm{e}^{-\pi i x\cdot\omega}\sum_{n\in\mathbb{N}}\langle e_n,\psi\rangle\,\langle\pi(-z)\varphi, e_n\rangle\\ &= \mathrm{e}^{-\pi i x\cdot\omega}\big\langle\pi(-z)\varphi,\psi\big\rangle = \mathrm{e}^{-\pi i x\cdot\omega}\big\langle \mathrm{e}^{2\pi i x\cdot\omega}\pi(z)^*\varphi,\psi\big\rangle\\ &= \mathrm{e}^{\pi i x\cdot\omega}\big\langle\varphi,\pi(z)\psi\big\rangle = \mathrm{e}^{\pi i x\cdot\omega}\,V_\psi\varphi(z)\,.\end{aligned}$$

From now on, in case $\mathbf{S}=\{S_1,S_2,\dots,S_N\}$ is a *set of stable generators* for the $\Lambda$-*shift-invariant subspace* $V_{\mathbf{S}}^2 := \overline{\mathrm{span}}_{\mathcal{HS}}\big\{\alpha_\lambda(S_n)\big\}_{\lambda\in\Lambda;\,n=1,2,\dots,N}$, that is, the sequence $\{\alpha_\lambda(S_n)\}_{\lambda\in\Lambda;\,n=1,2,\dots,N}$ is a Riesz sequence for $\mathcal{HS}(\mathbb{R}^d)$, then $V_{\mathbf{S}}^2$ can be described by

$$V_{\mathbf{S}}^2=\Big\{\sum_{n=1}^{N}\sum_{\lambda\in\Lambda}c_n(\lambda)\,\alpha_\lambda(S_n)\ :\ \{c_n(\lambda)\}_{\lambda\in\Lambda}\in\ell^2(\Lambda)\,;\ n=1,2,\dots,N\Big\}\,.$$

### 7.3.3 The Isomorphism $\mathcal{T}_{\mathbf{S}}$ connecting $\ell^2_N(\Lambda)$ with $V^2_{\mathbf{S}}$

Our sampling results rely on the following isomorphism $\mathcal{T}_{\mathbf{S}}$ which involves the spaces $\ell^2_N(\Lambda)$, the shift-invariant subspace $V^2_{\sigma_{\mathbf{S}}}$ in $L^2(\mathbb{R}^{2d})$ generated by the Kohn-Nirenberg symbols $\sigma_{S_n}$ of $S_n$, $n=1,2,\dots,N$, and the $\Lambda$-shift-invariant subspace $V^2_{\mathbf{S}}$. Namely,

$$\begin{array}{llllll}\mathcal{T}_{\mathbf{S}}: & \ell^2_N(\Lambda) & \longrightarrow & V^2_{\sigma_{\mathbf{S}}}\subset L^2(\mathbb{R}^{2d}) & \longrightarrow & V^2_{\mathbf{S}}\subset\mathcal{HS}(\mathbb{R}^d)\\ & (c_1,c_2,\dots,c_N)^\top & \longmapsto & \displaystyle\sum_{n=1}^{N}\sum_{\lambda\in\Lambda}c_n(\lambda)\,T_\lambda\sigma_{S_n} & \longmapsto & \displaystyle\sum_{n=1}^{N}\sum_{\lambda\in\Lambda}c_n(\lambda)\,\alpha_\lambda(S_n)\,.\end{array} \tag{7.11}$$

The isomorphism $\mathcal{T}_{\mathbf{S}}$ is the composition of the isomorphism $\mathcal{T}_{\sigma_{\mathbf{S}}}:\ell^2_N(\Lambda)\to V^2_{\sigma_{\mathbf{S}}}$ which maps the standard orthonormal basis $\{\delta_\lambda\}_{\lambda\in\Lambda}$ for $\ell^2_N(\Lambda)$ onto the Riesz basis $\{T_\lambda\sigma_{S_n}\}_{\lambda\in\Lambda;\,n=1,2,\dots,N}$ for $V^2_{\sigma_{\mathbf{S}}}$, and the Kohn-Nirenberg transform between $V^2_{\sigma_{\mathbf{S}}}$ and $V^2_{\mathbf{S}}$.

The Kohn-Nirenberg transform $L^2(\mathbb{R}^{2d})\ni f\mapsto K_f\in\mathcal{HS}(\mathbb{R}^d)$ is a unitary operator which respects translations in the sense that $K_{T_z f}=\alpha_z(K_f)$ for $f\in$

$L^2(\mathbb{R}^{2d})$ and $z \in \mathbb{R}^{2d}$. For a proof, see Ref. [7, Lemma 7.5.3]; for the Weyl transform, see Ref. [21].

### 7.3.4 An Expression for the Diagonal Channel Samples

For each $T = \sum_{n=1}^{N} \sum_{\mu\in\Lambda} c_n(\mu)\, \alpha_\mu(S_n)$ in $V_{\mathbf{S}}^2$ we define a set of *diagonal channel samples* as

$$\mathbf{s}_T(\lambda) := \big(\langle \alpha_{-\lambda}(T)g_1, \widetilde{g}_1\rangle, \langle \alpha_{-\lambda}(T)g_2, \widetilde{g}_2\rangle, \\ \dots, \langle \alpha_{-\lambda}(T)g_M, \widetilde{g}_M\rangle\big)^\top, \quad \lambda \in \Lambda, \tag{7.12}$$

where $g_m, \widetilde{g}_m$, $m = 1, 2, \dots, M$, denote $2M$ fixed functions in $L^2(\mathbb{R}^d)$.

For $m = 1, 2, \dots, M$ the above samples can be expressed by

$$\begin{aligned} s_{T,m}(\lambda) &:= \big\langle \alpha_{-\lambda}(T)g_m, \widetilde{g}_m\big\rangle_{L^2(\mathbb{R}^d)} \\ &= \Big\langle \sum_{n=1}^{N}\sum_{\mu\in\Lambda} c_n(\mu)\alpha_{\mu-\lambda}(S_n)g_m, \widetilde{g}_m\Big\rangle_{L^2(\mathbb{R}^d)} \\ &= \sum_{n=1}^{N}\sum_{\mu\in\Lambda} c_n(\mu)\big\langle \alpha_{\mu-\lambda}(S_n)g_m, \widetilde{g}_m\big\rangle_{L^2(\mathbb{R}^d)} \\ &= \sum_{n=1}^{N}\big(a_{m,n} *_\Lambda c_n\big)(\lambda), \qquad \lambda \in \Lambda, \end{aligned} \tag{7.13}$$

where

$$a_{m,n}(\mu) = \big\langle \alpha_{-\mu}(S_n)g_m, \widetilde{g}_m\big\rangle_{L^2(\mathbb{R}^d)}, \quad \mu \in \Lambda,$$

for $m = 1, 2, \dots, M$ and $n = 1, 2, \dots, N$.

Observe that $a_{m,n}(\lambda)$, $\lambda \in \Lambda$, and $m = 1, 2, \dots, M$, are precisely the samples $\mathbf{s}_{S_n}(\lambda)$, $\lambda \in \Lambda$, of the generator $S_n$.

**Lemma 7.3** *Concerning the samples defined in Eq.* (7.13) *we have:*

*1. For $m = 1, 2, \dots, M$, these samples can be expressed as*

$$\begin{aligned} \big\langle \alpha_{-\lambda}(T)g_m, \widetilde{g}_m\big\rangle_{L^2(\mathbb{R}^d)} &= \big\langle T\pi(\lambda)g_m, \pi(\lambda)\widetilde{g}_m\big\rangle_{L^2(\mathbb{R}^d)} \\ &= \big\langle T, \alpha_\lambda(\widetilde{g}_m \otimes g_m)\big\rangle_{\mathcal{HS}}, \qquad \lambda \in \Lambda. \end{aligned} \tag{7.14}$$

2. *For* $m = 1, 2, \ldots, M$ *and* $n = 1, 2, \ldots, N$, *the sequences* $\{a_{m,n}(\lambda)\}_{\lambda\in\Lambda}$ *involved in Eq.* (7.13) *belong to* $\ell^2(\Lambda)$.

***Proof*** For the first equality in (7.14) we have that

$$\begin{aligned} s_{T,m}(\lambda) &= \big\langle \alpha_{-\lambda}(T)g_m, \widetilde{g}_m\big\rangle_{L^2(\mathbb{R}^d)} \\ &= \big\langle \pi(-\lambda)T\pi(-\lambda)^* g_m, \widetilde{g}_m\big\rangle_{L^2(\mathbb{R}^d)} \\ &= \big\langle \sigma_T, R\big(\pi(-\lambda)^*\widetilde{g}_m, \pi(-\lambda)^* g_m\big)\big\rangle_{L^2(\mathbb{R}^{2d})}, \qquad \lambda \in \Lambda . \end{aligned}$$

On the other hand, it is easy to check that for the Rihaczek distribution one gets

$$R\big(\pi(-\lambda)^*\widetilde{g}_m, \pi(-\lambda)^* g_m\big)(z) = R\big(\pi(\lambda)\widetilde{g}_m, \pi(\lambda)g_m\big)(z), \quad z \in \mathbb{R}^{2d} .$$

Hence, for each $\lambda \in \Lambda$ we obtain

$$\begin{aligned} \big\langle \alpha_{-\lambda}(T)g_m, \widetilde{g}_m\big\rangle_{L^2(\mathbb{R}^d)} &= \big\langle \sigma_T, R\big(\pi(\lambda)\widetilde{g}_m, \pi(\lambda)g_m\big)\big\rangle_{L^2(\mathbb{R}^{2d})} \\ &= \big\langle T\pi(\lambda)g_m, \pi(\lambda)\widetilde{g}_m\big\rangle_{L^2(\mathbb{R}^d)} . \end{aligned}$$

For the second equality we get

$$\begin{aligned} \big\langle T, \alpha_\lambda(\widetilde{g}_m \otimes g_m)\big\rangle_{\mathcal{HS}} &= \big\langle T, \pi(\lambda)\widetilde{g}_m \otimes \pi(\lambda)g_m\big\rangle_{\mathcal{HS}} \\ &= \big\langle \sigma_T, \sigma_{\pi(\lambda)\widetilde{g}_m\otimes\pi(\lambda)g_m}\big\rangle_{L^2(\mathbb{R}^{2d})} \\ &= \big\langle \sigma_T, R(\pi(\lambda)\widetilde{g}_m, \pi(\lambda)g_m)\big\rangle_{L^2(\mathbb{R}^{2d})} \\ &= \big\langle T\pi(\lambda)g_m, \pi(\lambda)\widetilde{g}_m\big\rangle_{L^2(\mathbb{R}^d)} . \end{aligned}$$

We have used that the Kohn-Nirenberg symbol of $\pi(\lambda)\widetilde{g}_m \otimes \pi(\lambda)g_m$ coincides with the Rihaczek distribution of the pair of functions $\pi(\lambda)\widetilde{g}_m$ and $\pi(\lambda)g_m$ in $L^2(\mathbb{R}^d)$. This follows from the equality

$$\begin{aligned} \big\langle R(h,k), R(g,f)\big\rangle_{L^2(\mathbb{R}^{2d})} &= \langle f, k\rangle_{L^2(\mathbb{R}^d)}\, \langle h, g\rangle_{L^2(\mathbb{R}^d)} \\ &= \big\langle (h\otimes k)f, g\big\rangle_{L^2(\mathbb{R}^d)} \qquad \text{for } f, g, h, k \in L^2(\mathbb{R}^d) . \end{aligned}$$

As a consequence, one deduces that $\sigma_{h\otimes k} = R(h, k)$.

In particular we have proved that

$$\begin{aligned} a_{m,n}(\lambda) &= \big\langle \alpha_{-\lambda}(S_n)g_m, \widetilde{g}_m\big\rangle_{L^2(\mathbb{R}^d)} \\ &= \big\langle S_n, \alpha_\lambda(\widetilde{g}_m \otimes g_m)\big\rangle_{\mathcal{HS}} \\ &= \big\langle \alpha_{-\lambda}(S_n), \widetilde{g}_m \otimes g_m\big\rangle_{\mathcal{HS}}, \quad \lambda \in \Lambda . \end{aligned}$$

Since $\{\alpha_\lambda(S_n)\}_{\lambda\in\Lambda;\, n=1,2,\ldots,N}$ is a Riesz sequence for $\mathcal{HS}(\mathbb{R}^d)$, it is in particular a Bessel sequence in $\mathcal{HS}(\mathbb{R}^d)$. Hence, the sequences

$$\big\{\big\langle \alpha_{-\lambda}(S_n),\, \widetilde{g}_m \otimes g_m\big\rangle_{\mathcal{HS}}\big\}_{\lambda\in\Lambda}$$

belongs to $\ell^2(\Lambda)$ for $m = 1, 2, \ldots, M$, and $n = 1, 2, \ldots, N$. □

### 7.3.5 The Associated Convolution System

Once we have that $a_{m,n} \in \ell^2(\Lambda)$ for each $m = 1, 2, \ldots, M$, and $n = 1, 2, \ldots, N$, and denoting $A = [a_{m,n}]$ the corresponding $M \times N$ matrix with entries in $\ell^2(\Lambda)$, the sampling process in Eq. (7.12) is described by means of the discrete convolution system

$$T = \sum_{n=1}^{N}\sum_{\lambda\in\Lambda} c_n(\lambda)\,\alpha_\lambda(S_n) \in V_{\mathbf{S}}^2 \longmapsto \mathbf{s}_T(\lambda) = \big(A *_\Lambda \mathbf{c}\big)(\lambda)$$
$$= \sum_{\mu\in\Lambda} A(\lambda-\mu)\,\mathbf{c}(\mu), \quad \lambda \in \Lambda\,, \tag{7.15}$$

where $\mathbf{c} = (c_1, c_2, \ldots, c_N)^\top \in \ell^2_N(\Lambda) := \ell^2(\Lambda) \times \cdots \times \ell^2(\Lambda)$ ($N$ times).

Note that the $m$-th entry of $A *_\Lambda \mathbf{c}$ is

$$\sum_{n=1}^{N}(a_{m,n} *_\Lambda c_n)\,.$$

First of all, the operator $\mathcal{A} : \ell^2_N(\Lambda) \longrightarrow \ell^2_M(\Lambda)$ which maps $\mathbf{c} \longmapsto A *_\Lambda \mathbf{c}$ is a well-defined bounded operator if and only if the $M \times N$ matrix-valued function

$$\widehat{A}(\xi) := \big[\mathcal{F}_s^\Lambda(a_{m,n})(\xi)\big], \quad \text{a.e. } \xi \in \widehat{\Lambda}\,,$$

has entries in $L^\infty(\widehat{\Lambda})$ (see Proposition 5.1).

The needed results on discrete convolution systems $\mathcal{A} : \ell^2_N(\Lambda) \longrightarrow \ell^2_M(\Lambda)$, and their relationship with frames of translates in $\ell^2_N(\Lambda)$ can be found in Chap. 5.

Notice that the $m$-th component of $A *_\Lambda \mathbf{c}$ is

$$[A * \mathbf{c}]_m(\lambda) = \sum_{n=1}^{N}(a_{m,n} *_\Lambda c_n)(\lambda) = \big\langle \mathbf{c}, T_\lambda\, \mathbf{a}_m^*\big\rangle_{\ell^2_N(\Lambda)}, \quad \lambda \in \Lambda\,,$$

where $\mathbf{a}_m = \big(a_{m,1}(\mu), a_{m,2}(\mu), \ldots, a_{m,N}(\mu)\big)^\top \in \ell^2_N(\Lambda)$, and $a^*_{m,n}$ denotes the involution

$$a^*_{m,n}(\mu) := \overline{a_{m,n}(-\mu)}\,, \qquad \mu \in \Lambda\,.$$

As a consequence, the operator $\mathcal{A}$ is the analysis operator of the sequence

$$\big\{T_\lambda \mathbf{a}^*_m\big\}_{\lambda\in\Lambda;\, m=1,2,\ldots,M} \text{ in } \ell^2_N(\Lambda)\,.$$

Since the sequence $\big\{T_\lambda\, \mathbf{a}^*_m\big\}_{\lambda\in\Lambda;\, m=1,2,\ldots,M}$ is a frame for $\ell^2_N(\Lambda)$ if and only if its bounded analysis operator is injective with a closed range (see Corollary 1.1), it will be a frame for $\ell^2_N(\Lambda)$ if and only if

$$\begin{aligned} 0 < \alpha_A &:= \operatorname*{ess\,inf}_{\xi\in\widehat{\Lambda}} \lambda_{\min}[\,\widehat{A}(\xi)^*\widehat{A}(\xi)] \\ &\le \beta_A := \operatorname*{ess\,sup}_{\xi\in\widehat{\Lambda}} \lambda_{\max}[\widehat{A}(\xi)^*\widehat{A}(\xi)] < +\infty\,, \end{aligned} \tag{7.16}$$

where $\lambda_{\min}$ (respectively, $\lambda_{\max}$) denotes the smallest (respectively, the largest) eigenvalue of the positive semidefinite matrix $\widehat{A}(\xi)^*\widehat{A}(\xi)$ (see Proposition 5.4).

Concerning the duals of $\big\{T_\lambda\, \mathbf{a}^*_m\big\}_{\lambda\in\Lambda;\, m=1,2,\ldots,M}$ having its same structure, consider two matrices $\widehat{A} \in \mathcal{M}_{M\times N}(L^\infty(\widehat{\Lambda}))$ and $\widehat{B} \in \mathcal{M}_{N\times M}(L^\infty(\widehat{\Lambda}))$, and let $\mathbf{b}_m$ denote the $m$-th column of the matrix $B$ associated with $\widehat{B}$. Then, the sequences

$$\big\{T_\lambda\, \mathbf{a}^*_m\big\}_{\lambda\in\Lambda;\, m=1,2,\ldots,M} \quad \text{and} \quad \big\{T_\lambda\, \mathbf{b}_m\big\}_{\lambda\in\Lambda;\, m=1,2,\ldots,M}$$

form a pair of dual frames for $\ell^2_N(\Lambda)$ if and only if $\widehat{B}(\xi)\,\widehat{A}(\xi) = \mathbb{I}_N$, a.e. $\xi \in \widehat{\Lambda}$ (see Proposition 5.5).

Equivalently, they are dual frames if and only if $\mathcal{B}\mathcal{A} = I_{\ell^2_N(\Lambda)}$, i.e., the convolution system $\mathcal{B}$ with matrix $B$ is a left-inverse of the convolution system $\mathcal{A}$ with matrix $A$.

Thus, we get the frame expansion in $\ell^2_N(\Lambda)$

$$\mathbf{c} = \sum_{m=1}^{M}\sum_{\lambda\in\Lambda} \big\langle \mathbf{c}, T_\lambda \mathbf{a}^*_m\big\rangle_{\ell^2_N(\Lambda)} T_\lambda \mathbf{b}_m \quad \text{for each } \mathbf{c} \in \ell^2_N(\Lambda)\,.$$

Observe that a possible left-inverse $\widehat{B}(\xi)$ of the matrix $\widehat{A}(\xi)$ is given by its Moore-Penrose pseudo-inverse

$$\widehat{A}(\xi)^\dagger = \big[\widehat{A}(\xi)^*\widehat{A}(\xi)\big]^{-1}\widehat{A}(\xi)^*\,, \quad \text{a.e. } \xi \in \widehat{\Lambda}\,.$$

## 7.4 The Sampling Results

Firstly, we prove a sampling theorem in a $\Lambda$-shift-invariant subspace $V_{\mathbf{S}}^2$ of Hilbert-Schmidt operators by using the diagonal channel samples at a lattice $\Lambda$ in $\mathbb{R}^{2d}$. Later on, we deal with the general case of average samples at a lattice $\Lambda$.

### 7.4.1 *Sampling by using the Diagonal Channel Samples at a Lattice*

Let $V_{\mathbf{S}}^2$ be a $\Lambda$-shift-invariant subspace, where $\Lambda$ denotes a full rank lattice in $\mathbb{R}^{2d}$. Next we prove a sampling result in $V_{\mathbf{S}}^2$ which is the main result in this chapter.

**Theorem 7.6 (Sampling Theorem in $V_{\mathbf{S}}^2$)** *Suppose that for each $T \in V_{\mathbf{S}}^2$ we consider the samples defined in Eq.* (7.12), *and that the matrix $A = [a_{m,n}]$ with entries*

$$a_{m,n}(\lambda) = \big\langle \alpha_{-\lambda}(S_n)g_m, \widetilde{g}_m \big\rangle_{L^2(\mathbb{R}^d)}, \quad \lambda \in \Lambda, \tag{7.17}$$

*satisfies conditions in Eq.* (7.16). *Then, there exist $M \geq N$ elements $H_m \in V_{\mathbf{S}}^2$, $m = 1, 2, \dots, M$, such that the sampling formula*

$$T = \sum_{m=1}^{M} \sum_{\lambda \in \Lambda} s_{T,m}(\lambda)\, \alpha_\lambda(H_m) \quad \text{in } \mathcal{HS}\text{-norm} \tag{7.18}$$

*holds for each $T \in V_{\mathbf{S}}^2$.*

*The sequence $\{\alpha_\lambda(H_m)\}_{\lambda\in\Lambda;\, m=1,2,\dots,M}$ is a frame for $V_{\mathbf{S}}^2$, and the convergence of the series is unconditional in Hilbert-Schmidt norm.*

*Moreover, the $\ell^2$-norm of the samples $\|\mathbf{s}_T\|_{\ell^2_M}$ defines an equivalent norm to $\|T\|_{\mathcal{HS}}$ in $V_{\mathbf{S}}^2$, and for each $f \in L^2(\mathbb{R}^d)$ we have the pointwise expansion*

$$Tf = \sum_{m=1}^{M} \sum_{\lambda \in \Lambda} s_{T,m}(\lambda)\, \alpha_\lambda(H_m) f \quad \text{in } L^2(\mathbb{R}^d).$$

***Proof*** Under the hypotheses of the theorem, the sequence $\big\{T_\lambda\, \mathbf{a}_m^*\big\}_{\lambda\in\Lambda;\, m=1,2,\dots,M}$ is a frame for $\ell^2_N(\Lambda)$, and we can consider a dual frame $\big\{T_\lambda\, \mathbf{b}_m\big\}_{\lambda\in\Lambda;\, m=1,2,\dots,M}$ with the same structure.

As a consequence, for each $T = \sum_{n=1}^{N} \sum_{\lambda \in \Lambda} c_n(\lambda)\, \alpha_\lambda(S_n)$ in $V_{\mathbf{S}}^2$ we have the expansion

$$\begin{aligned}\mathbf{c} &= \sum_{m=1}^{M} \sum_{\lambda \in \Lambda} \big\langle \mathbf{c}, T_\lambda \mathbf{a}_m^* \big\rangle_{\ell^2_N(\Lambda)} T_\lambda \mathbf{b}_m \\ &= \sum_{m=1}^{M} \sum_{\lambda \in \Lambda} s_{T,m}(\lambda)\, T_\lambda \mathbf{b}_m \qquad \text{in } \ell^2_N(\Lambda)\,,\end{aligned} \tag{7.19}$$

where $\mathbf{c} = (c_1, c_2, \ldots, c_N)^\top \in \ell^2_N(\Lambda)$.

The fact that $\big\{T_\lambda\, \mathbf{a}_m^*\big\}_{\lambda \in \Lambda;\, m=1,2,\ldots,M}$ is a frame for $\ell^2_N(\Lambda)$ and the isomorphism $\mathcal{T}_{\mathbf{S}}$ defined in Eq. (7.11) give the equivalence of the norms.

The isomorphism $\mathcal{T}_{\mathbf{S}}$ applied in Eq. (7.19) gives the sampling expansion (7.18), where $H_m = K_{h_m} \in V_{\mathbf{S}}^2$ with Kohn-Nirenberg symbol

$$h_m = \mathcal{T}_{\sigma_{\mathbf{S}}}(\mathbf{b}_m) \in V^2_{\sigma_{\mathbf{S}}}\,, \quad m = 1, 2, \ldots, M\,.$$

Furthermore, since $\{\alpha_\lambda(H_m)\}_{\lambda \in \Lambda;\, m=1,2,\ldots,M}$ is a frame for $V_{\mathbf{S}}^2$ the convergence of the series is unconditional in the Hilbert-Schmidt norm.

Notice that

$$\mathcal{T}_{\sigma_S}(T_\lambda \mathbf{b}_m) = T_\lambda(\mathcal{T}_{\sigma_S} \mathbf{b}_m) = T_\lambda(h_m)\,,$$

where the same symbol $T_\lambda$ denotes both the translation by $\lambda$ in $\ell^2_N(\Lambda)$ and in $L^2(\mathbb{R}^{2d})$ respectively.

Observe that if $\mathbf{b}_m = \big(b_{1,m}(\lambda), b_{2,m}(\lambda), \ldots, b_{N,m}(\lambda)\big)^\top$ is the $m$-th column of a matrix $B$ such that $\widehat{B}(\xi)\, \widehat{A}(\xi) = \mathbb{I}_N$, a.e. $\xi \in \widehat{\Lambda}$, then the sampling operator $H_m$ is given by

$$H_m = \sum_{n=1}^{N} \sum_{\lambda \in \Lambda} b_{n,m}(\lambda)\, \alpha_\lambda(S_n)\,, \quad m = 1, 2, \ldots, M\,.$$

Since convergence in $\mathcal{HS}$-norm implies convergence in operator norm we deduce the pointwise expansion for each $f \in L^2(\mathbb{R}^d)$. □

Due to conditions (7.16) in Theorem 7.6 we have necessarily $M \geq N$. Whenever $M > N$, there are infinite dual frames $\big\{T_\lambda\, \mathbf{b}_m\big\}_{\lambda \in \Lambda;\, m=1,2,\ldots,M}$ of the sequence $\big\{T_\lambda\, \mathbf{a}_m^*\big\}_{\lambda \in \Lambda;\, m=1,2,\ldots,M}$ given by the samples (7.13).

They are obtained from the left-inverses $\widehat{B}(\xi)$ of $\widehat{A}(\xi)$ which are deduced, from the Moore-Penrose pseudo-inverse $\widehat{A}(\xi)^{\dagger}$, as the $N \times M$ matrices

$$\widehat{B}(\xi) := \widehat{A}(\xi)^{\dagger} + C(\xi)\big[\mathbb{I}_M - \widehat{A}(\xi)\widehat{A}(\xi)^{\dagger}\big], \quad \text{a.e. } \xi \in \widehat{\Lambda},$$

where $C$ denotes any $N \times M$ matrix with entries in $L^{\infty}(\widehat{\Lambda})$ (see (3.16)).

More can be said whenever $M = N$:

**Corollary 7.1** *In case $M = N$, assume that the conditions*

$$0 < \operatorname*{ess\,inf}_{\xi\in\widehat{\Lambda}} \big|\det[\widehat{A}(\xi)]\big| \le \operatorname*{ess\,sup}_{\xi\in\widehat{\Lambda}} \big|\det[\widehat{A}(\xi)]\big| < +\infty \tag{7.20}$$

*hold. Then, there exist $N$ unique elements $H_n$ in $V_{\mathbf{S}}^2$, $n = 1, 2, \ldots, N$, such that the associated sequence $\big\{\alpha_\lambda(H_n)\big\}_{\lambda\in\Lambda;\, n=1,2,\ldots,N}$ is a Riesz basis for $V_{\mathbf{S}}^2$ and the sampling formula*

$$T = \sum_{n=1}^{N}\sum_{\lambda\in\Lambda} s_{T,n}(\lambda)\, \alpha_\lambda(H_n) \quad \textit{in } \mathcal{HS}\textit{-norm}$$

*holds for each $T \in V_{\mathbf{S}}^2$.*

*Moreover, the interpolation property*

$$\big\langle \alpha_{-\lambda}(H_m) g_n,\, \widetilde{g}_n \big\rangle = \delta_{m,n}\delta_{\lambda,0}\,,$$

*where $\lambda \in \Lambda$ and $m, n = 1, 2, \ldots, N$, holds.*

***Proof*** In this case, the square matrix $\widehat{A}(\xi)$ is invertible and the statement (7.20) in corollary is equivalent to condition $0 < \alpha_A \le \beta_A < +\infty$ in (7.16); besides, any Riesz basis has a unique dual basis.

The uniqueness of the coefficients in a Riesz basis expansion gives the interpolation property. □

In particular, for the case $N = M = 1$ we have:

**Corollary 7.2** *Assume that, for a fixed pair of functions $g, \widetilde{g} \in L^2(\mathbb{R}^d)$, the sequence $\mathbf{a} = \{a(\lambda)\}_{\lambda\in\Lambda}$, where*

$$a(\lambda) = \big\langle \alpha_{-\lambda}(S) g,\, \widetilde{g}\big\rangle_{L^2(\mathbb{R}^d)}\,, \quad \lambda \in \Lambda\,,$$

*satisfies the conditions*

$$0 < \operatorname*{ess\,inf}_{\xi\in\widehat{\Lambda}} |\mathcal{F}_s^{\Lambda}(\mathbf{a})(\xi)| \le \operatorname*{ess\,sup}_{\xi\in\widehat{\Lambda}} |\mathcal{F}_s^{\Lambda}(\mathbf{a})(\xi)| < \infty\,. \tag{7.21}$$

*Then, there exists a unique* $H \in V_S^2$ *such that the sequence* $\{\alpha_\lambda(H)\}_{\lambda\in\Lambda}$ *is a Riesz basis for* $V_S^2$ *and, for each* $T \in V_S^2$*, the sampling formula*

$$T = \sum_{\lambda\in\Lambda} \langle \alpha_{-\lambda}(T)g, \widetilde{g}\rangle_{L^2(\mathbb{R}^d)} \alpha_\lambda(H) \quad \text{in } \mathcal{HS}\text{-norm}$$

*holds. Moreover, the interpolation property* $\langle \alpha_{-\lambda}(H)g, \widetilde{g}\rangle = \delta_{\lambda,0}$, $\lambda \in \Lambda$*, holds; in particular,* $\langle Hg, \widetilde{g}\rangle = 1$.

Concerning the above corollary, it is worth remarking that it is not necessary that the operators $T$ in $V_S^2$ have a bandlimited Kohn-Nirenberg symbol $\sigma_T$. In fact, the bandlimited setting is obtained from the above corollary as a particular case.

Namely, let $\Lambda = a\mathbb{Z}^d \times b\mathbb{Z}^d$ be a lattice in $\mathbb{R}^{2d}$ with $a, b > 0$. Assume that the generator $S$ of $V_S^2$ is a bandlimited operator to $Q := [\frac{-1}{2a}, \frac{1}{2a}]^d \times [\frac{-1}{2b}, \frac{1}{2b}]^d$, i.e., it belongs to

$$OPW^2(Q) := \left\{T \in \mathcal{HS}(\mathbb{R}^d) \,:\, \operatorname{supp}\widehat{\sigma}_T \subseteq Q\right\}.$$

Then any $T \in V_S^2$ also belongs to $OPW^2(Q)$. In case conditions (7.21) are satisfied, any $T \in V_S^2$ can be recovered from its diagonal channel samples as

$$T = \sum_{\lambda\in\Lambda} \langle T\pi(\lambda)g, \pi(\lambda)\widetilde{g}\rangle_{L^2(\mathbb{R}^d)} \alpha_\lambda(H) \quad \text{in } \mathcal{HS}\text{-norm},$$

where $H = \sum_{\lambda\in\Lambda} b(\lambda)\,\alpha_\lambda(S)$ in $V_S^2$ is obtained from the sequence $\mathbf{b} = \{b(\lambda)\}_{\lambda\in\Lambda}$ in $\ell^2(\Lambda)$ such that

$$\mathcal{F}_S^\Lambda(\mathbf{b})(\xi)\mathcal{F}_S^\Lambda(\mathbf{a})(\xi) = 1, \quad \text{a.e. } \xi \in \widehat{\Lambda}.$$

In Ref. [18] the reconstruction of pseudodifferential operators with a bandlimited Kohn-Nirenberg symbol is considered. In particular, Theorem 2 of the same reference proves that, under some appropriate assumptions, for any $T \in OPW^2(Q)$ we have

$$\sigma_T = \frac{1}{(ab)^d} \sum_{\lambda\in\Lambda} \langle T\pi(\lambda)g, \pi(\lambda)g\rangle_{L^2(\mathbb{R}^d)} T_\lambda\big(\operatorname{sinc}_{a,b} * k\big) \quad \text{in } L^2(\mathbb{R}^{2d}),$$

where the function $k$, independent of $T$, belongs to $L^1(\mathbb{R}^{2d})$ and $\mathrm{sinc}_{\mathrm{a,b}}$ denotes the *sinc function* adapted to the lattice $\Lambda = a\mathbb{Z}^d \times b\mathbb{Z}^d$, namely

$$\operatorname{sinc}_{a,b}(x) = \prod_{j=1}^{d} \frac{\sin \pi a x_j}{\pi a x_j} \prod_{j=d+1}^{2d} \frac{\sin \pi b x_j}{\pi b x_j}, \quad x \in \mathbb{R}^{2d}.$$

Using the Kohn-Nirenberg transform, the above sampling formula for $\sigma_T$ can be written as

$$T = \frac{1}{(ab)^d} \sum_{\lambda \in \Lambda} \langle T\pi(\lambda)g,\, \pi(\lambda)g \rangle_{L^2(\mathbb{R}^d)} \, \alpha_\lambda(k * K_{\text{sinc}_{a,b}}) \quad \text{in } \mathcal{HS}\text{-norm}\,,$$

where $k * K_{\text{sinc}_{a,b}}$ denotes the Hilbert-Schmidt operator obtained from the convolution of the function $k$ and the operator $K_{\text{sinc}_{a,b}}$. We have also used the following result:

**Lemma 7.4** *Let $K_f$ be an operator in $\mathcal{HS}(\mathbb{R}^d)$ with Kohn-Nirenberg symbol $f \in L^2(\mathbb{R}^{2d})$, and let $g$ be a function in $L^1(\mathbb{R}^{2d})$. Then we have that $K_{g*f} = g * K_f$.*

***Proof*** Recall that the convolution $g * K_f$ is the operator in $\mathcal{HS}(\mathbb{R}^d)$ defined by the operator-valued integral (in weak sense) as

$$g * K_f = \int_{\mathbb{R}^{2d}} g(z)\, \alpha_z(K_f)\, dz\,,$$

that is,

$$\Big\langle \Big( \int_{\mathbb{R}^{2d}} g(z)\, \alpha_z(K_f)\, dz \Big)\varphi,\, \psi \Big\rangle_{L^2(\mathbb{R}^d)} = \int_{\mathbb{R}^{2d}} g(z)\, \langle \alpha_z(K_f)\varphi,\, \psi \rangle\, dz\,,$$

for every $\varphi, \psi \in L^2(\mathbb{R}^d)$. See the details in Refs. [21, 25, 27].

Since the Kohn-Nirenberg map $\mathcal{K} : \mathcal{HS}(\mathbb{R}^d) \to L^2(\mathbb{R}^{2d})$ such that $\mathcal{K}(K_f) = f$ is a unitary operator and bounded operators commute with convergent integrals (see Ref. [21, Proposition 2.4]) we get

$$\begin{aligned} \mathcal{K}(g * K_f) &= \int_{\mathbb{R}^{2d}} g(z)\, \mathcal{K}\big(\alpha_z(K_f)\big) dz \\ &= \int_{\mathbb{R}^{2d}} g(z)\, \mathcal{K}\big(K_{T_z f}\big) dz \\ &= \int_{\mathbb{R}^{2d}} g(z)\, f(\cdot - z) dz = g * f\,, \end{aligned}$$

that is, $K_{g*f} = g * K_f$. □

**Example 7.4 (An Illustrative Example Involving Gabor Multipliers)** Assume that we have $N$ stable generators in $V_{\mathbf{S}}^2$ of the form $S_n = \varphi_n \otimes \widetilde{\varphi}_n$ with $\varphi_n, \widetilde{\varphi}_n \in \mathcal{S}_0(\mathbb{R}^d)$, $n = 1, 2, \ldots, N$. In this regard, note that in order to apply Theorem 7.5 we have (see Eq. (7.10))

$$\mathcal{F}_W(\varphi_n \otimes \widetilde{\varphi}_n)(z) = e^{\pi i x \cdot \omega}\, V_{\widetilde{\varphi}_n}\varphi_n(z)\,, \quad z = (x, \omega) \in \mathbb{R}^{2d}\,.$$

Next, for each $T \in V_{\mathbf{S}}^2$ we consider its diagonal channel samples

$$\big\langle \alpha_{-\lambda}(T) g_m, \widetilde{g}_m \big\rangle_{L^2(\mathbb{R}^d)}, \quad \lambda \in \Lambda \text{ and } m = 1, 2, \ldots M\,,$$

with $g_m, \widetilde{g}_m \in \mathcal{S}_0(\mathbb{R}^d)$. In this case, for $m = 1, 2, \ldots, M$ and $n = 1, 2, \ldots, N$, we get

$$\begin{aligned}
a_{m,n}(\lambda) &= \big\langle \alpha_{-\lambda}(\varphi_n \otimes \widetilde{\varphi}_n) g_m, \widetilde{g}_m \big\rangle_{L^2(\mathbb{R}^d)} \\
&= \big\langle (\varphi_n \otimes \widetilde{\varphi}_n) \pi(\lambda) g_m, \pi(\lambda) \widetilde{g}_m \big\rangle_{L^2(\mathbb{R}^d)} \\
&= \big\langle \langle \pi(\lambda) g_m, \widetilde{\varphi}_n \rangle \varphi_n, \pi(\lambda) \widetilde{g}_m \big\rangle_{L^2(\mathbb{R}^d)} \\
&= \overline{V_{g_m} \widetilde{\varphi}_n(\lambda)}\, V_{\widetilde{g}_m} \varphi_n(\lambda)\,, \quad \lambda \in \Lambda\,.
\end{aligned}$$

From Proposition 4.1 in Ref. [25] we deduce that the sequences $\big\{a_{m,n}(\lambda)\big\}_{\lambda \in \Lambda}$ belong to $\ell^1(\Lambda)$ and, as a consequence, the entries in the transfer matrix $\widehat{A}$ are continuous functions on the compact $\widehat{\Lambda}$.

In order to apply Theorem 7.6, conditions in Eq. (7.16) reduce to

$$\det[\widehat{A}(\xi)^* \widehat{A}(\xi)] \neq 0 \quad \text{for all } \xi \in \widehat{\Lambda}\,.$$

Under the above circumstances, any $T \in V_{\mathbf{S}}^2$, which is nothing but the finite sum of Gabor multipliers (see Eq. (7.9)),

$$T = \sum_{n=1}^{N} \mathcal{G}_{\mathbf{c}_n}^{\widetilde{\varphi}_n, \varphi_n}\,,$$

can be recovered, in a stable way, from its diagonal channel samples

$$\big\langle \alpha_{-\lambda}(T) g_m, \widetilde{g}_m \big\rangle_{L^2(\mathbb{R}^d)}\,, \quad \lambda \in \Lambda \text{ and } m = 1, 2, \ldots, M\,.$$

### 7.4.2 The General Case of Average Sampling

It is worth noticing that the proof of Theorem 7.6 depends on whether the matrix $A = [a_{m,n}]$, with entries given by (7.17), satisfies the frame conditions in Eq. (7.16). Thus, the following result holds:

**Theorem 7.7** *Theorem 7.6 is also valid for any sampling procedure like* (7.15) *such that the corresponding matrix A satisfies the conditions in Eq.* (7.16).

For instance, we can consider *average sampling* in $V_{\mathbf{S}}^2$. Namely, for any $T \in V_{\mathbf{S}}^2$, its *average samples* at $\Lambda$ are defined by

$$\big\langle T, \alpha_\lambda(Q_m)\big\rangle_{\mathcal{HS}}, \quad \lambda \in \Lambda, \ m = 1, 2, \dots, M,$$

from $M$ fixed operators $Q_1, Q_2, \dots, Q_M$ in $\mathcal{HS}(\mathbb{R}^d)$, not necessarily in $V_{\mathbf{S}}^2$.

Having in mind Eq. (7.14) in Lemma 7.3, observe that the diagonal channel samples defined in Eq. (7.12) are a particular case of average sampling where the average operators are $Q_m = \widetilde{g}_m \otimes g_m$, $m = 1, 2, \dots, M$.

The average samples of any $T = \sum_{n=1}^N \sum_{\mu\in\Lambda} c_n(\mu)\, \alpha_\mu(S_n)$ can be also expressed as the output of a discrete convolution system in $\ell^2_N(\Lambda)$.

Indeed, for $m = 1, 2, \dots, M$, we have

$$\begin{aligned}
\big\langle T, \alpha_\lambda(Q_m)\big\rangle_{\mathcal{HS}} &= \big\langle \sigma_T, T_\lambda \sigma_{Q_m}\big\rangle_{L^2(\mathbb{R}^{2d})} \\
&= \Big\langle \sum_{n=1}^N \sum_{\mu\in\Lambda} c_n(\mu)\, T_\mu \sigma_{S_n}, T_\lambda \sigma_{Q_m}\Big\rangle_{L^2(\mathbb{R}^{2d})} \\
&= \sum_{n=1}^N \sum_{\mu\in\Lambda} c_n(\mu) \langle T_\mu \sigma_{S_n}, T_\lambda \sigma_{Q_m}\rangle_{L^2(\mathbb{R}^{2d})} \\
&= \sum_{n=1}^N \sum_{\mu\in\Lambda} c_n(\mu) \langle \sigma_{S_n}, T_{\lambda-\mu} \sigma_{Q_m}\rangle_{L^2(\mathbb{R}^{2d})} \\
&= \sum_{n=1}^N \big(a_{m,n} *_\Lambda c_n\big)(\lambda), \quad \lambda \in \Lambda,
\end{aligned}$$

where

$$a_{m,n}(\mu) := \big\langle \sigma_{S_n}, T_\mu \sigma_{Q_m}\big\rangle_{L^2(\mathbb{R}^{2d})} = \big\langle S_n, \alpha_\mu(Q_m)\big\rangle_{\mathcal{HS}}, \quad \mu \in \Lambda,$$

and $\sigma_{S_n}, \sigma_{Q_m}$ are the Kohn-Nirenberg symbols of $S_n, Q_m$, respectively.

Observe that, for each $m = 1, 2, \dots, M$, and $n = 1, 2, \dots, N$, the sequence $\{a_{m,n}(\lambda)\}_{\lambda\in\Lambda}$ belongs to $\ell^2(\Lambda)$ since, in particular, $\{T_\lambda \sigma_{S_n}\}_{\lambda\in\Lambda;\, n=1,2,\dots,N}$ is a Bessel sequence in $L^2(\mathbb{R}^{2d})$.

Thus we have:

**Corollary 7.3** *Assume that for the $M \times N$ matrix $A = [a_{m,n}]$ with the $\ell^2(\Lambda)$ entries*

$$a_{m,n}(\lambda) = \big\langle S_n, \alpha_\lambda(Q_m)\big\rangle_{\mathcal{HS}}, \quad \lambda \in \Lambda,$$

*the associated constants $\alpha_A$ and $\beta_A$ satisfy conditions in* (7.16)*. Then, there exist $M \geq N$ operators $H_m \in V^2_{\mathbf{S}}$, $m = 1, 2, \ldots, M$, such that the sampling formula*

$$T = \sum_{m=1}^{M} \sum_{\lambda \in \Lambda} \big\langle T, \alpha_\lambda(Q_m) \big\rangle_{\mathcal{HS}} \, \alpha_\lambda(H_m) \quad \textit{in } \mathcal{HS}\textit{-norm} \tag{7.22}$$

*holds for each $T \in V^2_{\mathbf{S}}$ where $\{\alpha_\lambda(H_m)\}_{\lambda \in \Lambda;\, m=1,2,\ldots,M}$ is a frame for $V^2_{\mathbf{S}}$. The convergence of the series is unconditional in Hilbert-Schmidt norm.*

*Furthermore, the $\ell^2_M$-norm of the average samples defines an equivalent norm in $V^2_{\mathbf{S}}$, and we have the pointwise expansion*

$$Tf = \sum_{m=1}^{M} \sum_{\lambda \in \Lambda} \big\langle T, \alpha_\lambda(Q_m) \big\rangle_{\mathcal{HS}} \, \alpha_\lambda(H_m) f \quad \textit{in } L^2(\mathbb{R}^d)\,,$$

*for each $f \in L^2(\mathbb{R}^d)$.*

Notice that the above formula can be also deduced by using the Weyl symbols of $S_n$ and $Q_m$ instead of their Kohn-Nirenberg symbols; see the details in Ref. [13].

Finally, it is worth mentioning that each sampling result in this section admits a kind of converse result. For instance, for Corollary 7.3 it reads:

**Proposition 7.1** *Assume that a sampling formula like* (7.22) *holds in $V^2_{\mathbf{S}}$ where the samples are obtained by means of an $M \times N$ matrix $A = [a_{m,n}]$ with entries*

$$a_{m,n}(\lambda) = \big\langle S_n, \alpha_\lambda(Q_m) \big\rangle_{\mathcal{HS}}, \quad \lambda \in \Lambda\,,$$

*and satisfying $\beta_A < \infty$. If the sequence $\{\alpha_\lambda(H_m)\}_{\lambda \in \Lambda;\, m=1,2,\ldots,M}$ is a frame for $V^2_{\mathbf{S}}$, then the condition $\alpha_A > 0$ also holds.*

***Proof*** For each $T = \sum_{n=1}^{N} \sum_{\lambda \in \Lambda} c_n(\lambda)\, \alpha_\lambda(S_n)$ in $V^2_{\mathbf{S}}$, let $\mathbf{c}$ denote the vector sequence of its coefficients $\mathbf{c} = (c_1, c_2, \ldots, c_N)^\top \in \ell^2_N(\Lambda)$.

We have that $\big\{\mathcal{T}_{\mathbf{S}}^{-1}[\alpha_\lambda(H_m)]\big\}_{\lambda \in \Lambda;\, m=1,2,\ldots,M}$ is a frame for $\ell^2_N(\Lambda)$. Applying the operator $\mathcal{T}_{\mathbf{S}}^{-1}$ in (7.22) we obtain

$$\begin{aligned}
\mathbf{c} &= \sum_{m=1}^{M} \sum_{\lambda \in \Lambda} \big\langle T, \alpha_\lambda(Q_m) \big\rangle \mathcal{T}_{\mathbf{S}}^{-1}[\alpha_\lambda(H_m)] \\
&= \sum_{m=1}^{M} \sum_{\lambda \in \Lambda} \big\langle \mathbf{c}, T_\lambda \mathbf{a}^*_m \big\rangle_{\ell^2_N(\Lambda)} \mathcal{T}_{\mathbf{S}}^{-1}[\alpha_\lambda(H_m)] \quad \text{in } \ell^2_N(\Lambda)\,,
\end{aligned}$$

where $a^*_{m,n}(\mu) := \overline{a_{m,n}(-\mu)}$, $\mu \in \Lambda$.

Since $\beta_A < +\infty$, the sequence $\big\{T_\lambda\, \mathbf{a}_m^*\big\}_{\lambda\in\Lambda;\, m=1,2,\dots,M}$ is a Bessel sequence for $\ell^2_N(\Lambda)$. According to Proposition 1.11, the sequence $\big\{T_\lambda\, \mathbf{a}_m^*\big\}_{\lambda\in\Lambda;\, m=1,2,\dots,M}$ will be a dual frame of $\{\mathcal{T}_{\mathbf{S}}^{-1}[\alpha_\lambda(H_m)]\}_{\lambda\in\Lambda;\, m=1,2,\dots,M}$. Therefore, from Proposition 5.4 we deduce that the constant $\alpha_A > 0$. □

### 7.4.3 *Sampling at a Sub-lattice*

Let $\Lambda'$ be a sub-lattice of $\Lambda$ with finite index $L$, i.e., the quotient group $\Lambda/\Lambda'$ has finite order $L$. We consider a set of representatives $\{\lambda_1, \lambda_2, \dots, \lambda_L\}$ of the cosets of $\Lambda'$. That is, the lattice $\Lambda$ can be decomposed as

$$\Lambda = \bigcup_{l=1}^{L} (\lambda_l + \Lambda') \quad \text{with} \quad (\lambda_l + \Lambda') \cap (\lambda_{l'} + \Lambda') = \varnothing \ \text{ for } l \neq l' \,.$$

Thus, the space $V^2_{\mathbf{S}} = \Big\{\sum_{n=1}^{N}\sum_{\lambda\in\Lambda} c_n(\lambda)\,\alpha_\lambda(S_n) \,:\, c_n \in \ell^2(\Lambda)\Big\}$ can be written as

$$\begin{aligned} V^2_{\mathbf{S}} &= \Big\{\sum_{n=1}^{N}\sum_{l=1}^{L}\sum_{\mu\in\Lambda'} c_n(\lambda_l+\mu)\,\alpha_{\lambda_l+\mu}(S_n)\Big\} \\ &= \Big\{\sum_{n=1}^{N}\sum_{l=1}^{L}\sum_{\mu\in\Lambda'} c_{nl}(\mu)\,\alpha_\mu(S_{nl}) \,:\, c_{nl} \in \ell^2(\Lambda')\Big\}, \end{aligned}$$

where $c_{nl}(\mu) := c_n(\lambda_l + \mu)$ and $S_{nl} := \alpha_{\lambda_l}(S_n)$, and the new index $nl$ goes from 11 to $NL$ following the lexicographic order.

As a consequence, the subspace $V^2_{\mathbf{S}}$ can be rewritten as $V^2_{\widetilde{\mathbf{S}}}$ with $NL$ generators $\widetilde{\mathbf{S}} = \{S_{nl}\}$ and coefficients $c_{nl}$ in $\ell^2(\Lambda')$, $nl = 11, 12, \dots, NL$.

Now, let $T = \sum_{n=1}^{N}\sum_{l=1}^{L}\sum_{\nu\in\Lambda'} c_{nl}(\nu)\,\alpha_\nu(S_{nl})$ be in $V^2_{\mathbf{S}}$. Its diagonal samples

$$\big\langle \alpha_{-\mu}(T)g_m, \widetilde{g}_m\big\rangle_{L^2(\mathbb{R}^d)}, \quad \mu \in \Lambda',$$

can be expressed by

$$\begin{aligned} s_{T,m}(\mu) &:= \big\langle \alpha_{-\mu}(T)g_m, \widetilde{g}_m\big\rangle_{L^2(\mathbb{R}^d)} \\ &= \Big\langle \sum_{n=1}^{N}\sum_{l=1}^{L}\sum_{\nu\in\Lambda'} c_{nl}(\nu)\,\alpha_{\nu-\mu}(S_{nl})g_m, \widetilde{g}_m\Big\rangle_{L^2(\mathbb{R}^d)} \end{aligned}$$

$$= \sum_{n=1}^{N} \sum_{l=1}^{L} \sum_{\nu \in \Lambda'} c_{nl}(\nu) \langle \alpha_{\nu-\mu}(S_{nl}) g_m, \widetilde{g}_m \rangle_{L^2(\mathbb{R}^d)}$$

$$= \sum_{n=1}^{N} \sum_{l=1}^{L} \big(a_{m,nl} *_{\Lambda'} c_{nl}\big)(\mu), \quad \mu \in \Lambda',$$

where

$$a_{m,nl}(\nu) := \langle \alpha_{-\nu}(S_{nl}) g_m, \widetilde{g}_m \rangle_{L^2(\mathbb{R}^d)}, \quad \nu \in \Lambda'.$$

Hence, Theorem 7.6 gives:

**Corollary 7.4** *Let $A = [a_{m,nl}]$ be the $M \times NL$ matrix with entries*

$$a_{m,nl}(\nu) = \langle \alpha_{-\nu}(S_{nl}) g_m, \widetilde{g}_m \rangle_{L^2(\mathbb{R}^d)}, \quad \nu \in \Lambda',$$

*where $m = 1, 2, \ldots, M$ and the double index $nl = 11, 12, \ldots, NL$ follows the lexicographic order. Assume that the matrix $A$ satisfies conditions in Eq. (7.16) with respect to the dual group $\widehat{\Lambda'}$ of $\Lambda'$. Then, there exist $M \geq NL$ operators $H_m \in V_{\mathbf{S}}^2$, $m = 1, 2, \ldots, M$, such that for each $T \in V_{\mathbf{S}}^2$ the sampling formula*

$$T = \sum_{m=1}^{M} \sum_{\mu \in \Lambda'} s_{T,m}(\mu)\, \alpha_\mu(H_m) \quad \text{in } \mathcal{HS}\text{-norm}$$

*holds. Moreover, the sequence $\{\alpha_\mu(H_m)\}_{\mu \in \Lambda';\, m=1,2,\ldots,M}$ is a frame for $V_{\mathbf{S}}^2$, and the convergence of the series is unconditional in Hilbert-Schmidt norm.*

Observe that the columns of the matrix $A$ are indexed by the double index $nl$ which follows the lexicographic order of the indexes $n$ and $l$.

When we are using general average samples in $V_{\mathbf{S}}^2$ involving $M$ Hilbert-Schmidt operators $Q_m \in$, $m = 1, 2, \ldots, M$, not necessarily in $V_{\mathbf{S}}^2$, we obtain:

**Corollary 7.5** *Assume that the $M \times NL$ matrix $A = [a_{m,nl}]$ with entries*

$$a_{m,nl}(\nu) = \langle S_{nl}, \alpha_\nu(Q_m) \rangle_{\mathcal{HS}}, \quad \nu \in \Lambda',$$

*where $m = 1, 2, \ldots, M$, and $nl = 11, 12, \ldots, NL$, satisfies the conditions in Eq. (7.16). Then, there exist $M \geq NL$ operators $H_m \in V_{\mathbf{S}}^2$, $m = 1, 2, \ldots, M$, such that for each $T \in V_{\mathbf{S}}^2$ the sampling formula*

$$T = \sum_{m=1}^{M} \sum_{\lambda \in \Lambda} \langle T, \alpha_\lambda(Q_m) \rangle_{\mathcal{HS}}\, \alpha_\lambda(H_m) \quad \text{in } \mathcal{HS}\text{-norm}$$

*holds, where* $\{\alpha_\lambda(H_m)\}_{\lambda\in\Lambda;\, m=1,2,\ldots,M}$ *is a frame for* $V_S^2$*. The convergence of the series is unconditional in Hilbert-Schmidt norm.*

## 7.5 Some Closing Remarks

The contents of this chapter are a small contribution to the wider topic of time-frequency analysis. The results and proofs in this chapter concerning with sampling are mainly stated in terms of Hilbert space and frame theory. However, they can be rephrased in terms of operators following the results in Refs. [6, 21, 25].

Thus, by considering a rank-one operator as generator $S$, one can relate the elements in $V_S^2$ to *Gabor multipliers*, and by considering $Q = \varphi\otimes\varphi$, also a rank-one operator, one can relate their average samples to their *lower symbols* with respect to $\varphi$ and $\Lambda$. As a consequence, some comments are pertinent in order to help the sampling reader who is not familiar with time-frequency techniques and vice-versa, to help the time-frequency reader who is not familiar with sampling procedures:

(*a*) The necessary and sufficient condition on $\{\alpha_\lambda(S)\}_{\lambda\in\Lambda}$ to be a Riesz sequence for $\mathcal{HS}(\mathbb{R}^d)$ can be also expressed in terms of the Fourier transform in $\ell^2(\Lambda)$ of the convolution of two operators. It reads:
*The function* $\mathcal{F}_s^\Lambda(S *_\Lambda \check{S}^*) = P_{\Lambda^\circ}(|\mathcal{F}_W(S)|^2)$ *has no zeros in* $\widehat{\Lambda}$.
The *convolution of two operators* $S, T$ is formally defined as the function

$$S * T(z) := \operatorname{tr}[S\alpha_z(\check{T})]\,, \quad z \in \mathbb{R}^{2d}\,,$$

where $\check{T} = PTP$ and $P$ denotes the *parity operator* $(P\varphi)(t) = \varphi(-t)$ for $\varphi$ in $L^2(\mathbb{R}^d)$.

Replacing $\mathbb{R}^{2d}$ by a lattice $\Lambda \subset \mathbb{R}^{2d}$ we obtain the convolution of two operators $S, T$ at the lattice $\Lambda$ as the sequence $S *_\Lambda T(\lambda) := S * T(\lambda)$ for $\lambda \in \Lambda$.

(*b*) The *convolution of a function* $f$ *and an operator* $S$ is formally defined by the operator-valued integral

$$f * S = \int_{\mathbb{R}^{2d}} f(z)\,\alpha_z(S)dz\,.$$

Replacing $\mathbb{R}^{2d}$ by a lattice $\Lambda \subset \mathbb{R}^{2d}$ we get the definition

$$\mathbf{c} *_\Lambda S = S *_\Lambda \mathbf{c} := \sum_{\lambda\in\Lambda} c(\lambda)\,\alpha_\lambda(S)\,.$$

As a consequence, the subspace $V_S^2$ can be also expressed as the convolution

$$V_S^2 = \ell^2(\Lambda) *_\Lambda S\,.$$

(*c*) Under appropriate hypotheses (see, for instance, Ref. [25]), the average samples $s_T(\lambda) = \langle T, \alpha_\lambda(Q)\rangle_{\mathcal{HS}}$, $\lambda \in \Lambda$, can be expressed as a convolution like in the classical shift-invariant case.

Remind that, for each $f \in L^2(\mathbb{R}^d)$ its average samples are

$$\langle f, \psi(\cdot - n)\rangle_{L^2(\mathbb{R})} = f * \widetilde{\psi}(n), \quad n \in \mathbb{Z},$$

where $\widetilde{\psi}(t) = \overline{\psi(-t)}$ is the average function.

Now, in the case treated here, an easy computation gives

$$\begin{aligned}\langle T, \alpha_\lambda(Q)\rangle_{\mathcal{HS}} &= \mathrm{tr}[T\alpha_\lambda(Q)^*] = \mathrm{tr}[T\alpha_\lambda(Q^*)] \\ &= T *_\Lambda \check{Q}^*(\lambda) = T *_\Lambda \widetilde{Q}(\lambda), \quad \lambda \in \Lambda,\end{aligned}$$

where $\widetilde{Q} = \check{Q}^*$.

By using the appropriate convolution notation, the sampling formula in (7.22), i.e.,

$$T = \sum_{m=1}^{M} \sum_{\lambda \in \Lambda} \langle T, \alpha_\lambda(Q_m)\rangle_{\mathcal{HS}}\, \alpha_\lambda(H_m)$$

can be expressed by

$$T = (T *_\Lambda \widetilde{Q}) *_\Lambda H \quad \text{for each } T \in V_S^2 .$$

(*d*) Whenever $S = \varphi \otimes \psi$ is a rank-one operator, one gets that the $\Lambda$-shift-invariant subspace $V_S^2 = \ell^2(\Lambda) *_\Lambda S$ is a subspace of *Gabor multipliers* in $\mathcal{HS}(\mathbb{R}^d)$. Namely, as proved in Sect. 7.3, $\mathbf{c} *_\Lambda S = \mathcal{G}_{\mathbf{c}}^{\psi,\varphi}$, the Gabor multiplier with windows $\psi, \varphi$ and mask $\mathbf{c}$ (in $\ell^2(\Lambda)$) used in time-frequency analysis (see Refs. [8, 25]).

In Theorem 5.17 of Ref. [6] one can find the properties of the subspace $V_S^2$ for $S = \varphi \otimes \varphi$ with $\varphi$ in the Feichtinger algebra $\mathcal{S}_0(\mathbb{R}^d)$.

Theorem 5.20 of the same reference establishes the relationship between $V_S^2$ and the shift-invariant subspace of $L^2(\mathbb{R}^{2d})$ generated by the Kohn-Nirenberg symbol of $\varphi \otimes \varphi$.

In the present chapter we work in a more general context where the generalization of the alluded Theorem 5.17 has been proved in Ref. [25]. Thus, Theorem 7.6 in this chapter is a sampling result in a subspace of Gabor multipliers in $\mathcal{HS}(\mathbb{R}^d)$ and it can be thought as a generalization of results in Ref. [6].

A difference from the setup of Ref. [6] is that in sampling formula (7.18) the average operator $Q$ does not necessarily belong to $V_S^2$.

Observe that in this case the mapping

$$\ell^2(\Lambda) \ni \mathbf{c} \longmapsto \mathbf{c} *_\Lambda (\varphi \otimes \psi) = \mathcal{G}_{\mathbf{c}}^{\psi,\varphi} \in V_S^2$$

is precisely the isomorphism $\mathcal{T}_S$ defined in (7.11).

(*e*) Whenever both $S = \varphi \otimes \varphi$ and $Q = \psi \otimes \psi$ are rank-one operators Theorem 7.6 gives a necessary and sufficient condition to recover any Gabor multiplier of the form

$$T = \sum_{\lambda \in \Lambda} c(\lambda)\, \alpha_\lambda(\varphi \otimes \varphi)$$

from its lower symbol with respect to $\psi$ on $\Lambda$

$$\begin{aligned} s_T(\lambda) &= \big\langle T, \alpha_\lambda(Q)\big\rangle_{\mathcal{HS}} = \big\langle T, \pi(\lambda)\psi \otimes \pi(\lambda)\psi\big\rangle_{\mathcal{HS}} \\ &= \big\langle T\pi(\lambda)\psi, \pi(\lambda)\psi\big\rangle_{L^2(\mathbb{R}^d)}, \qquad \lambda \in \Lambda . \end{aligned}$$

Of course we are assuming that $\{\alpha_\lambda(\varphi \otimes \varphi)\}_{\lambda \in \Lambda}$ is a Riesz sequence in $\mathcal{HS}(\mathbb{R}^d)$; in this regard, remember that the Fourier-Wigner transform of $S$ is (see Eq. (7.10))

$$\mathcal{F}_W(\varphi \otimes \varphi)(z) = e^{\pi i x\cdot\omega} V_\varphi \varphi(z), \quad z = (x, \omega) \in \mathbb{R}^{2d} .$$

In order to apply Corollary 7.3, an easy computation gives

$$\begin{aligned} q(\lambda) &= \big\langle \varphi \otimes \varphi, \alpha_\lambda(\psi \otimes \psi)\big\rangle_{\mathcal{HS}} \\ &= \big\langle (\varphi \otimes \varphi)\pi(\lambda)\psi, \pi(\lambda)\psi\big\rangle_{L^2(\mathbb{R}^d)} \\ &= \big\langle \langle \pi(\lambda)\psi, \varphi\rangle\, \varphi, \pi(\lambda)\psi\big\rangle_{L^2(\mathbb{R}^d)} \\ &= |V_\psi \varphi(\lambda)|^2, \qquad \lambda \in \Lambda . \end{aligned}$$

In other words, in this particular case the corresponding $\mathbf{q} = \{q(\lambda)\}_{\lambda \in \Lambda}$ is the sequence of samples at $\Lambda$ of the *spectrogram* of $\varphi$ with window $\psi$.

If the functions $\varphi, \psi$ belong to Feichtinger's algebra $S_0(\mathbb{R}^d)$, by using Proposition 4.1 in Ref. [25] it is easy to deduce that the sequence $\mathbf{q} = \big\{|V_\psi \varphi(\lambda)|^2\big\}_{\lambda \in \Lambda}$ belongs to $\ell^1(\Lambda)$.

As a consequence, the thesis in Corollary 7.3 (for $N = M = 1$) holds if and only if the continuous function $\mathcal{F}_s^\Lambda(\mathbf{q})$ has no zeros on the compact $\widehat{\Lambda}$.

(*f*) Finally, we include a brief comment on the relationship between the Fourier-Wigner transform of a Hilbert-Schmidt operator and the symplectic Fourier transform of its Weyl symbol. To this end, consider the *integrated Schrödinger representation* which is defined in $L^1(\mathbb{R}^{2d})$ as

$$\begin{aligned} \rho : L^1(\mathbb{R}^{2d}) &\longrightarrow B\big(L^2(\mathbb{R}^d)\big) \\ f &\longmapsto \rho(f) = \iint_{\mathbb{R}^{2d}} f(z)\, \mathrm{e}^{-\pi i x\cdot\omega}\, \pi(z)\, dz\,, \end{aligned}$$

where the integral is defined in weak and pointwise sense.

This operator can be extended to a unitary operator from $L^2(\mathbb{R}^{2d})$ onto $\mathcal{HS}(\mathbb{R}^d)$ such that $L_f = \rho(\mathcal{F}_s f)$, where $L_f$ denotes the Hilbert-Schmidt operator with Weyl symbol $f$, and $\mathcal{F}_s f$ denotes the symplectic Fourier transform of $f$.

The last equality, $L_f = \rho(\mathcal{F}_s f)$, can be justified as follows: Having in mind the integral representation of $L_f$ (see Ref. [16]) we have:

$$\begin{aligned} L_f &= \iint_{\mathbb{R}^{2d}} \widehat{f}(\xi, u)\, \mathrm{e}^{-\pi i \xi\cdot u}\, T_{-u} M_\xi \, du d\xi \\ &= \iint_{\mathbb{R}^{2d}} \mathcal{F}_s f(-u, \xi)\, \mathrm{e}^{-\pi i \xi\cdot u}\, T_{-u} M_\xi \, du d\xi \\ &= \iint_{\mathbb{R}^{2d}} \mathcal{F}_s f(u, \xi)\, \mathrm{e}^{\pi i \xi\cdot u}\, T_u M_\xi \, du d\xi = \iint_{\mathbb{R}^{2d}} \mathcal{F}_s f(u, \xi)\, \mathrm{e}^{-\pi i u\cdot \xi}\, M_\xi T_u \, du d\xi \\ &= \rho(\mathcal{F}_s f)\,. \end{aligned}$$

Concerning the above integral representation for $L_f$, following Ref. [16, p. 312], for any function $f \in \mathcal{S}(\mathbb{R}^{2d})$ and any function $g \in L^2(\mathbb{R}^d)$ it can be easily proved that

$$\iint_{\mathbb{R}^{2d}} \widehat{f}(\xi, u)\, \mathrm{e}^{-\pi i \xi\cdot u}\, T_{-u} M_\xi \, \overline{g(x)}\, du d\xi = \int_{\mathbb{R}^d} k(x, y)\, \overline{g(y)}\, dy = L_f g(x)\,,$$

since the integral kernel $k(x, y) = \mathcal{T}_s^{-1} \mathcal{F}_2^{-1} f(x, y)$ is precisely the kernel of the Hilbert-Schmidt operator $L_f$ (see Sect. 7.2). By using a density argument, the result is true for any $f \in L^2(\mathbb{R}^{2d})$.

On the other hand, the Fourier-Wigner transform defined in (7.8) extends to a unitary operator $\mathcal{F}_W : \mathcal{HS}(\mathbb{R}^d) \longrightarrow L^2(\mathbb{R}^{2d})$ which is the inverse operator of the integrated Schrödinger representation $\rho$. Hence, we have $\mathcal{F}_W(L_f) = \mathcal{F}_s f$. See the details in Refs. [10, 21].

# References


1. Aldroubi, A., Sun, Q., Tang, W.S.: Convolution, average sampling, and a Calderon resolution of the identity for shift-invariant spaces. J. Fourier Anal. Appl. **11**(2), 215–244 (2005)
2. Benedetto, J.J., Pfander, G.E.: Frame expansions for Gabor multipliers. Appl. Comput. Harmon. Anal. **20**(1), 26–40 (2006)
3. Conway, J.B.: A Course in Functional Analysis, 2nd edn. Springer, New York (1990)
4. Conway, J.B.: A Course in Operator Theory. AMS, Providence RI (2000)
5. Deitmar, A., Echterhoff, S.: Principles of Harmonic Analysis, 2nd edn. Universitext. Springer, Berlin (2014)
6. Feichtinger, H.G.: Spline-type spaces in Gabor analysis. In: Zhou, D.-X. (ed.) Wavelet Analysis, pp. 100–122. World Sci. Publ., River Edge, NJ (2002)
7. Feichtinger, H.G., Kozek, W.: Quantization of TF lattice-invariant operators on elementary LCA groups. In: Feichtinger, H., Strohmer, T. (eds.) Gabor Analysis and Algorithms. Theory and Applications, pp. 233–266. Birkhäuser, Boston (1998)
8. Feichtinger, H.G., Nowak, K.: A first survey of Gabor multipliers. In: Feichtinger, H.G., Strohmer, T. (eds.) Advances in Gabor Analysis, pp. 99–128. Birkhäuser, Boston (2003)
9. Feichtinger, H.G., Luef, F., Wherter, T.: A guided tour from linear algebra to the foundations of Gabor analysis. In: Goh, S.S., et al. (eds.) Gabor and Wavelet Frames, pp. 1–49. Lecture Notes Series, IMS, National University of Singapore: Vol. 10. World Scientific, Singapore (2007)
10. Folland, G.B.: Harmonic Analysis in Phase Space, Princeton University Press, Princeton (1989)
11. Folland, G.B.: A Course in Abstract Harmonic Analysis. CRC Press, Boca Raton, FL (1995)
12. Führ, H.: Abstract Harmonic Analysis of Continuous Wavelet Transform. Springer, Berlin-Heidelberg (2005)
13. García, A.G.: Average sampling in certain subspaces of Hilbert-Schmidt operators on $L^2(\mathbb{R}^d)$. Sampl. Theory Signal Process. Data Anal. **19**, 10 (2021)
14. García, A.G.: Sampling in $\Lambda$-shift-invariant subspaces of Hilbert-Schmidt operators on $L^2(\mathbb{R}^d)$. Math. Found. Comput. **4**(4), 281–297 (2021)
15. García, A.G., Pérez-Villalón, G.: Dual frames in $L^2(0, 1)$ connected with generalized sampling in shift-invariant spaces. Appl. Comput. Harmon. Anal. **20**(3), 422–433 (2006)
16. Gröchenig, K.: Foundations of Time-Frequency Analysis. Birkhäuser, Basel (2001)
17. Gröchenig, K., Heil, C.: Modulation spaces and pseudodifferential operators. Integral Equations Oper. Theory **34**(4), 439–457 (1999)
18. Gröchenig, K., Pauwels, E.: Uniqueness and reconstruction theorems for pseudodifferential operators with a bandlimited Kohn-Nirenberg symbol. Adv. Comput. Math. **40**, 49–63 (2014)
19. Jakobsen, M.S.: On a (no longer) new Segal Algebra: a review of the Feichtinger Algebra. J. Fourier Anal. Appl. **24**, 1579–1660 (2018)
20. Krahmer, F., Pfander, G.E.: Local sampling and approximation of operators. Construct. Approx. **39**(3), 541–572 (2014)
21. Luef, F., Skrettingland, E.: Convolutions for localizations operators. J. Math. Pures Appl. **118**, 288–316 (2018)
22. Pfander, G.E.: Sampling of operators. J. Fourier Anal. Appl. **19**, 612–650 (2013)
23. Pfander, G.E., Walnut, D.F.: Sampling and reconstruction of operators. IEEE Trans. Inform. Theory **62**(1), 435–458 (2016)
24. Simon, B.: Trace Ideals and Their Applications. Cambridge University Press, Cambridge (1979)
25. Skrettingland, E.: Quantum harmonic analysis on lattices and Gabor multipliers. J. Fourier Anal. Appl. **26**, 48 (2020)
26. Strohmer, T.: Pseudodifferential operators and Banach algebras in mobile communications. Appl. Comput. Harmon. Anal. **20**, 237–249 (2006)
27. Werner, R.F.: Quantum harmonic analysis on phase space. J. Math. Phys. **25**(5), 1404–1411 (1984)

# Index





A. García García, *The Use of Frames in Sampling Theory*,
RSME Springer Series 14, https://doi.org/10.1007/978-3-031-63242-6

The manufacturer's authorised representative in the EU is Springer Nature Customer Service Centre GmbH, Europaplatz 3, 69115 Heidelberg, Germany. If you have any concerns regarding our products, please contact ProductSafety@springernature.com

Printed and bound by CPI Group (UK) Ltd, Croydon, CR0 4YY
01/10/2026
02225970-0009